GEOPHYSICAL INVERSE THEORY

GEOPHYSICAL INVERSE THEORY

ROBERT L. PARKER

PRINCETON UNIVERSITY PRESS
PRINCETON, NEW JERSEY

Copyright © 1994 by Princeton University Press
Published by Princeton University Press, 41 William Street,
Princeton, New Jersey 08540
In the United Kingdom: Princeton University Press,
Chichester, West Sussex

Library of Congress Cataloging-in-Publication Data

Parker, Robert L. (Robert Ladislav), 1942–
    Geophysical Inverse Theory / by Robert L. Parker.
        p. cm. — (Princeton series in geophysics)
    Includes bibliographical references and index.
    ISBN 0-691-03634-9
    1. Inversion (Geophysics). 2. Geophysics—measurement.
    3. Functional analysis. 4. Mathematical optimization.
    I. Title. II. Series.
QC808.5.P37 1994
550'.1'515—dc20   93-44915
                                                CIP

Printed on recycled paper

Princeton University books are printed on
acid-free paper and meet the guidelines for
permanence and durability of the Committee on
Production Guidelines for Book Longevity
of the Council on Library Resources

Printed in the United States of America

2 4 6 8 10 9 7 5 3 1

CONTENTS

Chapter 5: Nonlinear Problems

PREFACE

Geophysical Inverse Theory is an expanded version of my lecture notes for a one-quarter graduate-level class, which I have been teaching for twenty years at the Institute for Geophysics and Planetary Physics at La Jolla, California. I have organized the subject around a central idea: almost every problem of geophysical inverse theory can be posed as an optimization problem, where a function must be minimized subject to various constraints. The emphasis throughout is on mathematically sound results whenever possible. This has meant that the level of discussion must be raised above that of undergraduate linear algebra, a level on which the material is often presented. The theory can be set out quite simply in the setting of abstract linear spaces but to do this it is desirable to provide the students with an introduction to elementary functional analysis. In this book, as in the class, the introduction is an informal survey of essential definitions and results, mostly without proof. Fortunately for the students and the author, the classic text by Luenberger, *Optimization by Vector Space Methods*, covers all the necessary technical material and much more besides. Not only in the introduction, but throughout the book, I have normally omitted the proofs of standard mathematical results, unless the proof is short.

In geophysics, inverse theory must provide the tools for learning about the distribution in space of material properties within the earth, based on measurements made at the surface. The notion that optimization lies at the heart of much geophysical inverse theory arises from the observation that problems based on actual measurements can never have unique solutions—there just is not enough information in finitely many measurements to prescribe an unknown function of position. Nonetheless, we may elect to construct a particular model, which means choosing one from the infinite set of candidates. Alternatively, we can seek common properties valid for every member of the set. In the first case the choice should be deliberately made and designed to avoid misleading artificial features: this leads to the minimization of complexity, an optimization problem involving a suitable function, such as root-mean-square gradient. The second course also points to an optimization: if we wish to find an upper or lower bound on a property of the earth, like the average value in a given region, we should calculate the maximum or minimum value over the set of all models agreeing with the measurements. These two approaches will occupy most of our attention.

I believe in the value of concrete and realistic illustrations for learning. This is particularly important when we must translate a result from an abstract setting to a practical computation. Therefore, I have made a practice of working through a few examples in some detail; the examples here are all rather small computationally, but realistic in the sense that they are, with one exception, based on actual measurements in geomagnetism, seismology, gravity, and electromagnetic sounding. The observations used in the examples are tabulated, unless the number of data is too great. The same problem, for instance, one on seismic attenuation, appears in several different contexts, illustrating numerical instability, the application of smooth model construction, resolution studies, and the bounding of linear functionals. And of course, students learn best by practice so there are lots of exercises, some of them involving computer work. Computer calculations, when I started this work, required the students to program in Fortran which for most of them resulted in nothing but frustration and missed deadlines. Therefore I wrote a numerical linear-algebra computer language to alleviate these difficulties but, in the last few years, personal computers and wonderful software packages like Matlab and Mathematica have completely solved the problem and I have been able to retire my program. In fact, all my own calculations for chapter 5 were performed in Matlab.

The scope of the book is certainly not intended to be comprehensive. I hope to give my readers the principles from which they can solve the problems they encounter. I have avoided analytic theories that treat arbitrarily accurate and extensive data, and concentrated instead on the more practical side of inverse theory. I have devoted only a small amount of space to strongly statistical approaches, first because I am unconvinced that one can place much confidence in results that depend on so many unverifiable assumptions. Second, Albert Tarantola's book *Inverse Problem Theory: Methods for Fitting and Model Parameter Estimation* (1987) covers these methods in great detail and it would be redundant to attempt to reproduce it even in part.

It is time for a brief tour of the contents. There are five fairly long chapters, each divided into named sections, which vary in length from one to seventen pages. As I have already noted, we must have a survey of the results and language of abstract linear vector spaces and some optimization theory. This is chapter 1. It also includes a section on numerical linear algebra, mainly to introduce the QR factorization. The material covers the standard ideas of vector spaces up to Hilbert spaces; the Decomposition and Projection Theorems and the Riesz Representation Theorem each play a part later on because Hilbert space is the normal setting for the theory. Some of the material here, for example sections 1.08 and 1.09 on completion of normed spaces, is perhaps peripheral to geophysical interests, but useful if the student is ever to dip into the applied mathematical literature.

Chapter 2 discusses the simplified situation in which the data are error free and the problem is linear. We define linear inverse problems to be those in which the functional describing the measurement is linear in the model. Linear problems are much more fully understood than nonlinear ones; we see in chapter 5 that the only general theory for the nonlinear systems involves approximating them linearly. Thus most of the book is about linear inverse problems. In chapter 2 two general problems are discussed: finding the smallest model in the norm and constructing the smallest model when a finite-dimensional subspace is excluded from penalization (the latter is an example of seminorm minimization). We introduce a highly simplified, one-dimensional marine magnetization problem as the first illustration. More realistic examples follow, including interpolation of irregularly spaced data on a sphere.

In chapter 3 we continue to examine the issue of building model solutions, but now from uncertain data. Once more we cast the problem as one in optimization: finding the smallest model subject to the constraint that the data and the predictions of the linear model are sufficiently close in the sense of the norm. The calculations for the minimum norm model are identical with those arising from what applied mathematicians call regularization of unstable problems; our perspective provides a different and, in my view, clearer motivation for these calculations.

So far we have focused on the question of finding a simple model compatible with the observations; in chapter 4 we turn to the matter of drawing conclusions independent of any particular model. We briefly discuss the idea of resolution—that the incompleteness of our measurements smears out the true solution to an extent that can be calculated. To obtain more quantitative inferences we find it is necessary to add to the observations additional constraints in the model space, like a bound on a norm, or the condition of model positivity. These cases are examined, and to enable calculations to be made with positivity conditions, linear and quadratic programming methods are described. The theory of ideal bodies is introduced as an illustration of the use of a non-Hilbert space norm, in this case the uniform norm. Finally, we give our only example of an application a strong statistical theory, where the model is itself the realization of a statistical process, in this case sea-floor magnetization treated as a stationary stochastic process.

The last chapter deals with nonlinear problems, those with nonlinear relationships between observation and unknown model. As mentioned earlier, general results are essentially nonexistent: we must form approximations that draw upon the linear theory. To do this we need an analog of differentiation for abstract vector spaces: we describe Gateaux and Fréchet differentiation. The general idea continues to be that of minimization of the complexity function subject to the constraint of a good match to observation. Now the search for the minimum is iterative and numerical. We describe a once-popular scheme given by Backus and Gilbert, and a more recent and efficient method called Occam's process. By changing the parameterization we can convert the linear gravity ideal body problem considered in chapter 4 into a nonlinear problem, which we solve by the iterative methods. The other major example of this chapter is that of magnetotelluric sounding, which has the distinction of being one of the few for which the question of existence of solutions can be rigorously decided. In nonlinear systems it is by no means guaranteed that there are any models capable of fitting the data and in general the only way to proceed is to allow the iterative minimization process to search. At the end of the chapter we give the analytic theory for existence and construction of models from magnetotelluric data and compare the results with those found by iteration.

This book has been over ten years in the writing. I have received continuous encouragement from my friends and colleagues at IGPP, particularly Cathy and Steve Constable and Guy Masters. I am also most grateful to the many students who have read early versions and offered valuable advice, notably Loren Shure, Philip Stark, and Mary Kappus. When she was a student Loren Shure and I wrote the graphics computer program *plotxy* which generated all but one of the figures in this book; unlike my matrix arithmetic program, *plotxy* has not yet retired. John Booker provided numerous helpful suggestions and corrections, for which I am grateful. Sara Van Rheenen at Princeton University Press has been ever sympathetic and helpful to this neophyte printer. I would like to thank the Office of Naval Research, which provided a small grant to get this project started. Finally, I wish to express my heart-felt appreciation to my wife, Joan, for her support and understanding and for giving me the final impetus to complete this task.

La Jolla, California

GEOPHYSICAL INVERSE THEORY

CHAPTER 1

MATHEMATICAL PRECURSOR

1.01 Linear Vector Spaces

Most readers will have already encountered linear vector spaces before picking up this book: they are fundamental to much of mathematical analysis. In particular they supply the foundations for optimization theory, which is the branch of mathematics that meets the needs of geophysical inverse theory most closely. Vector spaces (as they are often called for brevity) allow us to visualize abstract relations in simple geometrical terms; this is because a vector space is an abstraction of ordinary space and the members of it can be loosely regarded as ordinary vectors. In a completely general vector space the notion of length has no place, nor does that of the angle between two vectors. We must supply additional structure to the space in order to have these useful properties. We follow the usual practice in building up the required machinery starting from the most general concept, the linear vector space.

The definition of a linear vector space involves two types of object: the *elements* of the space and the *scalars*. Usually the scalars will be the real numbers but quite often complex scalars will prove useful; the reader may assume real scalars are intended unless it is specifically stated otherwise. The elements of the space are much more diverse as we shall see in a moment when we give a few examples. First we lay out the rules that define a *real linear vector space* ("real" because the scalars are the real numbers): it is a set V containing elements which can be related by two operations, addition and scalar multiplication; the operations are written

$$f + g \quad \text{and} \quad \alpha f$$

where $f, g \in V$ and $\alpha \in \mathbb{R}$. For any $f, g, h \in V$ and any scalars α and β, the following set of nine relations must be valid:

$$f + g \ \in V \tag{V1}$$

$$\alpha f \ \in V \tag{V2}$$

$$f + g = g + f \tag{V3}$$

$$f + (g + h) = (f + g) + h \tag{V4}$$

$$f + g = f + h, \quad \text{if and only if } g = h \qquad (V5)$$

$$\alpha(f + g) = \alpha f + \alpha g \qquad (V6)$$

$$(\alpha + \beta)f = \alpha f + \beta f \qquad (V7)$$

$$\alpha(\beta f) = (\alpha\beta)f \qquad (V8)$$

$$1f = f \ . \qquad (V9)$$

In $(V9)$ we mean that scalar multiplication by the number *one* results in the same element. The notation $-f$ means *minus one* times f and the relation $f - g$ denotes $f + (-g)$. These nine "axioms" are only one characterization; other equivalent definitions are possible. An important consequence of these laws (so important, some authors elevate it to axiom status and eliminate one of the others), is that every vector space contains a unique zero element 0 with the properties that

$$f + 0 = f, \quad f \in V \qquad (1)$$

and whenever

$$\alpha f = 0 \qquad (2)$$

either $\alpha = 0$ or $f = 0$. The reader who has not seen it may like to supply the proof of this assertion. A most important real linear vector space is $\mathbb{R}^N$. Members of this set consist of ordered sequences of N real numbers (called N-vectors or N-tuples):

$$\mathbf{x} = (x_1, x_2, x_3, \ \cdots \ x_N) \ . \qquad (3)$$

Scalar multiplication is defined by

$$\alpha\mathbf{x} = (\alpha x_1, \alpha x_2, \alpha x_3, \ \cdots \ \alpha x_N) \qquad (4)$$

and addition by

$$\mathbf{x} + \mathbf{y} = (x_1 + y_1, x_2 + y_2, x_3 + y_3, \ \cdots \ x_N + y_N) \ . \qquad (5)$$

It is easy to verify that, with these rules, $\mathbb{R}^N$ is a vector space. When $N = 3$ we have $\mathbb{R}^3$, the familiar Cartesian representation of ordinary "three-dimensional space." The set of observations pertaining to any geophysical system is usually in the form of an N-tuple of real numbers; therefore the data set will be an element in $\mathbb{R}^N$.

With the normal rules of addition and multiplication by constant real factors, the set of all real-valued functions on an interval forms a real vector space. For most purposes this space is too "large"; this means it includes too many things, and so mathematicians have little use for it. Instead further conditions are introduced and each collection of functions obeying common restrictions is given its own name. For example, $C^n[a,b]$ is the vector space of real-valued functions on the closed interval $[a,b]$ which possess continuous derivatives of all orders up to and including order n. The ordinary definitions of what $f+g$ and αf mean insure the validity of the axioms $(V2)$ through $(V9)$; axiom $(V1)$ follows from the fact that the sum of two functions each of which is continuously differentiable n times is itself differentiable equally often. We denote the space of merely continuous functions on $[a,b]$ by $C^0[a,b]$ and the one containing functions that are differentiable to all orders by $C^\infty[a,b]$. Note that in any of these spaces the element 0 is the constant function that is equal to zero at every point in the interval.

A collection of functions each of which obeys a particular linear differential equation with homogeneous boundary conditions, forms a linear vector space, by virtue of the fact that the sum of two such solutions also satisfies the equation, as does a scalar multiple of any solution. An important geophysical example is the space of *harmonic* functions, that is, those obeying Laplace's equation $\nabla^2 f = 0$. Bounded functions which are harmonic outside the Earth define a space containing the potentials for magnetic and gravitational fields with their sources inside the Earth; the study of harmonic functions is an essential part of geomagnetism and terrestrial gravity. Curiously, this space seems not to have acquired a generally accepted name.

Table 1.01A is a list of linear vector spaces, many of which will be used in this book. A few of the names (like CL_2) will not be found in the standard mathematical literature, because the spaces are not widely used; in naming these spaces I followed some conventions employed by Jacob Korevaar in his little-known but fine introduction to functional analysis misleadingly titled *Mathematical Methods* (1968). At this point the reader may be unfamiliar with most of the terms in the table but all of them will be defined later in the chapter. I hope that the reader may find this table convenient when at some point he or she comes across a named space and cannot quite remember its definition.

Table 1.01A: Some Linear Vector Spaces

Symbol	Description	Remarks
$\mathbb{R}^N$	The set of ordered real N-tuples $(x_1, x_2, \cdots x_N)$	The flagship of the finite-dimensional linear vector spaces
E^N	$\mathbb{R}^N$ equipped with any norm	–
$C^n[a,b]$	The set of functions, continuously differentiable to order n on the real interval $[a,b]$	Not a normed space
$C[a,b]$	$C^0[a,b]$ equipped with the uniform norm $\|f\|_\infty = \max\limits_{a \le x \le b} |f(x)|$	A Banach (complete normed) space
$CL_1[a,b]$	$C^0[a,b]$ equipped with the L_1-norm $\|f\|_1 = \int_a^b |f(x)|\, dx$	An incomplete normed space
$CL_2[a,b]$	$C^0[a,b]$ equipped with the L_2-norm $\|f\|_2 = [\int_a^b f(x)^2\, dx]^{\frac{1}{2}}$; implied inner product $(f,g) = \int_a^b f(x)g(x)\, dx$	An inner product or pre-Hilbert space; this is an incomplete space
$C^n L_2[a,b]$	$C^n[a,b]$ equipped with a 2-norm that penalizes $d^n f/dx^n$	Another pre-Hilbert space
$L_2[a,b]$	The completion of $CL_2[a,b]$; each element is an equivalence class	The flagship Hilbert (complete inner product) space
l_2	The set of infinite ordered real sequences $(x_1, x_2, x_3 \cdots)$ normed by $\|x\| = [\sum\limits_j x_j^2]^{\frac{1}{2}}$	Another Hilbert space
$W_2^n[a,b]$	The completion of $C^n L_2[a,b]$	A Sobolev space (the norm acts on a derivative)

Exercise

1.01(i) Which of the following sets is a linear vector space?
The set of complex numbers with the real numbers as scalars.
The set of real numbers with the rational numbers as scalars.
The set of bounded, continuous functions on the real line with real numbers for scalars.

The set of continuous functions on the real line such that $|f(x)| < M$ with real scalars.

1.02 Subspaces, Linear Combinations, and Linear Independence

Let us now define some useful terminology, doubtless known to most readers. A *linear subspace* of V is a subset of V that forms a linear vector space under the rules of addition and scalar multiplication defined for V. We shall usually drop the adjective and refer simply to a subspace, although there are in mathematics kinds other than linear subspaces, for example, metric subspaces. Examples are easily constructed, for instance, $C^{n+1}[a,b]$ is a subspace of $C^n[a,b]$. The simplest mental picture of a subspace is the one suggested by the example of a straight line or a plane through the origin in ordinary space.

A *linear combination* of elements $f_1, f_2, \cdots f_n$ is any vector of the form

$$\alpha_1 f_1 + \alpha_2 f_2 + \cdots + \alpha_n f_n . \tag{1}$$

The axioms $1.01(V1)$–$(V9)$ and definitions deal only with the addition of elements two at a time, but, because of axiom $1.01(V4)$, the order in which a finite linear combination is assembled has no effect on the final answer; thus the notation is unambiguous. The set of all linear combinations formed from a fixed collection of elements is a subspace of the original space; the fixed elements are said to *span* the subspace.

Elements $f_1, f_2, \cdots f_n$ are *linearly dependent* if it is possible to find a linear combination of them whose value is the zero element and not all the scalars of the combination are zero. Equivalently, when a set of elements is linearly dependent it is then possible to express one of the elements as a linear combination of the others. The converse of linear dependence is of course linear independence; the elements $f_1, f_2, \cdots f_n$ are *linearly independent* if the only linear combination of them that equals 0 is the one in which all the scalars vanish. It follows from this that if $f_1, f_2, \cdots f_n$ are linearly independent and

$$\sum_{j=1}^{n} \alpha_j f_j = \sum_{j=1}^{n} \beta_j f_j \tag{2}$$

then $\alpha_1 = \beta_1, \alpha_2 = \beta_2, \cdots \alpha_n = \beta_n$; in other words, the coefficients are unique in the expansion of an element in terms of a linear combination of linearly independent elements. A useful example of fundamental importance is the linear independence of the powers of x. Consider the space $C^\infty[a,b]$; the powers $1, x, x^2, \cdots x^{n-1}$ are elements of this space. They are linearly independent. To prove this, assume the contrary, that the powers are linearly dependent. Then there are fixed coefficients α_j, not all zero, for which it is true that with every x in $[a,b]$

$$\sum_{j=1}^{n} \alpha_j x^{j-1} = 0 . \tag{3}$$

Suppose the largest power with a nonvanishing coefficient in the above sum is x^k. If $k=0$, the result is obvious nonsense, so we assume $k > 0$; then we can rearrange the sum to give

$$x^k = \sum_{j=1}^{k} (\alpha_j/\alpha_{k+1}) x^{j-1} \tag{4}$$

for all x in $[a,b]$. When this expression is differentiated k times we find $k!=0$, which for $k>0$, is impossible. The contradiction proves that the powers cannot be linearly dependent.

At this point, we should note that the expression of an element as a linear combination of other linearly independent elements in a vector space is one of the most widely used techniques for solving problems in mathematical physics. Expansions in spherical harmonics, Fourier series, normal modes, and power series are all examples of the approach familiar to every geophysicist.

1.03 Bases and Dimension

One obvious but clearly basic subspace of V is V itself. We have already seen that the set of elements formed from all linear combinations of a fixed collection of elements is a subspace; suppose now that the fixed elements are linearly independent and that the subspace which they span is V, the whole space; then those elements form a *basis* for V. If a basis for V can be found and it consists of a finite number of elements, the space is said to be *finite dimensional*; otherwise it is *infinite dimensional*. For a given finite-dimensional space there is more than one basis, but all of them possess the same number of elements; that number is called the *dimension* of the space. The proof of this can be found in any book on linear algebra. Other standard results are that in an N-dimensional space V, every set of $N+1$ or more elements is linearly dependent; that every subspace has dimension less than N, unless the subspace is V itself.

The classical example of an N-dimensional vector space is $\mathbb{R}^N$. This is shown by direct construction of a basis from the N elements

$$\mathbf{e}_1 = (1, 0, 0, \, \cdots \, 0)$$
$$\mathbf{e}_2 = (0, 1, 0, \, \cdots \, 0) \qquad\qquad (1)$$
$$\vdots$$
$$\mathbf{e}_N = (0, 0, 0, \, \cdots \, 1) \, .$$

The linear independence of this set of elements is immediate. They span the space because we can give the value of the scalars needed in the expansion of $\mathbf{x}$: if

$$\mathbf{x} = (x_1, x_2, x_3, \, \cdots \, x_N) \qquad\qquad (2)$$

and we assert that

$$\mathbf{x} = \sum_{j=1}^{N} \alpha_j \, \mathbf{e}_j \, , \qquad\qquad (3)$$

then $\alpha_j = x_j$, $j = 1, 2, 3, \, \cdots \, N$ for any $\mathbf{x} \in \mathbb{R}^N$.

Many of the spaces of functions that we shall meet are infinite dimensional. As an example consider $C^n[a,b]$; again we use the trick of temporarily assuming what is in fact untrue, that $C^n[a,b]$ is finite dimensional with dimension K. Recall that the $K+1$ powers $1, x, x^2, \, \cdots \, x^K$ are linearly independent and they are evidently valid elements of $C^n[a,b]$; but this is impossible since any $K+1$ elements of a K-dimensional space are linearly dependent. The contradiction shows that $C^n[a,b]$ is not K dimensional for any finite value of K.

The astute reader will have noticed that we have evaded the question of defining a basis for infinite-dimensional spaces. Aside from the fact that we shall not use the concept later, the question is more tricky than the one we have treated. While there is only one reasonable way to define a basis for finite-dimensional vector spaces, there are several in the infinite-dimensional case. One course is to define convergence of sequences in V (see 1.08) and to consider sets of elements (spanning sets), from linear combinations of which any element of V can be obtained as a limit. To weed out unnecessary elements in the spanning set one can demand that none of the basis elements is obtainable from the others by this limiting process; this defines a topologically free set. One finds, however, that representation by such a set is not necessarily unique as it is in the finite-dimensional case; an alternative is to define a basis by the uniqueness of the representation (the Schauder basis), but not every Schauder basis is topologically free. See Korevaar (1968) for more on this topic.

1.04 Functionals

We take the view in this book that the observations of a geophysical or geological system are in the form of a finite collection of real numbers; a single datum, that is, a measurement, will be a single real number from this set of observations. When complex quantities are estimated, for example in electromagnetic sounding, they can if necessary be treated as real number pairs. As the reader must have guessed, our mathematical model for the unknown parameters in the geophysical system will be an element in a vector space. Thus we need a notation to describe mathematically the process of taking a measurement. This is where the *functional* comes in: it is a rule that unambiguously assigns a single real number to an element in the space V. Strictly the value of a functional is a scalar; thus for complex spaces, functionals are complex-valued in general. Not every element in V need be connected with a real number by the functional: the subset of V for which values are defined is the *domain* of the functional. Moreover, not every real number may result from the action of a given functional; the subset of $\mathbb{R}$ where results may lie is termed the *range*. If the rule assigns something other than a scalar number to a given element of the space, for example an element of the space or even of a different space, then the terms *mapping, operator,* or *transformation* are appropriate rather than functional. (The word *function* is also used in this context, but if the elements of the space are functions, confusion seems unavoidable.) There is, as most readers will know, a compact way to write where a mapping finds its argument and where it puts the result:

$$F : D \subseteq V \to \mathbb{R} . \qquad (1)$$

Here we have defined a mapping from a subset D of the linear vector space V into the real numbers, and so F is a functional.

Usually we write a functional as an upper-case letter (possibly subscripted) with its argument in square brackets. Here are some simple functionals that will be encountered later on:

$$N_1[\mathbf{x}] = |x_1| + |x_2| \ \cdots \ + |x_N|, \quad \mathbf{x} \in \mathbb{R}^N \tag{2}$$

$$I_j[m] = \int_a^b g_j(x)m(x)\,dx, \quad m \in C^0[a,b] \tag{3}$$

$$D_2[f] = \left.\frac{d^2f}{dx^2}\right|_{x=0}, \qquad f \in C^2[0,1] \tag{4}$$

$$U[f] = \int_{S^2} (\nabla_s^2 f)^2\,d^2\hat{\mathbf{r}}, \qquad f \in C^2[S^2]. \tag{5}$$

In the last functional the set S^2 is the surface of the unit sphere in $\mathbb{R}^3$. Other functionals arising in geophysics may be too complicated to condense into a single simple formula. For example, the time taken for an impulse to pass from a seismic source to a receiver is a functional of the seismic velocity function in the intervening medium; if the velocity varies simply (changing only with depth, for instance) the time can be computed by a simple integral, but otherwise the calculations are very complex, involving the solution of differential equations.

Returning now to purely mathematical questions, we single out two types of functionals of special significance: linear functionals and norms. The second of these is so important to our development and to functional analysis in general that we will devote several sections to its exploration. A *linear functional L* is one which, for any elements f, g in D, its domain of definition, and for any scalars α, β, obeys the rule

$$L[\alpha f + \beta g] = \alpha L[f] + \beta L[g]. \tag{6}$$

In general we must have $\alpha f + \beta g \in D$, so that D is a linear subspace of V. The second example will recur again and again in the coming chapters; it is a linear functional. To show this just plug the definition into the left side of the equation above:

$$I[\alpha f + \beta g] = \int_a^b (\alpha f(x) + \beta g(x))w(x)\,dx \tag{7}$$

$$= \alpha \int_a^b f(x)w(x)\,dx + \beta \int_a^b g(x)w(x)\,dx \tag{8}$$

$$= \alpha I[f] + \beta I[g]. \tag{9}$$

In $\mathbb{R}^N$ the most general linear functional is

$$Y[\mathbf{x}] = y_1 x_1 + y_2 x_2 + \ \cdots \ + y_N x_N. \tag{10}$$

This may be construed as the dot product between two elements both in $\mathbb{R}^N$. It would be natural to write

$$Y[\mathbf{x}] = \mathbf{y} \cdot \mathbf{x}. \tag{11}$$

This functional is an example of an inner product, something we shall be treating in some detail later (section 1.10). The third example, equation (4), is also linear as the reader can easily confirm, but those in (2) and (5) are not. The functional of (2) is an example of a norm, which we study next.

1.05 Norms

The norm provides a means of attributing sizes to the elements of a vector space. The addition of this single property opens up almost unbelievably rich mathematical possibilities. It should be recognized at once that there are many ways to define the size of an element even within the context of a particular vector space. Each of the functions in figure 1.05a belongs to the space $C^\infty[-1, 1]$; which one is the largest? In fact, the point cannot be settled unambiguously because, with the appropriate choice of norm, any of the functions could be said to be the largest, except for f_0; see $(N2)$ below. This apparent arbitrariness is in fact a great virtue: we can often choose a norm with just the right behavior to suit a particular problem. It is time to define these ideas more precisely.

Because the definition of the norm hinges on properties of linear vector spaces, the proper way to introduce the idea is with a special kind of space. A space without a norm can be supplied with one, but then, strictly speaking, it becomes a different space. Pedantically one should give the space a new name if one "equips" it with a norm, although often this renaming is skipped. A *normed vector space* V is a linear vector space in which every element f has a *norm*, written $\|f\| \in \mathbb{R}$. A norm is a real-valued functional satisfying the following four conditions: for every $f, g, h \in$ V and $\alpha \in \mathbb{R}$

$$\|f\| \geq 0 \tag{$N1$}$$

$$\|\alpha f\| = |\alpha| \, \|f\| \tag{$N2$}$$

$$\|f + g\| \leq \|f\| + \|g\| \tag{$N3$}$$

$$\|h\| = 0 \text{ only if } h = 0. \tag{$N4$}$$

If we omit the last condition, the functional is called a *seminorm*.

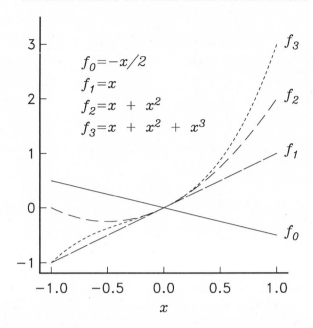

Figure 1.05a: Four continuously differentiable functions. Each may be considered to be an element of the infinite-dimensional linear vector space $C^\infty[-1, 1]$. Any of them can be made the largest by choice of a suitable norm for the space, except for f_0, which must always be smaller than f_1 by the axiom $(N2)$.

Each of the conditions corresponds to a property of length for ordinary vectors; $(N3)$ is called the triangle inequality because of its evident truth in the case of plane triangles. A normed vector space is automatically provided with a measure of the distance between two elements:

$$d(f, g) = \|f - g\|. \tag{1}$$

The distance between two elements, called the *metric*, is more fundamental mathematically than the norm, and can be defined on spaces without norms.

1.06 Some Norms on Finite-Dimensional Spaces

Three important norms for elements in $\mathbb{R}^N$ are

$$\|\mathbf{x}\|_1 = |x_1| + |x_2| + \cdots + |x_N| \tag{1}$$

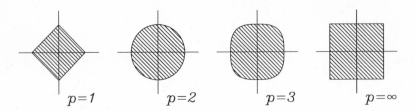

$p=1$ $p=2$ $p=3$ $p=\infty$

Figure 1.06a: The unit ball $\|\mathbf{x}\|_p \leq 1$ in the Euclidean space E^2 for various values of p.

$$\|\mathbf{x}\|_2 = (x_1^2 + x_2^2 + \cdots + x_N^2)^{1/2} \qquad (2)$$

$$\|\mathbf{x}\|_\infty = \max_i |x_i| . \qquad (3)$$

For $\|\mathbf{x}\|_1$ all the conditions $1.05(N1)$–$(N4)$ are immediately evident; the same is true of $\|\mathbf{x}\|_\infty$. The second norm, $\|\mathbf{x}\|_2$ corresponds to the ordinary length of a vector in the sense with which we are most familiar; it is sometimes called the *Euclidean* norm and written $\|\mathbf{x}\|_E$. Conditions $1.05(N1), (N2)$, and $(N4)$ are easily verified for $\|\mathbf{x}\|_2$ but the triangle inequality $1.05(N3)$ presents a little more difficulty. We shall not give the proof at this point because $\|\mathbf{x}\|_2$, like the majority of other norms used in the first three chapters, can be derived from an inner product; in section 1.10 we shall show the triangle inequality holds for all such norms. The three norms above are the most interesting members of a family of p-norms on $\mathrm{I\!R}^N$:

$$\|\mathbf{x}\|_p = (|x_1|^p + |x_2|^p + \cdots + |x_N|^p)^{1/p} . \qquad (4)$$

The result that the triangle inequality is valid when $p \geq 1$ is called *Minkowski's inequality* and appears in most texts on linear vector spaces. It is not hard to show that

$$\|\mathbf{x}\|_\infty = \lim_{p \to \infty} \|\mathbf{x}\|_p . \qquad (5)$$

The normed vector space derived from $\mathrm{I\!R}^N$ by supplying one of these norms is called E^N, irrespective of the particular one used. (In fact, conventional usage is rather sloppy and $\mathrm{I\!R}^N$ is often associated with a norm.)

The set of points defined by $\|\mathbf{x}\| \leq 1$, that is, the points not more than unit distance from the origin, is called the *unit ball* in E^N. The shape of the region depends on the norm in use, of course. Figure 1.06a illustrates the shape of the unit ball under various p-norms in the plane E^2.

The reader will be assumed to be familiar with the elementary properties of matrices, including the simple eigenvalue properties; useful accounts at the appropriate level include those of Strang (1980) and Noble and Daniel (1977). We therefore take this opportunity to discuss matrix norms. The usual way to regard a matrix is as a linear mapping:

$$A : \mathbb{R}^n \to \mathbb{R}^m \tag{6}$$

is a real matrix that maps vectors in $\mathbb{R}^n$ into vectors in $\mathbb{R}^m$ and can be written as an array of real numbers of n columns each one m in length. The set of all such matrices with m rows and n columns can also be seen as a linear vector space which we shall call $M(m \times n)$ of dimension mn using the rules of matrix arithmetic to define addition and scalar multiplication. An obvious class of norms is the one that treats the elements as a collection of numbers ignoring the special ordering; then any of the norms for a vector of length mn would work. An example of such a norm is the *Frobenius* norm, $\|A\|_F$, which is simply the Euclidean norm of the matrix entries.

A more powerful way of equipping the space of matrices with a norm is to give norms to the range and domain spaces of the mapping and to rely upon them for a measure of size of A. Suppose we say now

$$A : E^n \to E^m . \tag{7}$$

We define a *subordinate matrix norm* by

$$\|A\| = \max_{\|\mathbf{x}\|=1} \|A\mathbf{x}\| . \tag{8}$$

Thus we calculate the maximum possible magnification of the length of a vector after the action of A. (To be rigorous here, we should replace the maximum with the least upper bound in case the greatest value cannot be achieved with any vector; there is no need for the distinction with any of the norms that we consider. We remark in passing that this is a method used to assign a norm to linear mappings in infinite-dimensional spaces too, where the supremum rather than the maximum would be essential.) All the conditions 1.05(N1)–(N4) are automatically satisfied as well as

$$\|AB\| \le \|A\| \, \|B\| , \tag{9}$$

a property conventionally demanded of all matrix norms. We assume the same type of norm is in use in both E^m and E^n; then the matrix norm simply takes its name from the one in the spaces of vectors—thus $\| x \|_1$ gives rise to $\| A \|_1$, etc. Explicit calculation of $\| A \|_1$ is simple: one finds

$$\| A \|_1 = \max_j \| \mathbf{a}_j \|_1 \tag{10}$$

where $\mathbf{a}_j$ is the jth column of the matrix considered as an m-vector. Similarly for $\| A \|_\infty$ with the maximum 1-norm of the *rows* of A instead. The 2-norm is much more difficult to evaluate: it is given by

$$\| A \|_2 = [\text{ maximum eigenvalue of } (A^T A)]^{\frac{1}{2}} . \tag{11}$$

This prescription follows quite simply from the extremal property of eigenvalues. The 2-norm is also called the *spectral norm*. For more on spectral representations of matrices see 3.03.

The chief interest in matrix norms stems from their applications in numerical analysis. These days all serious data analysis makes use of digital computers, which carry out their calculations with only limited precision. Matrix manipulation is central to many calculations, including those of geophysical inverse theory, and therefore an understanding of the effects of small errors in the representation of matrix problems and in their numerical solution is vital. As a simple illustration, we briefly mention the concept of the *condition* of a system of linear equations. Consider the solution for **x** of the system

$$A \mathbf{x} = \mathbf{b} \tag{12}$$

where $\mathbf{x}, \mathbf{b} \in E^n$, $A \in M(n \times n)$, and $A, \mathbf{b}$ are both known and A is nonsingular. Computers cannot store an arbitrary real number exactly, so that (unless A and **b** are exact binary fractions, with sufficiently small magnitudes and denominators) a small error is committed just representing the problem internally in the machine: the actual problem being solved is then

$$A' \mathbf{x}' = \mathbf{b}' \tag{13}$$

where A' and $\mathbf{b}'$ are the computer's approximations for the original matrix and the right side of (12). Even if subsequent arithmetic is done exactly, the perturbations in A and **b** cause $\mathbf{x}'$, the exact solution to (13) to be different from **x**, the exact solution to (12) and presumably what is sought. In most scientific computers the arithmetic operations are carried out to high accuracy then the result is rounded or truncated before it is stored, so that the precision of storage, not of the arithmetic, limits the reliability of the final result. Suppose, for a moment that $\mathbf{b}'=\mathbf{b}$; then it can be shown that

$$\frac{\|\mathbf{x}'-\mathbf{x}\|}{\|\mathbf{x}'\|} \leq \|A\| \cdot \|A^{-1}\| \cdot \frac{\|A'-A\|}{\|A\|} = \kappa \frac{\|A'-A\|}{\|A\|} \tag{14}$$

where

$$\kappa = \text{cond}(A) = \|A\| \cdot \|A^{-1}\|. \tag{15}$$

The number κ is called the *condition number* of the matrix A. This inequality shows how far in the norm the computer solution can depart from the true one just because of small errors of representation of the numbers. The corresponding result for errors in $\mathbf{b}$ is

$$\frac{\|\mathbf{x}'-\mathbf{x}\|}{\|\mathbf{x}\|} \leq \kappa \frac{\|\mathbf{b}'-\mathbf{b}\|}{\|\mathbf{b}\|}. \tag{16}$$

It is easily shown that $\kappa \geq 1$, and as we shall see later κ may be very large for some linear systems. Suppose $\kappa = 10^7$ and the relative accuracy of the computer's floating-point word is a few parts in 10^7 (a typical figure for a 32-bit machine); then there is no way of guaranteeing that $\mathbf{x}'$ will bear any relationship to the true $\mathbf{x}$.

Exercises

1.06(i) Show that the functional on E^N with $N > 1$ defined by the relation

$$F[\mathbf{x}] = (|x_1|^{1/2} + |x_2|^{1/2} + \cdots |x_N|^{1/2})^2$$

does *not* define a norm for the space. Sketch the region $F[\mathbf{x}] \leq 1$ for $N = 2$ and compare it with the unit balls of figure 1.06a.

1.06(ii) Show from the definition (8) that whenever $\|\cdot\|$ is a norm on the space of vectors E^n, E^m, that the corresponding subordinate matrix norm is a norm for $M(m \times n)$; that is, it obeys 1.05(N1)–(N4).
Calculate the matrix subordinate norms for $p = 1, 2, \infty$ of the matrix

$$A = \begin{bmatrix} 2.8 & -0.4 & -1 \\ -0.4 & 2.2 & -2 \\ -1 & -2 & -2 \end{bmatrix}.$$

Find the condition number under the different norms. Give a unit vector that causes the linear transformation to achieve the greatest magnification under each of the norms. See also exercise 1.14(i).

1.06(iii) If $A \in M(N \times N)$ is symmetric and positive definite, show that under the spectral norm

$$\text{cond}(A) = \lambda_{max}/\lambda_{min}$$

where λ_{max} and λ_{min} are respectively the largest and smallest eigenvalues of A.

1.07 Some Norms on Infinite-Dimensional Spaces

Consider the vector space $C^0[a, b]$, the space of continuous functions on the closed interval $[a, b]$. Just as in $\mathbb{R}^N$, we may define the following functionals all of which exist for an arbitrary element of the space:

$$\|f\|_1 = \int_a^b |f(x)| \, dx \tag{1}$$

$$\|f\|_2 = [\int_a^b f(x)^2 \, dx]^{1/2} \tag{2}$$

$$\|f\|_\infty = \max_{a \le x \le b} |f(x)|. \tag{3}$$

Verification of the norm conditions is trivial except for the triangle inequality with $\|f\|_2$; again we defer the proof to 1.10 because this norm derives from an inner product. Each of the three norms generates a normed vector space from $C^0[a, b]$. The third is the most important and is called $C[a, b]$, the space of bounded continuous functions on $[a, b]$. The norm $\|f\|_\infty$ is referred to as the *uniform norm* for the following reason: if we are given $g, f_1, f_2, \cdots \in C[a, b]$ and the sequence of distances $\|f_k - g\|_\infty$ tends to zero as $k \to \infty$, then the sequence of functions f_k is said to converge to g *uniformly*; uniform convergence is the backbone of classical analysis. We shall shortly be discussing convergence in more general terms.

While the two norms $\|f\|_1$ and $\|f\|_2$ are of the utmost importance, the normed spaces they create from $C^0[a, b]$, called $CL_1[a, b]$ and $CL_2[a, b]$ are of only minor significance: they are actually subspaces of the more fundamental normed spaces $L_1[a, b]$ and $L_2[a, b]$. The more important spaces can be derived from $CL_1[a, b]$ and $CL_2[a, b]$ by an enlargement process called completion; see section 1.09.

All three norms above measure the size of f fairly directly, either by seeking its peak magnitude or by averaging in some way. Provided that we start in a space of smooth functions, say $C^2[a,b]$, we may incorporate derivative information into the norm; then the norm is influenced by the roughness of the function. Denoting derivatives of f by $f'(x)$, $f''(x)$, we can define

$$\|f\|_s = [\int_a^b (w_0(x)f(x)^2 + w_1(x)f'(x)^2)\, dx\,]^{\frac{1}{2}} \tag{4}$$

$$\|f\|'' = [f(a)^2 + f'(a)^2 + \int_a^b f''(x)^2\, dx\,]^{\frac{1}{2}}. \tag{5}$$

Provided that $w_0 > 0$ and $w_1 \geq 0$ in $\|f\|_s$ the norm condition $1.05(N4)$ is satisfied; again both norms may be derived from appropriate inner products so that the triangle inequality is obeyed. The first norm is a simple example of a Sobolev norm which in general contains a weighted sum of squares of derivatives. The functional is associated with a differential equation

$$(w_1 f')' - w_0 f = 0 \tag{6}$$

for which it provides a variational principle. Generalized to functions in more than one independent variable, the norm and the variational principle play a central role in the theory of partial differential equations. At first glance, $\|f\|''$ presents difficulties in complying with $(N4)$. One version of Taylor's Theorem for functions in $C^2[a,b]$ states for each x on $[a,b]$ there is a y on the interval such that

$$f(x) = f(a) + (x-a)f'(a) + \frac{(x-a)^2}{2}f''(y). \tag{7}$$

Thus if f'' vanishes everywhere on $[a,b]$ and $f(a)=f'(a)=0$, it follows that $f(x)$ vanishes identically too; so whenever $\|f\|''$ is zero, f is the zero function, exactly as $(N4)$ requires.

Even more than $\|f\|_s$, the norm $\|f\|''$ is increased by wiggliness of the function f. Later on we shall be seeking solutions to problems using functions that minimize a specific norm; the function minimizing $\|f\|''$ tends to be very smooth. For example, the minimization of $\|f\|''$ with certain modifications leads to a solution of the one-dimensional interpolation problem via cubic splines.

Our final example is another inner product norm: for $f \in C^0[a,b]$ define

$$\|f\|_I = [\int_a^b \{\int_a^x f(t)dt\}^2 dx]^{1/2}. \tag{8}$$

Clearly this functional is very insensitive to the "high frequency" behavior of f because we have integrated twice. For future reference the normed space connected with $C^0[a,b]$ and $\|f\|_I$ will be named $CI[a,b]$; when we need this kind of norm (in section 4.02) our interest will center on the bigger space obtained from $C^0[a,b]$ by completion.

Exercises

1.07(i) Consider the three elements $f_1, f_2, f_3 \in C^\infty[-1, 1]$ defined by

$$f_1(x) = x$$
$$f_2(x) = x + x^2$$
$$f_3(x) = x + x^2 + x^3.$$

These functions are depicted in figure 1.05a. Find three norms for the space with these properties

$$\|f_1\|_a > \|f_2\|_a, \|f_3\|_a$$
$$\|f_2\|_b > \|f_1\|_b, \|f_3\|_b$$
$$\|f_3\|_c > \|f_1\|_c, \|f_2\|_c$$

1.07(ii) Equip the space $C^\infty[0, 1]$ with a norm such that the function e^x is larger in the norm than e^{2x}.

1.07(iii) Let L define a *linear mapping* of the normed space V onto itself; thus for every $f, g \in V$ and $\alpha, \beta \in \mathbb{R}$, $L(\alpha f + \beta g) = \alpha Lf + \beta Lg$. Show $\|Lf\|$ is always a seminorm of f. When is it a true norm?

1.08 Convergence, Cauchy Sequences, and Completeness

In a normed vector space an infinite sequence of elements $f_1, f_2, f_3, \cdots$ is said to *converge* to the element g if as $k \to \infty$, $\|f_k - g\| \to 0$. Precisely in the same way that norms measure different aspects of the size of f, a given sequence may be convergent under one norm, but not under another. If a sequence converges, the members f_k tend to get closer and closer together; by 1.05($N3$)

$$\|f_k - f_j\| = \|f_k - g - f_j + g\| \tag{1}$$

$$\leq \| f_k - g \| + \| f_j - g \| \to 0 \tag{2}$$

as j and k both grow large. Any sequence in which $\| f_k - f_j \| \to 0$ as $k, j \to \infty$ is called a *Cauchy sequence*, and we have just seen that every convergent sequence is Cauchy. Is it true that every Cauchy sequence converges to an element in the normed space? Somewhat surprisingly, at least on first encounter, the answer is no. A normed vector space is said to be *incomplete* if there are Cauchy sequences in it that do not converge to an element. Conversely, of course, a space is *complete* if every Cauchy sequence converges to an element of the space; complete normed linear vector spaces are called *Banach spaces*.

The most familiar example of a complete normed space is $\mathbb{R}$, the real numbers (with themselves as scalars!). It is a well known but deep property of the real number system that every Cauchy sequence in it tends to a real limit. This fact makes it easy to prove that the norms of the elements of a Cauchy sequence in any normed vector space always tend to a definite limit. While this is a necessary condition for a sequence to be Cauchy, it is certainly not sufficient.

Let us examine some Cauchy sequences that illustrate incompleteness. Recall $CL_1[a, b]$, the space of continuous functions normed by $\| f \|_1$; we take the interval to be $[-1,1]$ and define $f_1, f_2, \cdots$ by

$$f_k(x) = \begin{cases} 0, & x \leq 0 \\ k^2 x^2, & 0 < x \leq 1/k \\ 1, & 1/k < x \ . \end{cases} \tag{3}$$

These are continuous functions (see figure 1.08a). We calculate the separation between two elements f_j, f_k; assume $j \leq k$, then $f_j(x) \leq f_k(x)$ and

$$\| f_k - f_j \|_1 = \int_{-1}^{1} |f_k(x) - f_j(x)| \, dx \tag{4}$$

$$= \int_{-1}^{1} [f_k(x) - f_j(x)] \, dx \tag{5}$$

$$= \int_{-1}^{0} + \int_{0}^{1/k} + \int_{1/k}^{1/j} + \int_{1/j}^{1} \tag{6}$$

$$= 0 + \int_{0}^{1/k} (k^2 x^2 - j^2 x^2) dx + \int_{1/k}^{1/j} (1 - j^2 x^2) \, dx + 0 \tag{7}$$

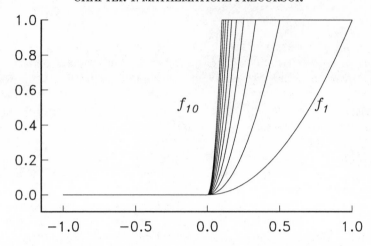

Figure 1.08a: The first ten members of an infinite sequence $f_1, f_2, f_3, \cdots$ of continuous functions belonging to the normed space $CL_1[-1, 1]$. The sequence is Cauchy under the norm of the space but does not converge to an element of the space because the limiting member is evidently discontinuous. Hence the space is incomplete.

$$= \frac{2}{3} \left(\frac{1}{j} - \frac{1}{k} \right). \tag{8}$$

Obviously if $j > k$ we may interchange them in the formula, and so in general

$$\|f_k - f_j\|_1 = \frac{2}{3} \left| \frac{1}{j} - \frac{1}{k} \right|. \tag{9}$$

Thus, as j and k grow large, the distance between f_j and f_k shrinks in the 1-norm, and the sequence is Cauchy in $CL_1[-1,1]$. However, the sequence is plainly converging to an end member whose value is zero on the closed interval $[-1,0]$ and unity on $(0,1]$; this cannot be a continuous function. The Cauchy sequence is not converging to an element in $CL_1[-1,1]$ which contains only continuous functions. The space is incomplete.

We remark that in the space $C[-1,1]$ (recall this is the space of continuous functions normed by the uniform norm) the same sequence is not Cauchy. The reader may easily verify that

$$\|f_k - f_j\|_\infty = 1 - j^2/k^2, \quad j \le k. \tag{10}$$

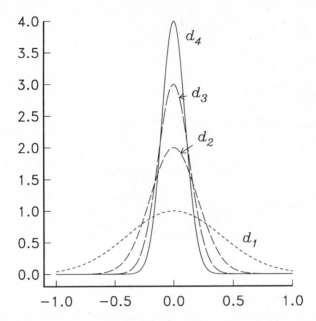

Figure 1.08b: Four members of an infinite sequence $d_1, d_2, d_3, \cdots$ of elements in the normed space $CI[-1, 1]$. Again the sequence is Cauchy but does not converge to any conventional function; once more this implies the space is incomplete. The limiting element in a completed space corresponds to the infamous Dirac delta function.

Note that in this case $\|f_k\| \to 1$ but the existence of a limit for the norm of the sequence is not by itself reason for the sequence to be Cauchy.

The second example has bearing upon a question we shall see again in chapter 4, that of finding normed spaces which admit delta functions as elements. On the space $CI[-1,1]$, defined in section 1.07, consider the family of continuous functions $d_1, d_2, d_3, \cdots$ given by

$$d_k(x) = k\, e^{-\pi k^2 x^2}, \quad k = 1, 2, 3, \cdots . \tag{11}$$

These functions become narrower and taller with increasing k (see figure 1.08b); more precisely, as $k \to \infty$ for any fixed $x \neq 0$, $d_k(x) \to 0$; but $d_k(0) = k$. It can be established that, if $g \in C^1[-1,1]$ (so that g is a lot smoother than the general element of CI)

$$\int_{-1}^{1} g(x)\, d_k(x)\, dx \;\to\; g(0)\,, \qquad k \to \infty\,. \tag{12}$$

As a special case of this result we have

$$\int_{-1}^{1} d_k(x)\, dx \;\to\; 1\,. \tag{13}$$

No continuous function (nor even any kind of ordinary function) can exhibit the properties of the putative end member of the sequence. However, in CI[−1, 1] the sequence is a Cauchy sequence. To show this we calculate

$$\|d_k - d_j\|_I^2 = \int_{-1}^{1} \Big[\int_{-1}^{x} (d_k(t) - d_j(t))\, dt \Big]^2 dx\,. \tag{14}$$

The evaluation is easier if the range of integration is extended; clearly

$$\|d_k - d_j\|_I^2 \leq \int_{-\infty}^{\infty} \Big[\int_{-\infty}^{x} (d_k(t) - d_j(t))\, dt \Big]^2 dx\,. \tag{15}$$

Using integration by parts twice, we can evaluate the expression exactly but the reader is spared the irksome algebra:

$$\|d_k - d_j\|_I^2 \leq \frac{1}{\pi\sqrt{2}} \Big[\Big\{ \frac{2}{k^2} + \frac{2}{j^2} \Big\}^{\frac{1}{2}} - \frac{1}{k} - \frac{1}{j} \Big]\,. \tag{16}$$

This inequality proves that $\|d_k - d_j\|_I$ tends to zero if j and k both grow without bound. Therefore $d_1, d_2, d_3, \cdots$ is a Cauchy sequence. This fact, together with the previously noted one that no element of CI[−1, 1] can be the limiting member, demonstrates that the space is incomplete.

The sequence $d_1, d_2, d_3, \cdots$ is not a Cauchy sequence in C[−1, 1] since $\|d_k - d_j\|_\infty = |k - j|$; nor is it in CL$_1$[−1, 1] but the calculation is fairly messy.

Some spaces are of course complete. The following results are to be found in any book on functional analysis. Every finite-dimensional space is complete; thus the idea of incompleteness has no counterpart in ordinary space. The space C[a, b], that is, the set of continuous functions under the uniform norm, is also complete. Associated with every incomplete space is a complete one, obtained by "filling in the gaps"; this is the topic of the next section.

Exercises

1.08(i) The three elements of $C^\infty[-1, 1]$ in exercise 1.07(i) may be seen as the first three of an infinite sequence

$$f_k(x) = \sum_{n=1}^{k} x^n, \quad k = 1, 2, 3, \cdots .$$

Give $C^\infty[-1, 1]$ a norm under which this sequence is Cauchy.
Hint: What function does the sequence "converge" to?

1.08(ii) Consider the sequence d_j of figure 1.08b in the $CL_2(-\infty, \infty)$. Show that $\|d_j\|_2 \to \infty$ as j increases. Can the sequence $d_1, d_2, d_3, \cdots$ be Cauchy?

1.09 Completion

Incomplete spaces are awkward places to perform mathematical analysis: often the solution to a problem can be formulated as a (Cauchy) convergent sequence (for example, by an "infinite sum") and it is troublesome to discover the result does not exist in the space where one started. The dilemma is sidestepped by the idea of *completion*: for a normed vector space N we may define a complete normed vector space $\hat{N}$ in which every Cauchy sequence of N has a limit element in $\hat{N}$. In this way we enlarge the space to capture formerly inadmissible objects. Indeed, the enlargement process is a powerful way to define new abstract kinds of elements which cannot easily be handled with classical analysis (e.g., delta functions). The mathematical ideas are as follows. We write a Cauchy sequence $f_1, f_2, f_3, \cdots$ in N as $\{f\}$; two Cauchy sequences $\{f_k\}$, $\{\tilde{f}_k\}$ are said to be *equivalent* if

$$\|f_k - \tilde{f}_k\|_N \to 0 \quad \text{as} \quad k \to \infty \tag{1}$$

where $\|\cdot\|_N$ denotes the norm of N. The completion $\hat{N}$ of N consists of elements $\hat{f}$ each of which is the set of all Cauchy sequences equivalent to a particular sequence in N. The elements are called *equivalence classes*. Having elements of $\hat{N}$ that are *sets of sequences* in N seems a little abstract at first, but we have all the machinery we need to handle it. If $\{f_k\} \in \hat{f} \in \hat{N}$, then $\hat{f}$ is the set of all Cauchy sequences in N equivalent to $\{f_k\}$. We see that $\hat{N}$ is a vector space if we define addition and scalar multiplication in the natural way:

$$\{f_k + g_k\} \in \hat{f}_k + \hat{g}_k , \quad \alpha f_k \in \alpha \hat{f} \tag{2}$$

where

$$\{f_k\} \in \hat{f} \in \hat{N}, \quad \{g_k\} \in \hat{g} \in \hat{N}, \quad \alpha \in \mathbb{R}. \tag{3}$$

It is also a normed space if we define

$$\|\hat{f}\|_{\hat{N}} = \lim_{k \to \infty} \|f_k\|_N \tag{4}$$

where $\{f_k\} \in \hat{f}$. It is possible to show that $\hat{N}$ is a complete space from these definitions.

Obviously every element of N generates a corresponding element in $\hat{N}$ by sequences like $f_1, f_2, f_3, \cdots$, where $f_k = f$ for all k, and we can think of N as a subspace of $\hat{N}$. There is a certain amount of circumlocution involved in referring to elements of $\hat{N}$ as sets of equivalent sequences (not to mention the mental strain of thinking this way!); in ordinary usage, one fastens upon a representative element of N (if one exists) for a reference. To make this point clearer let us discuss $L_1[a,b]$ the completion of $CL_1[a,b]$. We know from 1.08 that sequences leading to discontinuous end members are admitted, and indeed so are those tending to unbounded functions provided they are integrable. The reason why one cannot rigorously regard the *discontinuous function*

$$g(x) = \begin{cases} 1, & x \geq 0 \\ 0, & x < 0 \end{cases} \tag{5}$$

as an element of $L_1[-1,1]$ is that other mathematically different functions are equivalent to it, for example,

$$\tilde{g}(x) = \begin{cases} 1, & x > 0 \\ 0, & x \leq 0, \end{cases} \tag{6}$$

or

$$\tilde{\tilde{g}}(x) = \begin{cases} 1, & x > 0 \\ \frac{1}{2}, & x = 0 \\ 0, & x < 0, \end{cases} \tag{7}$$

and so on. Their Cauchy sequences are equivalent since $\|g_k - \tilde{g}_k\|_1 \to 0$; the 1-norm cannot "see" the difference between g_k and $\tilde{g}_k$ because discrepancies at the single point $x = 0$ are lost by the integration. Every "function" in $L_1[a,b]$ is joined by a family of others, differing from it at disconnected points in such a way that

$$\| f - \tilde{f} \|_1 = 0 . \tag{8}$$

Nonetheless, we usually just refer to a function $f \in L_1[a, b]$; the other members of the ghostly horde are there by implication.

A different mathematical approach to the definition of L_1 is the one using the theory of measure and Lebesgue integration. In that development elements are indeed sets of functions rather than sets of sequences of functions, which on the surface makes it more appealing. An element of L_1 is defined to be the set of all Lebesgue integrable functions differing from each other at a set of argument points with measure zero (a different example of an equivalence class). No attempt will be made here to define these terms, which form the foundation of real analysis. Functions differing on a set of zero measure are said to be identical *almost everywhere*; it will be understood that the 1-norm of the difference of two such functions vanishes.

For most purposes, and certainly for all the purposes of this book, one may think of $\hat{N}$ as N with something added to make it complete. This mode of thinking is facilitated, and indeed made rigorous by setting up mathematical problems in a *pre-space*, a normed vector space whose completeness is not examined but whose elements are familiar functions with desirable properties; only when the stage is set is the space completed to $\hat{N}$.

1.10 Inner Products

The norm is the mathematical abstraction of the length of a vector; the inner product is the generalization of the dot product. We can give meaning with this concept to the angle between two elements and to orthogonality, the importance of which will be seen in the next section. A natural norm comes with the inner product, the kind that sums or integrates over the square of the element; in this section we demonstrate the validity of the triangle inequality for this whole class of norms.

A real *inner product space* I (or more grandly, a real *pre-Hilbert space*) is a real linear vector space in which there is defined for every pair of elements f, g a functional, the *inner product* (f, g), with these properties: if $f, g, h \in$ I and $\alpha \in \mathbb{R}$:

$$(f,g) = (g,f) \tag{I1}$$

$$(f+g,h) = (f,h) + (g,h) \tag{I2}$$

$$(\alpha f, g) = \alpha(f,g) \tag{I3}$$

$$(f,f) > 0, \quad \text{if } f \neq 0. \tag{I4}$$

Condition $(I3)$ with $g = \alpha f$ and $\alpha = 0$ shows that $(0,0) = 0$, so that from $(I4)$ we have $(f,f) = 0$ if and only if $f = 0$.

We remark in passing that in a complex inner product space (f,g) is a complex number and that $(I1)$ reads $(f,g) = (g,f)^*$. Complex inner product spaces are sometimes called *unitary spaces*. Several other notations appear in the literature for (f,g), including $<f,g>$, $(f \mid g)$, and in quantum mechanics $<f \mid g>$.

For simplicity we shall treat real inner product spaces. It follows directly from the axioms that *Schwarz's inequality* is obeyed:

$$|(f,g)| \leq (f,f)^{1/2}(g,g)^{1/2} \tag{1}$$

and equality holds if and only if $g = 0$ or $f = \alpha g$. Here is the proof. When $g = 0$ equality is obvious; now examine the case $g \neq 0$. We have seen that

$$(f - \alpha g, f - \alpha g) \geq 0, \tag{2}$$

with equality if and only if $f = \alpha g$. From $(I1)$–$(I3)$ we can show

$$(f - \alpha g, f - \alpha g) = (f,f) - 2\alpha(f,g) + \alpha^2(g,g) \tag{3}$$

and therefore

$$0 \leq (f,f) - 2\alpha(f,g) + \alpha^2(g,g). \tag{4}$$

Choose $\alpha = (f,g)/(g,g)$, which exists because $(g,g) > 0$; then direct substitution yields

$$0 \leq (f,f) - (f,g)^2/(g,g) \tag{5}$$

$$(g,g)(f,f) \geq (f,g)^2 \tag{6}$$

$$\|g\|^2 \|f\|^2 \geq (f,g)^2 \tag{7}$$

and Schwarz's inequality follows at once.

Every inner product space comes equipped with the norm

$$\|f\| = (f,f)^{1/2}. \tag{8}$$

A check of 1.05$(N1)$–$(N4)$ quickly reveals that only 1.05$(N3)$ needs any work; the proof of the triangle inequality for these norms is something we have been waiting for. Consider

$$\|f + g\|^2 = (f + g, f + g) \tag{9}$$

$$= (f, f) + 2(f, g) + (g, g) \tag{10}$$

$$= \|f\|^2 + 2(f, g) + \|g\|^2 \tag{11}$$

$$\leq \|f\|^2 + 2|(f, g)| + \|g\|^2 . \tag{12}$$

By Schwarz's inequality

$$\|f + g\|^2 \leq \|f\|^2 + 2\|f\| \, \|g\| + \|g\|^2 \tag{13}$$

or

$$\|f + g\|^2 \leq (\|f\| + \|g\|)^2 . \tag{14}$$

The square root yields the triangle inequality for norms based upon inner products.

Examples of inner product spaces have already been given earlier in the discussion of normed spaces. Whenever the square $\|f\|^2$ is some sort of quadratic functional of f we can easily see the inner product lurking there. For instance, in 1.06, the inner product associated with $\|\mathbf{x}\|_2$ on E^N is the *dot product*

$$(\mathbf{x}, \mathbf{y}) = \mathbf{x} \cdot \mathbf{y} \tag{15}$$

$$= x_1 y_1 + x_2 y_2 + \cdots + x_N y_N . \tag{16}$$

A useful generalization of this is an inner product connected with $A \in \mathrm{M}(N \times N)$ where A is a positive definite and symmetric matrix:

$$(\mathbf{x}, \mathbf{y})_A = \mathbf{x} \cdot (A\mathbf{y}) . \tag{17}$$

The reader should know that a *symmetric matrix* is one with the property $A^T = A$ and a *positive definite matrix* satisfies $\mathbf{x} \cdot A\mathbf{x} > 0$ for all nonzero $\mathbf{x}$. Let us first show that $(I1)$, the commutative property for inner products on real pre-Hilbert spaces, is satisfied; this gives us the chance to remind the reader of the isomorphism between vectors and column matrices and a number of other elementary rules of matrix algebra. An ordered N-tuple of real numbers can be thought of as an element of a vector space in several ways: as an ordinary vector, $\mathbf{x} \in \mathrm{E}^N$; as a single-column matrix, $x \in \mathrm{M}(N \times 1)$; or as a row matrix, which we shall write as a transposed column matrix, $x^T \in \mathrm{M}(1 \times N)$. It is very valuable to move between these alternative forms when performing matrix and vector manipulations. For instance, it is easily seen by expansion into components that

$$\mathbf{x} \cdot \mathbf{z} = x^T z = z^T x \ . \tag{18}$$

We mention two other (it is to be hoped) familiar rules of matrix algebra. For $A \in M(M \times N)$ and $B \in M(N \times K)$, so that the product AB is defined, $(A B)^T = B^T A^T$; again this is easily verified by expansion into components. For any pair of square nonsingular (that is, invertible) matrices A and B, it is always true that $(A B)^{-1}$ exists and equals $B^{-1}A^{-1}$. With the first of these properties in mind, we can now verify (I 1) for the generalized inner product of $\mathbf{E}^N$ as follows:

$$(\mathbf{x}, \mathbf{y})_A = x^T (A \ y) = (A \ y)^T x \tag{19}$$

$$= y^T A^T x \tag{20}$$

$$= y^T A \ x = (\mathbf{y}, \mathbf{x})_A \tag{21}$$

where we have recognized that $A = A^T$, in other words that A is symmetric. To obtain (I 4), the other nontrivial property of an inner product, we observe that it is just the requirement that the matrix A be positive definite.

Moving next to infinite-dimensional spaces, we recall from 1.07 the most famous perhaps of all inner products of functions comes from $CL_2[a, b]$ and $\|f\|_2$

$$(f, g) = \int_a^b f(x)g(x) \ dx \ . \tag{22}$$

It should be obvious to the reader by now how to write the inner products for $\|f\|_s$ and the other norms of section 1.07. The inner product associated with $\|f\|''$ will be of considerable interest to us in section 2.07.

Exercises

1.10(i) In the following functional, obviously suggested by Schwarz's inequality, g is a fixed element of an inner product space I. Show that the functional $G^{\frac{1}{2}}$ is a seminorm on I.

$$G[f] = (g, g)(f, f) - (g, f)^2$$

1.10(ii) Prove Schwarz's inequality for complex inner product spaces. Notice, for example, that it is no longer true that $(f, \alpha g) = \alpha(f, g)$ and modify the proof accordingly.

1.11 Hilbert Space

A *Hilbert space* H is an inner product space that is complete under the norm $(f, f)^{1/2}$. An incomplete inner product space can always be completed in the regular way; the inner products in the Hilbert space may be defined by sequences which can be shown to converge; hence every pair of elements in the completed space has a valid inner product. For the applications in this book the question of whether the inner product space is complete (that is, Hilbert) or not can usually be avoided because most of the results we need do not depend upon completeness.

Every finite-dimensional normed space is complete, and therefore E^N under $\|\mathbf{x}\|_2$ is a Hilbert space. The simplest infinite-dimensional Hilbert space is l_2 whose elements are the infinite sequences $x = (x_1, x_2, x_3, \cdots)$ such that $\sum x_j^2$ converges. The norm is the obvious extension of the 2-norm of E^N to the case with $N = \infty$. Another classic example of a Hilbert space is $L_2[a, b]$ which we have arrived at by completing $CL_2[a, b]$, and invested it with the inner product

$$(f, g) = \int_a^b f(x) g(x) \, dx . \tag{1}$$

We may usually regard elements in $L_2[a, b]$ as functions with bounded 2-norms, or *square-integrable* functions; but from our discussion in 1.09 we should keep in mind that each element is a set of functions (an equivalence class). The difference between any two of the functions in the equivalence class is a function that vanishes at almost every point and indeed the 2-norm of the difference is zero. The space $L_2[a, b]$ often seems to be a very natural space in which to perform the operations of geophysical inverse theory but this appearance can be deceptive. One reason is that the space is too "large": it admits functions that may be arbitrarily discontinuous or unbounded but it is usually wise to deny solutions of physical problems such freedom. Almost always there is a better Hilbert space comprised of smoother solutions in which the identical mathematical machinery applies and in this space more satisfactory results are frequently obtained. Our favorite spaces for building models will be W_2^1 and W_2^2 which are the completed versions of C^1L_2 and C^2L_2, spaces comprising relatively smooth functions because derivatives enter the norm. The diversity of Hilbert spaces and the variety of possible norms is something the reader should remember.

We present several theorems for Hilbert spaces which will not be proved. First is the *Riesz Representation Theorem*. Consider a fixed element l of a Hilbert space H; then the inner product of (l,f) from 1.10(I2) and 1.10(I3) is a linear functional of f for any $f \in$ H (see 1.04). Let us write

$$L[f] = (l, f). \tag{2}$$

Moreover, L is a *bounded linear functional*, which means that

$$|L[f]| < c \|f\| \tag{3}$$

where $c \in \mathbb{R}$ and $0 \le c < \infty$. This follows from Schwarz's inequality:

$$|L[f]| = |(l, f)| \le \|l\| \|f\|. \tag{4}$$

Recall that we often use functionals to represent measurements made on physical systems and we shall rarely be concerned with unbounded functionals. The converse of the above elementary result is Riesz's Theorem for Hilbert spaces: every bounded linear functional $L[f]$ on a Hilbert space may be written (l, f) and l is uniquely determined by L. The element l is called the *representer* for the linear functional. The complete characterization of linear functionals can be achieved on other Banach spaces too, but then the generating elements (corresponding to l here) lie in a different space (called the *dual space*) from the one containing f.

Another idea, deriving its appeal from geometrical analogy, is orthogonality; two elements f and g are *orthogonal* if $(f, g) = 0$. For a given set of vectors S, there is another set called the *orthogonal complement* of S and written $S^{\perp}$, such that, if $f \in$ S and $g \in S^{\perp}$, $(f, g) = 0$; $S^{\perp}$ is always a complete subspace of the Hilbert space. A *complete subspace* C is one in which Cauchy sequences in C converge to elements in C; thus C is really a "sub-Hilbert-space." Usually, C is called a *closed* subspace, but in complete linear vector spaces closed and complete subspaces are the same thing and we have no need to make a distinction.

This terminology allows us to introduce the *Decomposition Theorem*. Given a complete subspace C $\subset$ H, any element $f \in$ H can be written as the sum of a part in C and a part in $C^{\perp}$:

$$f = g + h, \quad g \in C, \quad h \in C^{\perp}, \tag{5}$$

and the decomposition is unique. The element g is said to be the *orthogonal projection* of f onto the subspace C. It is helpful to

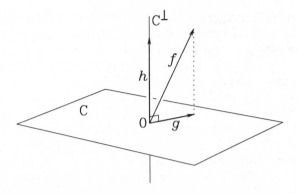

Figure 1.11a: The decomposition of the vector $f \in E^3$ into a part g lying in the two-dimension subspace C (a plane containing the origin) and h which lies in the orthogonal complement $C^\perp$ (a line perpendicular to the plane).

have a name for the function that maps an arbitrary element of a Hilbert space into its unique orthogonal projection in a fixed closed subspace; the function is called the *orthogonal projection operator* associated with the subspace and it is written P_C. Thus

$$P_C : H \to C . \tag{6}$$

In the decomposition above we have

$$g = P_C f . \tag{7}$$

Intuitively, we may picture a vector in ordinary space projected perpendicularly onto a plane C through the origin; the orthogonal complement $C^\perp$, is just the straight line through the origin perpendicular to the plane. An obvious property of any projection operator is that it is *idempotent*, which means that application twice must yield the same answer as one application; naturally, this may be written $P^2 = P$. It should be mentioned that there are projection operators that are not orthogonal so that the unique decomposition does not result in parts whose inner product always vanishes; we have no use for these in this book. At this point even orthogonal projection operators may look like very abstract objects. In the next section we shall see how to construct them explicitly for projections into finite-dimensional subspaces, which will meet nearly all our practical needs.

Figure 1.11a suggests an important question whose answer is in fact contained in the proof of the Decomposition Theorem: of all the elements g in the closed subspace C, which is the one that lies closest to f, that is, minimizes $\|f - g\|$? Indeed, there always exists a unique element minimizing the distance between an arbitrary element f and elements in a closed subspace C and it is given by the orthogonal projection of f onto C, that is, $P_C f$. This is a statement of the *Projection Theorem*. While the existence and uniqueness of the minimizing element may seem trivially obvious to the reader, they are in fact special properties of Hilbert space not shared by other spaces associated with different norms; the Projection Theorem is what makes Hilbert space so convenient for problems of minimization and optimization.

To give some life to these definitions and theorems, we turn in the next section to two applications that will be central to our approach to the questions of inverse theory.

1.12 Two Simple Minimization Problems

Many problems in inverse theory can be reduced to the minimization of the norm of a function that is constrained in some way. The simplest solutions for this type of problem are obtained with norms derived from inner products; in other words, the natural setting for these considerations is Hilbert space or pre-Hilbert space.

A frequently occurring example takes this form: we are given a set of linearly independent elements $g_1, g_2, \cdots g_N$ of a Hilbert space H, and an unknown element $m \in$ H obeying the conditions

$$(g_1, m) = d_1$$
$$(g_2, m) = d_2 \qquad\qquad (1)$$
$$\vdots$$
$$(g_N, m) = d_N$$

which we shall call the constraint equations. When the dimension of H is greater than N (it must be at least N; why?), the element m is not uniquely determined by these constraint equations (see 2.04). What is the solution satisfying these conditions which possesses the smallest norm? Assume for the moment there is at least one solution. We may write it using the Decomposition Theorem

$$m = p + q, \quad p \in \text{G}, q \in \text{G}^{\perp} \tag{2}$$

where G is the (finite-dimensional and hence closed) subspace of H spanned by the elements g_j, and $\text{G}^{\perp}$ is its orthogonal complement. By definition

$$(g_j, q) = 0, \quad j = 1, 2, 3, \cdots N . \tag{3}$$

Substituting the decomposition into the constraint equations (1), we have

$$(g_j, p) = d_j, \quad j = 1, 2, 3, \cdots N \tag{4}$$

so that only the element p determines the fit to the constraints; the element q can be chosen freely without influencing the agreement. Let us now calculate the size of the solution m:

$$\|m\|^2 = (m, m) \tag{5}$$

$$= (p + q, p + q) \tag{6}$$

$$= \|p\|^2 + 2(p, q) + \|q\|^2 . \tag{7}$$

$$= \|p\|^2 + \|q\|^2 . \tag{8}$$

Therefore the choice $q = 0$ yields the solution with the smallest norm. To find which element of G satisfies the constraints, notice that the elements g_j form a basis for the subspace (because we assumed that they were linearly independent) and so the solution with the smallest norm m_0 can be written

$$m_0 = p = \sum_{j=1}^{N} \alpha_j g_j . \tag{9}$$

Now substitute (9) into the constraint equations (1)

$$(g_i, \sum_j \alpha_j g_j) = d_i, \quad i = 1, 2, 3, \cdots N \tag{10}$$

$$\sum_{j=1}^{N} (g_i, g_j) \alpha_j = d_i, \quad i = 1, 2, 3, \cdots N . \tag{11}$$

This is a system of linear equations for the unknown expansion coefficients α_j in the expression for m_0. We may write the system more compactly as

$$\Gamma \alpha = \mathbf{d} \tag{12}$$

where $\Gamma \in \text{M}(N \times N)$ and $\alpha, \mathbf{d} \in \mathbb{R}^N$. Explicitly

$$\Gamma_{ij} = (g_i, g_j) = \Gamma_{ji} \tag{13}$$

$$\boldsymbol{\alpha} = (\alpha_1, \alpha_2, \cdots \alpha_N) \tag{14}$$

$$\mathbf{d} = (d_1, d_2, \cdots d_N). \tag{15}$$

The symmetric matrix Γ is called the *Gram matrix*. If the Gram matrix is nonsingular a solution to (12) for $\boldsymbol{\alpha}$ always exists for any $\mathbf{d}$:

$$\boldsymbol{\alpha} = \Gamma^{-1}\mathbf{d} . \tag{16}$$

We demonstrate now that the linear independence of the elements g_i guarantees that Γ is nonsingular.

Assume that $\det \Gamma = 0$, while the elements g_i are linearly independent; we shall see that this leads to a contradiction. If $\det \Gamma = 0$, an elementary property of determinants states that a linear combination of rows can be found yielding a row that vanishes identically; in other words, the rows of the matrix, treated as vectors in $\mathbb{R}^N$, must be linearly dependent. For Γ, this means that there are constants β_i, not all zero, such that

$$0 = \sum_{i=1}^{N} \beta_i \Gamma_{ij} , \qquad j = 1, 2, 3, \cdots N \tag{17}$$

$$= (\sum_i \beta_i g_i , g_j) , \qquad j = 1, 2, 3, \cdots N . \tag{18}$$

Weight each of these equations by β_j and sum:

$$0 = \sum_j \beta_j (\sum_i \beta_i g_i , g_j) \tag{19}$$

$$= (\sum_i \beta_i g_i , \sum_j \beta_j g_j) \tag{20}$$

$$= \|\sum_i \beta_i g_i \|^2 . \tag{21}$$

From the norm property 1.05(N4) this implies

$$\sum_i \beta_i g_i = 0 \tag{22}$$

and therefore the elements g_i are linearly dependent, contrary to the initial assumption. We must conclude that $\det \Gamma \neq 0$ and it follows at once that Γ^{-1} always exists. It is also easy to see that $\det \Gamma = 0$ when the elements g_j actually are linearly dependent. Incidentally, a demonstration following almost the same plan as the one just given can be used to show that Γ is positive definite so that, among other things, det $\Gamma > 0$.

The fact that Γ^{-1} exists establishes something assumed earlier, namely, that there is always at least one solution to the original set of constraint equations; it is m_0, the one in the form $m_0 = \Sigma\, \alpha_j g_j$. The norm-minimization property of m_0 remains valid even when H is a pre-Hilbert space, but our proof used the Decomposition Theorem which relies on the completeness of H. Therefore we should show directly that m_0 really is the smallest norm solution, if we want the more general result. We suppose there is another $n \in$ H satisfying these constraints. Then

$$(g_j, n) = d_j \, , \qquad j = 1, 2, 3, \cdots N \tag{23}$$

and so subtracting (1)

$$(g_j, n - m_0) = 0 \, , \qquad j = 1, 2, 3 \cdots N \, . \tag{24}$$

Consider

$$\| n \|^2 = \| n - m_0 + m_0 \|^2 \tag{25}$$

$$= \| n - m_0 \|^2 + \| m_0 \|^2 + 2(n - m_0, m_0) \tag{26}$$

$$= \| n - m_0 \|^2 + \| m_0 \|^2 + 2 \sum_j \alpha_j (n - m_0, g_j) \tag{27}$$

$$= \| n - m_0 \|^2 + \| m_0 \|^2 \tag{28}$$

$$\geq \| m_0 \|^2 \, . \tag{29}$$

Clearly equality holds if and only if $n = m_0$, which proves that m_0 is unique in minimizing the norm.

The second minimization problem is the one arising from the Projection Theorem when the subspace is finite dimensional. We wish to discover the element $\tilde{g}$ of a finite-dimensional subspace G that lies closest to f, a fixed element of a Hilbert space H. The Projection Theorem tells us the answer is simply

$$\tilde{g} = P_G f \tag{30}$$

but now we want an explicit solution when a basis for G is known. From the Decomposition Theorem we can see

$$f - \tilde{g} = h \in G^\perp \, . \tag{31}$$

Let $g_1, g_2, \cdots g_N$ be a basis for G; since $g_j \in$ G it follows that

$$0 = (g_j, h) \tag{32}$$

$$= (g_j, f - \tilde{g}), \quad j = 1, 2, 3, \cdots N . \tag{33}$$

Introducing the expansion of $\tilde{g}$ in the basis elements

$$(g_j, f - \sum_k \beta_k g_k) = 0 . \tag{34}$$

Rearranging as we did in the first problem we find the following set of linear equations for the unknown coefficients β_j

$$\sum_{k=1}^{N} \Gamma_{jk} \beta_k = (g_j, f), \quad j = 1, 2, 3, \cdots N \tag{35}$$

where Γ is of course the Gram matrix. This is a linear system just like (12) but with a different right side vector. The elements g_j are linearly independent (why?) and so Γ is nonsingular; therefore the linear system (35) can always be solved for the coefficients. In this way we have found an explicit expression for the orthogonal projection of an element onto a finite-dimensional subspace and hence the best approximation to f made from elements of G:

$$\tilde{g} = \sum_{j=1}^{N} \beta_j g_j . \tag{36}$$

If the space of approximating elements is not finite dimensional, the calculation of the projection is not always so simple. One case is easy, however. When the orthogonal complement to the space G is finite dimensional and a basis for it is known, we write

$$P_G f = f - P_{G^\perp} f \tag{37}$$

and the projection operator on the right side can be evaluated by the methods we have just described.

1.13 Numerical Aspects of Hilbert Space

In order to obtain a numerical solution to a geophysical inverse problem it is often necessary to approximate an element in an infinite-dimensional Hilbert space by a vector in a finite-dimensional space. Therefore we shall find it useful to discuss the best numerical techniques for performing some of the operations in the Hilbert space E^M. The inner product on E^M will be the ordinary dot product: with $\mathbf{x}, \mathbf{y} \in E^M$

$$(\mathbf{x}, \mathbf{y}) = \mathbf{x} \cdot \mathbf{y} = \sum_{i=1}^{M} x_i y_i \ . \tag{1}$$

The more general inner product

$$(\mathbf{x}, \mathbf{y})_A = \mathbf{x} \cdot A \mathbf{y} \tag{2}$$

can be reduced to the simpler one as we shall show later.

With the ordinary dot product, the finite-dimensional version of the minimum norm problem discussed in the last section is this: given N linearly independent $\mathbf{g}_j \in E^M$ with $N < M$, what vector $\mathbf{x} \in E^M$ satisfies

$$\mathbf{g}_j \cdot \mathbf{x}_j = d_j, \quad j = 1, 2, 3, \cdots N \tag{3}$$

and has the smallest Euclidean norm? In matrix notation (3) reads

$$G\mathbf{x} = \mathbf{d} \tag{4}$$

where $G \in M(N \times M)$ so that the N rows of G are vectors $\mathbf{g}_j$ and $\mathbf{d} \in E^N$ is the vector of constants d_j. This minimization problem is usually called an *underdetermined linear least squares problem*. Following the recipe of 1.12 we first calculate

$$\Gamma_{jk} = (\mathbf{g}_j, \mathbf{g}_k) = \mathbf{g}_j \cdot \mathbf{g}_k \tag{5}$$

or in matrix notation

$$\Gamma = G\, G^T \ . \tag{6}$$

Then the vector of coefficients $\boldsymbol{\alpha} \in E^N$ that expands the solution $\mathbf{x}_0$ in the basis vectors $\mathbf{g}_j$ is

$$\boldsymbol{\alpha} = \Gamma^{-1} \mathbf{d} \tag{7}$$

$$= (G\, G^T)^{-1}\mathbf{d} \ . \tag{8}$$

The explicit solution from 1.12(9) is

$$\mathbf{x}_0 = \sum_{j=1}^{N} \alpha_j \mathbf{g}_j \tag{9}$$

$$= G^T \boldsymbol{\alpha} \tag{10}$$

$$= G^T (G\, G^T)^{-1}\mathbf{d} \ . \tag{11}$$

Algebraically this answer is of course correct, but from a

numerical analysis viewpoint it is expressed in a quite unsatisfactory way. We shall show that the condition number of $\Gamma = GG^T$ is unnecessarily large, which, as we saw in 1.06, may lead to serious computational trouble. Fortunately a more stable alternative exists. The classic reference to this material is by Lawson and Hanson (1974). Another accessible treatment is given by Strang (1980).

We introduce the QR factorization. Suppose that we can write G^T as the product of two matrices $Q \in M(M \times M)$ and $R \in M(M \times N)$:

$$G^T = Q R \tag{12}$$

where Q is an *orthogonal matrix*, which means $Q^T = Q^{-1}$, and R has a special form

$$R = \begin{bmatrix} R_1 \\ O \end{bmatrix} . \tag{13}$$

Here the submatrix $O \in M(M-N \times N)$ consists entirely of zeros, R_1 is a square $N \times N$ right *triangular matrix*, that is, all the elements below the diagonal are zero. The matrix R is rather empty: only $\frac{1}{2}N(N+1)$ of its MN elements need not be zero. This factorization divides action of G^T into two parts: first a linear transformation is formed on $\mathbf{x}$ with R_1 leaving it in E^N; the resultant vector is elevated into E^M by adding zeros at the end; the factor Q rotates the augmented vector in E^M. Multiplication by an orthogonal matrix is thought of as rotation (with possible reflection) because the new vector has the same Euclidean length (2-norm) as the original:

$$\|Q\mathbf{y}\|^2 = (Q\mathbf{y}) \cdot (Q\mathbf{y}) = \mathbf{y} \cdot Q^T Q \mathbf{y} = \mathbf{y} \cdot Q^{-1} Q \mathbf{y} = \mathbf{y} \cdot \mathbf{y} = \|\mathbf{y}\|^2 . \tag{14}$$

If the series of steps above appears mysterious, refer to the matrix algebra properties mentioned in 1.10; we shall draw heavily upon them in the next few paragraphs. Before describing how the factors Q and R are actually constructed, let us see how they affect the solution of the minimization problem.

Substitute the factored form of G^T into equation (11):

$$\mathbf{x}_0 = Q R ((Q R)^T Q R)^{-1} \mathbf{d} \tag{15}$$

$$= Q R (R^T Q^T Q R)^{-1} \mathbf{d} \tag{16}$$

$$= Q R (R^T R)^{-1} \mathbf{d} . \tag{17}$$

Referring to (13) we can see that

$$R^T R = R_1^T R_1.$$ (18)

Also from (6) and (12) note that $\Gamma = R^T R = R_1^T R_1$. The assumption that the vectors $\mathbf{g}_j$ are linearly independent implies that Γ is non-singular and therefore R_1 must also be nonsingular. Thus, following the rules of matrix manipulation

$$\mathbf{x}_0 = Q\, R\, R_1^{-1}\, (R_1^T)^{-1}\, \mathbf{d}$$ (19)

$$= Q \begin{bmatrix} R_1 \\ O \end{bmatrix} R_1^{-1}\, (R_1^T)^{-1}\, \mathbf{d}$$ (20)

$$= Q \begin{bmatrix} (R_1^T)^{-1}\mathbf{d} \\ O \end{bmatrix}.$$ (21)

(There is a more elegant path to this result starting from the QR factors without using the equation in Γ; the reader may wish to discover it.) To find $\mathbf{x}_0$ we must therefore first solve the linear system

$$R_1^T \mathbf{y} = \mathbf{d}$$ (22)

then augment with zeros and rotate by applying Q. The solution of this linear system is very easy, as can be seen by writing out the first few lines

$$\begin{aligned} r_{11}y_1 &= d_1 \\ r_{12}y_1 + r_{22}y_2 &= d_2 \\ r_{13}y_1 + r_{23}y_2 + r_{33}y_3 &= d_3 \\ &\cdots \end{aligned}$$ (23)

Starting at the top and working down one has at the kth line exactly enough information to calculate y_k. To understand the numerical superiority of this way of computing $\mathbf{x}_0$, consider the condition number of R_1 in the matrix 2-norm. Recall from 1.06 that

$$\|R_1\|_2 = [\lambda_{\max}(R_1^T R_1)]^{\frac{1}{2}}$$ (24)

$$= [\lambda_{\max}(\Gamma)]^{\frac{1}{2}}$$ (25)

and that the condition number is

$$c = \text{cond}(R_1)$$ (26)

$$= \|R_1\|_2 \|R_1^{-1}\|_2 \tag{27}$$

$$= [\lambda_{\max}(\Gamma)\lambda_{\max}(\Gamma^{-1})]^{\frac{1}{2}} \tag{28}$$

where $\lambda_{\max}[A]$ is the functional giving the largest eigenvalue of the matrix A. Since Γ is positive definite, $\lambda_{\max}$ is real and positive. Now calculate the condition number of Γ:

$$\text{cond}(\Gamma) = \|\Gamma\|_2 \|\Gamma^{-1}\|_2 \tag{29}$$

$$= [\lambda_{\max}(\Gamma^T \Gamma)\lambda_{\max}(\{\Gamma^T \Gamma\}^{-1})]^{\frac{1}{2}} \tag{30}$$

$$= [\lambda_{\max}(\Gamma^2)\lambda_{\max}(\Gamma^{-2})]^{\frac{1}{2}} . \tag{31}$$

The last equation follows from the symmetry of Γ, namely, that $\Gamma^T = \Gamma$. It is elementary that the eigenvalues of Γ^2 are the squares of those belonging to Γ and so

$$\text{cond}(\Gamma) = \lambda_{\max}(\Gamma)\lambda_{\max}(\Gamma^{-1}) \tag{32}$$

$$= c^2 . \tag{33}$$

We saw in 1.06 that when the condition number of a matrix is large, the accuracy of a numerical solution of a linear system based on that matrix may deteriorate, sometimes to the point of rendering the numbers worthless. Obviously the system invoking R_1 is much less prone to this defect than the one for α involving Γ. An example illustrating this point is given in section 2.08 of the next chapter. A more detailed error analysis accounting for the effects of applying Q and constructing the factors from G^T supports the assertion that QR is numerically superior to the approach based upon the normal equations. See Lawson and Hanson (1974) or Wilkinson (1965).

We return to the question of obtaining the factors Q and R from G^T. The principle is to build Q from a series of elementary rotations:

$$Q = Q_N Q_{N-1} \cdots Q_3 Q_2 Q_1 . \tag{34}$$

The first rotation Q_1 is designed to act on the first column vector of G^T so that only the first element of the transformed column is nonzero; Q_2 is constructed to annihilate all but the first two elements of the second column of $Q_1 G^T$ (without undoing the work of Q_1!); Q_3 attacks the third column of the matrix $Q_2 Q_1 G^T$, making all but the top three elements vanish; and so on. At every stage the rotations are performed so that at the end the matrix R has to be explicitly constructed. The elementary rotations, often called *Householder rotations* after their inventor, are interesting objects; each is in the form

$$I - 2\mathbf{uu} \tag{35}$$

where I is the unit $M \times M$ matrix and $\mathbf{uu}$ denotes the *dyad* matrix with components $(\mathbf{uu})_{ij} = u_i u_j$; alternatively we can express $\mathbf{uu}$ as $u\, u^T$, where u is the column matrix corresponding to the vector $\mathbf{u}$. In addition, $\mathbf{u}$ must have unit Euclidean length. It is easily verified that every such Householder matrix is orthogonal and symmetric and therefore it is its own inverse. To transform the last $K-1$ components of a vector $\mathbf{g} \in E^M$ to zero, we calculate $\mathbf{u}$ as follows. Denote the last K elements of $\mathbf{g}$ by the vector $\mathbf{v}$. Then the first $M-K-1$ elements of $\mathbf{u}$ are zeros and

$$\mathbf{u} = \gamma\,(0,\, 0,\, \cdots\, v_1 \pm \|\mathbf{v}\|,\, v_2,\, v_3,\, \cdots\, v_K) \tag{36}$$

where γ is chosen to give $\mathbf{u}$ unit length and the sign of $\|\mathbf{v}\|$ in element number $M-K$ is the same as that of v_1. The process is in fact less complicated than it may appear at first sight. In practical computer codes, the matrix Q is not constructed; only the vectors $\mathbf{u}$ are saved in the space occupied by the original matrix G^T. In terms of amount of computational effort, Householder triangularization requires about $MN^2 - N^3/3$ operations (treating addition, multiplication, and division as equivalent); the formation of GG^T is somewhat more efficient—it can be performed in about $MN^2/2 + N^3/6$ operations. Bounds on the accuracy of solutions found via the Householder procedure are provided in Lawson and Hanson (1974).

Another problem profitably inspected with QR factorization in mind is the familiar overdetermined linear least squares problem. One seeks to minimize by suitable choice of $\beta \in E^N$ the Euclidean norm

$$\|G^T \beta - \mathbf{d}\| \tag{37}$$

where $\mathbf{d} \in E^M$ and G is the matrix of N row vectors $\mathbf{g}_j \in E^M$. This is a typical modeling problem in which only a few parameters β_k are thought to be important in determining a vector of observations $\mathbf{d}$. A Hilbert space interpretation is as follows: the vectors $\mathbf{g}_j$, when linearly independent, form the basis of a (closed) subspace; the linear combination $\Sigma\, \beta_j \mathbf{g}_j$ is an element of that subspace which we shall call G. Thus we find the finite-dimensional realization of the second problem addressed in section 1.12. Equation 1.12(35)

$$\sum_{k=1}^{N} \Gamma_{jk} \beta_k = (g_j, f), \quad j = 1, 2, 3, \cdots N \tag{38}$$

must now be rewritten in our matrix notation as

$$G\,G^T\beta = G\,\mathbf{d}\,. \tag{39}$$

This is the matrix form of the famous *normal equations* for least squares problems (looking slightly strange because we have persisted in using G^T in the original statement of the problem). The numerical solution of this linear system is inferior to one based upon QR factorization because of course the condition of $GG^T = \Gamma$ is unnecessarily large. Without going into detail, we state that when the QR factors of G^T are introduced, the equation for β becomes

$$R_1\beta = P\,Q\,\mathbf{d} \tag{40}$$

where $P \in \mathrm{M}\,(N \times M)$ is a projection matrix that maps a vector in E^M into E^N simply by zeroing the bottom $M - N$ components. A detailed error analysis confirms the remarkably stable numerical performance of this solution to least squares systems; see Lawson and Hanson (1974).

There are circumstances in which the coefficient vector β is not needed explicitly and what is required is the best approximating vector to $\mathbf{d}$ given by

$$\tilde{\mathbf{d}} = G^T\beta\,. \tag{41}$$

An example of this situation is found in the preparation of long period seismic data which can be contaminated by tidal signals; other uses are found in the next chapter when we consider seminorm minimization. The usual procedure is to fit by least squares a series of tidal constituents whose frequencies are well known, but whose amplitudes are not; the best-fitting tidal model is then subtracted from the record. Here the coefficients β_k are unimportant—all that is needed is $\tilde{\mathbf{d}}$, the best approximation to the data. We shall meet other examples when we examine seminorm minimization in section 2.05. Substituting in the QR factors and the solution found by Householder triangularization, we obtain

$$\tilde{\mathbf{d}} = Q^T P\,Q\,\mathbf{d}\,. \tag{42}$$

Notice that R_1 does not appear in the equation; there is no linear system to solve, so that the condition number of R_1 or Γ is irrelevant. Calculated this way, the values of $\tilde{\mathbf{d}}$ are numerically even more robust than those found from the vector β obtained by QR: the projected vector is virtually immune to problems of arithmetic error. The matrix $Q^T P Q$ is the orthogonal projection operator P_G mapping vectors in E^M into their orthogonal projections in the subspace G; the matrix is usually written $G^T(GG^T)^{-1}G$, and though it is algebraically identical to our expression, this matrix is quite unsuited to practical evaluation when poor conditioning may arise.

At the beginning of this section we mentioned a more general inner product for E^M based on the matrix $A \in M\ (M \times M)$:

$$(\mathbf{x, y})_A = \mathbf{x} \cdot A \mathbf{y} . \tag{43}$$

Let us call this inner product space E_A^M. Operations in E_A^M can be transformed into those in E^M with the ordinary inner product as follows. We find for every element $\mathbf{x} \in E_A^M$ a transformed vector $\mathbf{x}' \in E^M$ using

$$\mathbf{x}' = \bar{R}\mathbf{x} \tag{44}$$

where $\bar{R}$ is a right triangular matrix. Suppose

$$A = \bar{R}^T \bar{R} \tag{45}$$

then

$$\mathbf{x}' \cdot \mathbf{y}' = (\bar{R}\mathbf{x}) \cdot (\bar{R}\mathbf{y}) \tag{46}$$

$$= \mathbf{x} \cdot \bar{R}^T \bar{R}\mathbf{y} \tag{47}$$

$$= (\mathbf{x, y})_A . \tag{48}$$

Thus inner product calculations on the transformed elements in E^M correspond to those with the original vector in E_A^M. It remains to be shown that a right triangular matrix $\bar{R}$ always exists for every A. While this is untrue for a general matrix, A has certain properties whenever $(\mathbf{x, y})_A$ is to be an inner product: A must be symmetric and positive definite. These are precisely the conditions needed for the factorization

$$A = \bar{R}^T \bar{R} \tag{49}$$

to exist. This is called the *Cholesky factorization* of A; details can be found in any book on numerical analysis (see e.g., Golub and Van Loan, 1983) but the reader should be aware that for some reason the conventional notation takes $L = \bar{R}^T$, and then

$$A = L L^T \tag{50}$$

Exercise

1.13(i) The elementary transformation

$$Q = I - 2\mathbf{u}\mathbf{u}$$

is an example of a matrix that is simultaneously symmetric and orthogonal. One is naturally curious to know what other matrices have both these properties. Show that $\mathbf{u}\mathbf{u}$ is an orthogonal projection in E^N with the Euclidean norm and that in general the matrix

$$T = I - 2P$$

is orthogonal and symmetric when P is any orthogonal projection matrix in $M(N \times N)$. Show that the eigenvalues of every orthogonal symmetric matrix are either +1 or −1 and hence from its spectral factorization (see 3.03) that any such matrix can be written like T.

1.14 Lagrange Multipliers

The approach to inverse theory adopted in this book often converts the mathematical work into an optimization problem—the maximization or minimization of a functional. On the whole we shall not rely heavily upon the methods of simple calculus or calculus of variations for our results. Nonetheless, these techniques provide powerful ways of finding candidate models which, in a truly rigorous analysis, must be subjected to further scrutiny (see 3.02 for an example). To be useful a method must be able to solve *constrained* optimizations, those in which the argument of the functional is not free, but is held to one or more conditions and this generalization is what the method of Lagrange Multipliers supplies.

We treat a functional F on the linear vector space E^N with the ordinary Euclidean length as norm. We seek a local minimum (or any other type of stationary point) of F when the argument is subject to additional conditions in the form

$$S_i[\mathbf{x}] = s_i, \quad i = 1, 2, 3, \cdots M \tag{1}$$

where S_i are also functionals on E^N. As we shall assume $N > M$, equation (1) does not generally define a single point $\mathbf{x}$, but a set of points which will be presumed not to be empty. Suppose that a local minimum of the constrained system occurs at $\mathbf{x}_*$ and that in an open neighborhood around this point F and all the S_i are continuously differentiable; then we have the following necessary condition:

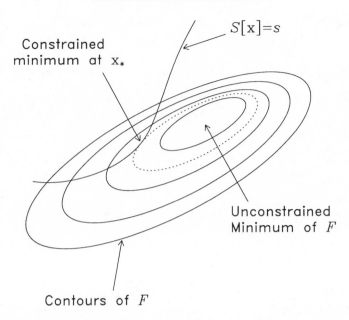

Figure 1.14a: Constrained minimization of the functional F on the space E^2. The closed solid lines are contours of F depicting a local minimum. An additional condition to be applied is that $\mathbf{x}$ must satisfy the equation $S[\mathbf{x}]=s$ shown as the sinuous solid curve. It will be clear that the constrained minimum occurs where $S=s$ is tangent to a contour of F, which is the condition (6).

$$\nabla F[\mathbf{x}_*] = \sum_{i=1}^{M} \lambda_i \nabla S_i[\mathbf{x}_*] \tag{2}$$

if the $M < N$ gradient vectors ∇S_i are linearly independent at $\mathbf{x}_*$. A common way to summarize these results is to introduce an *unconstrained* functional on E^{M+N} given by

$$U[\mathbf{x}, \lambda_1, \lambda_2, \cdots \lambda_M] = F[\mathbf{x}] - \sum_{i=1}^{M} \lambda_i (S_i[\mathbf{x}] - s_i). \tag{3}$$

Here the unknown real parameters λ_i are called *Lagrange multipliers*. The stationary points of U, those where $\nabla U = 0$ with respect to all its arguments, correspond to the constrained stationary points of F. This is easily seen by differentiating U: the differentials with respect to the Lagrange multipliers λ_i return the M constraint equations (1); the gradient with respect to $\mathbf{x}$ gives us (2). A stationary point of U is not necessarily a local minimum or maximum of F but may be a saddle point; only a study of the higher order (usually quadratic) behavior at the point can decide. Also a saddle in U can correspond to a constrained maximum or minimum of F. Remember, the linear independence of the ∇S_i is required if condition (2) or the equivalent stationarity of (3) is to be valid. If this condition fails, the blind application of (3) can lead to inconsistent results. Finally, the Lagrange multipliers have a useful interpretation—they give the rate of change of the stationary functional as the constraints vary:

$$\frac{\partial F[\mathbf{x}_*]}{\partial s_i} = \lambda_i \ . \tag{4}$$

To give the reader a flavor of the proof we shall derive the condition in the case of one constraint, that is, when $M = 1$; the situation is illustrated in figure 1.14a. For a full treatment including generalizations to inequalities see Bazaraa and Shetty (1979), and for functionals on infinite-dimensional spaces see Smith (1974). In the simple case (2) reduces to the requirement that the ∇F and ∇S_1 must be parallel at $\mathbf{x}_*$; This means the surface of constant F is tangent to the constraint surface $S_1[\mathbf{x}] = s_1$ there. The illustration in figure 1.14a shows a case when $N = 2$: the constraint here is just a line which is tangent to one of the contours of F at the constrained minimum. In the following we drop the subscript on S as there is only one constraint in the problem.

We shall assume that $\mathbf{x}_* \in E^N$ is a constrained local minimum of $F[\mathbf{x}]$ subject to

$$S[\mathbf{x}] = s \tag{5}$$

and that F and S are continuously differentiable in an open set around $\mathbf{x}_*$ and further that $\nabla S[\mathbf{x}_*] \neq 0$. We wish to show that

$$\nabla F[\mathbf{x}_*] = \lambda \nabla S[\mathbf{x}_*] \tag{6}$$

for some real λ. One of the difficulties with a proof is finding a rigorous way to write the condition for remaining in the constraint set; here we do this by writing a differential equation whose solution obeys the constraint. We shall prove (6) by showing that the assumption of the contrary leads to a contradiction. Suppose that there is a nonvanishing vector $\mathbf{t} \in E^N$ such that

$$\nabla F[\mathbf{x}_*] = \lambda \nabla S[\mathbf{x}_*] + \mathbf{t} \tag{7}$$

with $\mathbf{t} \cdot \nabla S[\mathbf{x}_*] = 0$. Define the vector-valued function $\mathbf{x}(\beta)$ of real

$\beta \geq 0$ as the solution of the ordinary differential equation

$$\frac{d\mathbf{x}(\beta)}{d\beta} = -P_\beta \mathbf{t} \tag{8}$$

with initial condition $\mathbf{x}(0)=\mathbf{x}_*$, where P_β is the orthogonal projection operator onto the subspace orthogonal to the vector $\nabla S[\mathbf{x}(\beta)]$; P_β is easily constructed (see 1.12(36)):

$$P_\beta \mathbf{t} = \mathbf{t} - \mathbf{g}(\beta)\mathbf{g}(\beta) \cdot \mathbf{t} \tag{9}$$

where $\mathbf{g}(\beta)$ is the unit normalized form of ∇S, that is, $\mathbf{g}(\beta)=\nabla S[\mathbf{x}(\beta)]/\|\nabla S[\mathbf{x}(\beta)]\|$. Notice that $d\mathbf{x}/d\beta=-\mathbf{t}$ at $\beta=0$. We shall show that $\mathbf{x}(\beta)$ is always a solution of (5). For some γ, $0 \leq \gamma \leq \beta$ by the Mean Value Theorem of elementary calculus

$$S[\mathbf{x}(\beta)] = S[\mathbf{x}(0)] + \beta \frac{dS[\mathbf{x}(\gamma)]}{d\gamma} \tag{10}$$

$$= s + \beta \frac{dS[\mathbf{x}(\gamma)]}{d\gamma} . \tag{11}$$

Use the Chain Rule of calculus and then substitute (8)

$$S[\mathbf{x}(\beta)] = s + \beta \frac{d\mathbf{x}}{d\gamma} \cdot \nabla S[\mathbf{x}(\gamma)] \tag{12}$$

$$= s + \beta(-P_\gamma \mathbf{t}) \cdot \nabla S[\mathbf{x}(\gamma)] \tag{13}$$

$$= s . \tag{14}$$

The vanishing of the second term follows from the way we constructed P. Thus $\mathbf{x}(\beta)$ satisfies (5) as we asserted. But now using the chain rule again and (7) we have

$$\frac{dF[\mathbf{x}(\beta)]}{d\beta}\Big|_{\beta=0} = \nabla F[\mathbf{x}_*] \cdot \frac{d\mathbf{x}(\beta)}{d\beta}\Big|_{\beta=0} \tag{15}$$

$$= (\lambda \nabla S[\mathbf{x}_*] + \mathbf{t}) \cdot (-\mathbf{t}) \tag{16}$$

$$= -\|\mathbf{t}\|^2 < 0 . \tag{17}$$

Thus for sufficiently small β we can find an $\mathbf{x}$ in the constraint set that yields a value of $F[\mathbf{x}]$ smaller than $F[\mathbf{x}_*]$, which is a contradiction. Thus we must have that $\mathbf{t}=0$ in (7) from which (6) follows. Notice how it is impossible to find $\mathbf{g}$ if $\nabla S[\mathbf{x}_*]$ vanishes.

As an example we solve again the first simple minimization problem of the previous section. For $\mathbf{x} \in E^N$ with the Euclidean norm, we minimize $F[\mathbf{x}] = \|\mathbf{x}\|^2$ subject to the K equality constraints

$$\mathbf{g}_k \cdot \mathbf{x} = d_k, \quad k = 1, 2, 3, \cdots K \qquad (18)$$

where $\mathbf{g}_k \in E^N$ are K known linearly independent vectors and d_k are K known real numbers. We follow the rule (3) of forming an unconstrained functional on E^{N+K}:

$$U[\mathbf{x}, \lambda_1, \lambda_2, \cdots \lambda_K] = \mathbf{x} \cdot \mathbf{x} - \sum_{k=1}^{K} \lambda_k (\mathbf{g}_k \cdot \mathbf{x} - d_k). \qquad (19)$$

Setting the $\mathbf{x}$ gradient of U to zero for an unconstrained stationary point gives

$$2\mathbf{x}_* - \sum_{k=1}^{K} \lambda_k \mathbf{g}_k = 0 \qquad (20)$$

which is just the familiar result 1.12(9) that the minimizing model is formed from a linear combination of the constraining elements. The expansion coefficients are (one half of) the Lagrange multipliers and they must be obtained by substituting (20) into (18) as usual. The linear independence of the vectors $\mathbf{g}_k$ gives the linear independence of the gradients. Without examining the higher derivatives at $\mathbf{x}_*$ we cannot tell if the stationary solution minimizes F or maximizes it or is merely stationary; nor do we know whether the locally stationary solution has any global significance. This last factor is the crippling disadvantage of all calculus methods. In fact we have shown elsewhere that the result gives the smallest possible value of the norm and so we know that the constrained F is locally and globally minimum. In the next section we describe a general class of functionals in which local optimization implies the global property.

Exercise

1.14(i) Recall from section 1.06 the definition of the matrix subordinate p-norm:

$$\|A\|_p = \max \|A\mathbf{x}\|_p$$

subject to the condition $\|\mathbf{x}\|_p = 1$. Write this optimization problem in an unconstrained form by using a Lagrange multiplier. Differentiate and write out the condition for a solution. Verify the result given in

1.06 for the spectral norm, the case when $p = 2$.
Show that $\|A\|_3 = 4.255$ when

$$A = \begin{bmatrix} 1 & 1 & 1 \\ 1 & 2 & 1 \\ 1 & 1 & 3 \end{bmatrix}.$$

You must solve the equations numerically; you may assume the optimizing vector has only positive components.

1.15 Convexity

As we discussed in the previous section, methods relying upon calculus and Lagrange multipliers to locate the minimum of a functional are open to the criticism that their solutions may be merely local, that is, the value is the smallest one in a neighborhood, but the functional is in fact smaller elsewhere. A property that confers global validity on a local minimum is the *convexity* of the functional.

To begin we need the notion of a convex set. In a linear vector space V, a set C of elements is said to be *convex* if for every $x, y \in$ C it is true that

$$\alpha x + (1 - \alpha)y \in C \tag{1}$$

when $\alpha \in \mathbb{R}$ and $0 \le \alpha \le 1$. The geometrical idea, illustrated in figure 1.15a, is that all the points on a chord joining any pair of

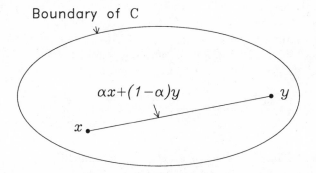

Boundary of C

$\alpha x + (1-\alpha)y$

y

x

Figure 1.15a: The interior and edge of the ellipse shown is an example of a convex set of points in the plane $\mathbb{R}^2$.

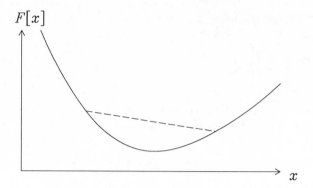

Figure 1.15b: A convex functional of one real argument. By equation (2) the functional value cannot exceed the straight-line interpolation of any two values on the curve.

elements in a convex set are also within the set. Obviously linear subspaces are always convex sets. It is also easily proved that the intersection of two convex sets is itself a convex set.

A real-valued functional F over V is convex in its convex domain C of definition if for all $x, y \in$ C we have

$$F[\alpha x + (1 - \alpha)y] \le \alpha F[x] + (1 - \alpha)F[y] \qquad (2)$$

and real α obeys the same condition as before. Figure 1.15b conveys the general idea. Again an almost trivial example of a convex functional is provided by any linear functional, since equality always obtains for them in the defining relation. A more interesting class of convex functionals is provided by the norms. If V is a normed vector space and $F[x] = \|x\|$, then by the norm conditions 1.05($N3$), the triangle inequality, and 1.05($N2$) it is easily shown that the norm is always a convex functional. The triangle inequality gives:

$$F[\alpha x + (1 - \alpha)y] \le F[\alpha x] + F[(1 - \alpha)y] \qquad (3)$$

$$\le |\alpha| F[x] + |1 - \alpha| F[y]. \qquad (4)$$

But since α lies between 0 and 1 the magnitude operations are immaterial and so (2) is satisfied for the norm. This result does not depend on 1.05($N4$) and therefore all seminorms are convex too.

It is obvious from the definitions that associated with a convex functional, the set of points C′ defined by those x with

$$F[x] \leq c \tag{5}$$

where c is a constant (and C' is not empty), is also a convex set.

We are now ready to state the useful theorem; for a proof see chapter 7 of Luenberger (1969). If C is a convex subset of a vector space V and F is a real-valued functional on C then if $x_0 \in C$ is a local minimum of F, then $F[x_0]$ is the smallest value attained by the functional anywhere in C. We have not defined the term local minimum carefully; this is simple enough on a normed space: x_0 is a *local minimum* of F if there is a positive real number ε such that for all y with $\|y\| < \varepsilon$ and $x_0 + y \in$ C

$$F[x_0] \leq F[x_0 + y] . \tag{6}$$

It should be noted that the element x_0 need not be the only one at which the global minimum is achieved; uniqueness of the minimizer requires stronger conditions.

Many of the functionals we shall minimize are norms or seminorms, and the constraint sets are convex; for example, 1.14(18) represents the intersection of a set of linear, and hence convex, sets. Thus convexity may provide a powerful tool for establishing the legitimacy of a local stationary point. On the whole, however, we shall use other, more direct methods for this purpose.

CHAPTER 2

LINEAR PROBLEMS WITH EXACT DATA

2.01 An Example

Figure 2.01a shows an idealized magnetic anomaly profile of a kind that might be observed during a marine geophysical survey of a mid-oceanic ridge system. The anomaly is the contribution to the Earth's magnetic field caused by magnetized rocks on the sea floor. Every geophysicist knows that there is a characteristic pattern of positive and negative anomalies wherever new oceanic crust is being formed, namely, in the submarine mountain chains of the ocean rises. The pattern is caused by the imprinting of geomagnetic field reversals in the volcanic rocks of the upper oceanic crust; to a first approximation the crust divides symmetrically at the ridge axis and is carried away horizontally at a steady rate, taking with it a record of the state of the geomagnetic field at the time when the rocks were new.

Suppose now we wish to find a mathematical model of the

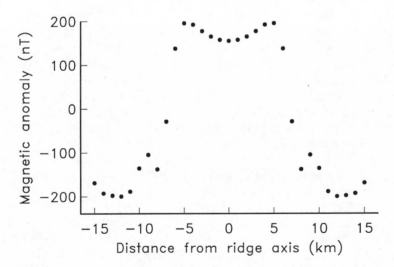

Figure 2.01a: An idealized marine magnetic anomaly at a mid-ocean ridge. The solid dots are the values of the vertical component of the magnetic field.

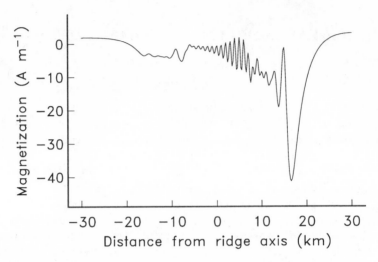

Figure 2.01b: A mathematically acceptable model for crustal magnetization exactly reproducing the magnetic field anomaly of figure
2.01a. Note the asymmetry and the highly oscillatory behavior.

magnetization of the sea-floor rocks based upon these anomaly
"observations," which are the solid dots in the figure. To keep
everything as simple as possible we will pretend the magnetization is confined to an extremely thin layer, that the direction of
the magnetization vector is constant, and that the intensity varies
only in the direction of spreading, perpendicular to the strike of
the ridge; all but the first of these idealizations is traditionally
made in the conventional interpretation of marine magnetic
anomalies. The model magnetization is just a real-valued function
of one variable, the distance of the point in the layer from the
ridge axis. Finding a function that can reproduce the anomaly
measurements is a task within the province of geophysical inverse
theory. Let us postpone the details until later in this chapter and
merely present a particular solution to the stated problem: it is
shown in figure 2.01b. The reader must accept on trust for the
moment that the magnetization given will account exactly for the
anomaly values indicated by the dots in figure 2.01a; that this is
indeed so is by no means obvious. First of all, despite the bilat-
eral symmetry of the anomaly, the magnetization is quite un-
symmetrical; second, there are wild oscillations that have no

counterparts in the magnetic field; finally, the actual intensities themselves, averaged over a one-kilometer-thick layer, are extraordinarily large, certainly outside the range of values found in samples recovered by dredging or drilling.

Something is drastically wrong with this magnetization as a solution to the geophysical problem, although we reassure the reader that it is a valid solution to the mathematical problem. One could be confident that if Vine and Mathews, the discoverers of sea-floor spreading, had calculated models like this one and believed in them, somebody else would have launched marine geology into its modern era. They did not find this model and they would certainly not have accepted it if they had: it is too complicated to be credible. The trouble arises from the unfortunate fact that more than one magnetization can match the magnetic anomaly values; indeed there are infinitely many widely different solutions. Many of these solutions are totally unacceptable geophysically because, like the one shown in figure 2.01b, they are too wiggly or too large; indeed our solution is fairly ordinary in comparison with some that might be constructed—for example, a magnetization that vanishes exactly everywhere within 20 km of the ridge axis, or one whose magnitude exceeds 10^{10} $A m^{-1}$ at every point.

This chapter is concerned with a theory for finding reasonable solutions to a particularly simple class of problems, the linear ones. We discover that there is in general an infinitely large family of solutions, each member of which is capable of accounting for the observations. Just as in this first example, however, not all of these solutions will be geophysically or geologically acceptable. Because the measurements themselves cannot distinguish between the different competing structures, we must impose our own preference by introducing a selection criterion. Our strategy in this chapter will be to seek functions emphasizing certain desirable attributes such as overall smoothness or closeness to an ideal form. Obviously this is not an objective exercise, but the nonuniqueness of solutions forces upon us the necessity of an arbitrary choice if we insist on representing the answer to the problem by a single model. An alternative approach, explored in later chapters, is examination of the common properties shared by the whole family of solutions. The mathematical setting for the theory will be Hilbert space, which is sufficiently flexible for most of our needs.

2.02 Linear Inverse Problems

The following is an informal prescription of a typical geophysical inverse problem. There is available a set of geophysical measurements whose values depend upon a physical parameter with unknown spatial distribution. Also available is a mathematical model which supplies a means of calculating the values of the observations if the function describing the parameter distribution is specified; this is called the solution of the *forward problem*. The corresponding *inverse problem* is to determine the unknown function from the given measurements, or if that is not possible, to learn as much as possible about it.

In our treatment we insist that the measurements comprise only a finite collection of numbers, either known exactly or known with limited precision. This assumption is clearly closer to practical reality than the one that requires infinite data sets. We represent the unknown as an element m in a suitable linear vector space V, which is normally an inner product space. (The model space will always be of infinite dimension; when only a finite number of parameters completely describe the model, their estimation is a problem primarily in statistics, not inverse theory. See, for example, Bard, 1974.) These arrangements allow us to write the solution to the forward problem symbolically as

$$d_j = F_j[m], \quad j = 1, 2, 3, \cdots N \tag{1}$$

where each functional F_j relates the model $m \in$ V to the measurement $d_j \in \mathbb{R}$. We remarked earlier that this notation can cover up some terribly complicated calculations. The simplest functionals, not only for the purposes of the forward problem but for the inverse problem too, are the linear ones. When all the functionals F_j are linear, the corresponding inverse problem is said to be a *linear inverse problem*.

Suppose now that V is a Hilbert space and that F_j is not only linear but a bounded functional as well. Then by the Riesz Representation Theorem discussed in section 1.11 we are assured that we may write in place of the general (1)

$$d_j = (g_j, m), \quad j = 1, 2, 3, \cdots N \tag{2}$$

with $g_j \in$ V. The form based upon the inner product is the one we use to develop a general theory.

As an illustration we first consider the problem of inverting magnetic anomalies for magnetization, of which the example in the previous section is a special case. Magnetic material is confined to a region $D \subset \mathbb{R}^3$ and the magnetization within D is represented by the vector-valued function of position $\mathbf{M}(\mathbf{s})$, $\mathbf{s} \in D$. The precise space of functions to which $\mathbf{M}$ belongs can be chosen at our convenience later. To an adequate approximation marine magnetic anomalies are magnetic field components in a fixed direction $\hat{\mathbf{u}}$; this direction is that of the background magnetic field which is generated in the core and hardly varies over the survey region. We measure magnetic anomalies $\hat{\mathbf{u}} \cdot \mathbf{B}$ at positions $\mathbf{r}_1, \mathbf{r}_2, \mathbf{r}_3, \cdots \mathbf{r}_N$, points outside D. If the magnetic anomaly at the jth observation point is our datum d_j:

$$d_j = \hat{\mathbf{u}} \cdot \mathbf{B}(\mathbf{r}_j), \quad j = 1, 2, 3, \cdots N . \tag{3}$$

The solution to the forward problem is found by integrating the contributions from all the elementary dipoles that constitute the magnetic source material: at an observer position $\mathbf{r}$ not within D

$$\hat{\mathbf{u}} \cdot \mathbf{B}(\mathbf{r}) = \int_D \mathbf{G}(\mathbf{r} - \mathbf{s}) \cdot \mathbf{M}(\mathbf{s}) \, d^3 \mathbf{s} \tag{4}$$

where $\mathbf{G}(\mathbf{x})$ is Green's function giving the field component at $\mathbf{x}$ in the direction of $\hat{\mathbf{u}}$ due to a point dipole at the origin with magnetic moment $\mathbf{M}$. Explicitly

$$\mathbf{G}(\mathbf{x}) = \frac{\mu_0}{4\pi} \hat{\mathbf{u}} \cdot \nabla \nabla \frac{1}{|\mathbf{x}|} \tag{5}$$

$$= \frac{\mu_0}{4\pi} \left[\frac{3\mathbf{x} \, \hat{\mathbf{u}} \cdot \mathbf{x}}{|\mathbf{x}|^5} - \frac{\hat{\mathbf{u}}}{|\mathbf{x}|^3} \right] . \tag{6}$$

We briefly digress here to remind the reader about the units used in electromagnetism. The quantity μ_0 is a defined constant of *Système International d'Unités* or SI set of measurement units that we follow. It is usually stated that $\mu_0 = 4\pi \times 10^{-7}$ H m^{-1}, but for our purposes a more convenient and equivalent way of expressing this is to write $\mu_0/4\pi = 100$ nT/(A m^{-1}) since in geophysics magnetic fields are commonly stated in nanotesla $(1\,\text{nT} = 10^{-9}\,\text{T})$ and intensity of magnetization, which is just magnetic dipole density, is measured in amps per meter (A m^{-1}). Everyone, including the author, refers to the vector $\mathbf{B}$ as the magnetic field, although officially its name is the magnetic induction or magnetic flux density.

Combining (3), (4), and (6) we write

$$d_j = \int_D \mathbf{G}_j(\mathbf{s}) \cdot \mathbf{M}(\mathbf{s}) \, d^3 \mathbf{s} \tag{7}$$

where $\mathbf{G}_j(\mathbf{s}) = \mathbf{G}(\mathbf{r}_j - \mathbf{s})$. The formula (7) for d_j is a linear functional of $\mathbf{M}$ and suggests an obvious inner product space in which magnetization distributions would be elements: it is the vector-valued function equivalent to L_2 (ignoring the distinction between Hilbert and pre-Hilbert spaces). The elements of this space are vector-valued functions of position $\mathbf{f}(\mathbf{s})$ where $\mathbf{s} \in D \subset \mathbb{R}^3$; the inner product is

$$(\mathbf{f}, \mathbf{g}) = \int_D \mathbf{f}(\mathbf{s}) \cdot \mathbf{g}(\mathbf{s}) \, d^3\mathbf{s} . \tag{8}$$

By picking this form of the inner product we have automatically accepted the norm

$$\|\mathbf{M}\| = [\int_D |\mathbf{M}(\mathbf{s})|^2 \, d^3\mathbf{s}]^{\frac{1}{2}} . \tag{9}$$

as the measure of model size in the problem. Provided the observer position $\mathbf{s}$ lies outside D, the function $\mathbf{G}_j$ has a bounded norm in this space. If the region D is bounded then we need only require that $|\mathbf{M}|$ be bounded to be certain $\mathbf{M}$ belongs to the space too. With this inner product (7) becomes an example of the general form (2).

It is important to note something that will be encountered repeatedly in later examples. The integral relating the magnetic field d_j to the internal magnetization represents the physical content of the forward problem, but this relationship may be written as an inner product in more than one way. The inner product corresponds to the norm used to measure the size of a model in the Hilbert space and as we have continually stressed there are many ways of choosing a norm even when we remain in the restricted class of normed spaces that are Hilbert spaces. If the observations are linearly related to the model, the solution to the forward problem will always be expressible as a linear functional no matter what norm is used. It is not always true, however, that the linear functional must be *bounded*, but if it is, Riesz's Theorem assures us that then we can find an inner product for it. As a trivial variant of the inverse problem for marine magnetism consider the alternative norm for the strength of magnetization:

$$\|\mathbf{M}\|_w = [\int_D |\mathbf{M}(\mathbf{s})|^2 \, w(\mathbf{s}) \, d^3\mathbf{s}]^{\frac{1}{2}} \tag{10}$$

where w is some arbitrary positive weight function inserted to give extra emphasis to the importance of magnetization in some region, say near the base of the crust. Then the inner product must become

$$(\mathbf{f}, \mathbf{g})_w = \int_D \mathbf{f}(\mathbf{s}) \cdot \mathbf{g}(\mathbf{s}) \, w(\mathbf{s}) \, d^3\mathbf{s} . \tag{11}$$

If we wish to write (7) as this kind of inner product, that is,

$$d_j = (\mathbf{g}_j, \mathbf{m})_w \tag{12}$$

and we want to retain magnetization as an element of the space, the representers $\mathbf{g}_j$ must be modified:

$$\mathbf{g}_j = \mathbf{G}_j(\mathbf{s})/w(\mathbf{s}) . \tag{13}$$

At first sight it may appear that multiplying and dividing by the same function under the integral can have no effect on the results we might obtain, but this is not the case. The weighted inner product assigns a different norm to each magnetization, and models that are small in the new norm might be very large in the original one. Because we shall be selecting our solutions on the basis of their norms, these differences have profound effects on the kinds of models we generate. We shall see that by modifying the norm, and hence the space in which we work, we can select different kinds of solutions with various desirable properties. In fact this is a major theme of the present chapter.

The general magnetization problem is too complicated to make an entirely suitable vehicle for illuminating the elementary principles of the theory. For the solution of this particular problem in all its glory, see Parker et al. (1987). Let us therefore return to the example in 2.01. For definiteness we set up a Cartesian axis system having its origin at the ridge axis, z vertically upwards, and y parallel to the ridge. We make the following sweeping simplifications: the sources lie in the plane $z = 0$, and are vertical dipoles whose intensities vary only with x; the observations are made at the sea surface (the plane $z = h$) and are z components only. Suppose the x coordinates for the magnetic field measurements are $x_1, x_2, x_3, \cdots x_N$; then by integrating the general expression (7) we find

$$d_j = \hat{z} \cdot \mathbf{B}(x_j, 0, h) = \int_{-\infty}^{\infty} g_j(x) m(x) \, dx \tag{14}$$

where

$$g_j(x) = \frac{-\mu_0}{2\pi} \frac{(x_j - x)^2 - h^2}{[(x_j - x)^2 + h^2]^2} \tag{15}$$

and $m(x)$ is the density per unit area of dipoles in the source plane. When actual models are computed, the magnetization is plotted scaled by a factor of 0.001 and given units of $A\,m^{-1}$; in this way we simulate the redistribution of the magnetization from a thin layer into one that is one kilometer thick and we retain familiar units and magnitudes for our solutions.

The Hilbert space $L_2(-\infty, \infty)$ with inner product

$$(f, g) = \int_{-\infty}^{\infty} f(x)g(x)\,dx \qquad (16)$$

seems a very natural choice, particularly since $g_j \in L_2(-\infty, \infty)$. There are, however, other ways of expressing d_j as an inner product involving m which may be useful (see 2.06).

2.03 Existence of a Solution

We continue to analyze the standard linear problem: the unknown is m, an element in a Hilbert space H; the measurements d_j are predicted in the mathematical model via the bounded linear functionals

$$d_j = (g_j, m), \quad j = 1, 2, 3, \cdots N . \qquad (1)$$

The elements $g_j \in H$ are called *data kernels* in the geophysical literature; in applied mathematics they are called *representers*, a name which the author prefers and which will be used exclusively. Let us assume at first that the values of d_j have been determined exactly. Can we always find an element m satisfying the N equations? The answer is yes, provided that the representers are linearly independent; this was shown in 1.12 by explicitly calculating expansion coefficients for a solution m_0 constructed by a linear combination of the representers. How can we tell if the representers g_j are linearly independent or not? It is possible in many geophysical problems to demonstrate linear independence of a general system by a careful analysis. Alternatively, if this should prove difficult, one need only evaluate the determinant of the Gram matrix Γ in any particular case; we saw in 1.12 that $\det \Gamma = 0$ only if the elements g_j are linearly dependent.

Linear independence of the representers is the normal situation in almost every geophysical inverse problem; nonetheless we shall discuss the question of existence of solutions should linear dependence be encountered. The following program can answer the question. By the definition of linear dependence, one representer of the set, say g_1 for definiteness, can be expressed as a linear combination of the others:

$$g_1 = \sum_{j=2}^{N} \beta_j g_j \ . \tag{2}$$

Using (1), the solution of the forward problem, we predict d_1:

$$d_1 = (g_1, m) \tag{3}$$

$$= \sum_{j=2}^{N} \beta_j (g_j, m) \tag{4}$$

$$= \sum_{j=2}^{N} \beta_j \, d_j \ . \tag{5}$$

If the observed value supplied for d_1 does not satisfy this equation, the mathematical model is inconsistent with the observations and no element m can exist that simultaneously obeys all the constraints. Consider the alternative, when the relation among the values of d_j is satisfied: we may remove the equation for d_1, because whenever the equations for $d_2, d_3, \cdots d_n$ are satisfied, the one for d_1 automatically is. Now we have $N-1$ data and $N-1$ constraints:

$$d_j = (g_j, m), \quad j = 2, 3, \cdots N \ . \tag{6}$$

This smaller system can be tested in exactly the same way: if the new system has linearly independent representers, there is a solution; if not, we check for consistency as before. Eventually the consistency of the original system will be demonstrated or refuted; the reader is invited to look at the extreme case $N=1$ which might be arrived at. When a solution does exist in the presence of linear dependence, the above procedure discovers a subset of the representers which form a basis for a subspace of H, which we usually call G.

Consider the problem of existence when the measurements are not exact. A criterion is needed to decide when the predictions of the mathematical model are in accord with the observations, since we must no longer demand perfect agreement. As might be imagined, the idea of a norm proves useful again. We regard the measurements as an N-vector $\mathbf{d}$ in the normed space E^N. The model m satisfactorily predicts the observations if the misfit as measured by the norm of E^N is small enough; that is,

$$\|\mathbf{d} - \mathbf{g}[m]\| \leq T \tag{7}$$

where

$$\mathbf{g} = ((g_1, m), (g_2, m), (g_3, m), \cdots (g_N, m)) \tag{8}$$

and T is a positive constant, the tolerance. For a statistical treatment based upon a Gaussian error law we find the following inner product norm

$$\|\mathbf{x}\| = [\mathbf{x} \cdot (C^{-1}\mathbf{x})]^{\frac{1}{2}} \tag{9}$$

where C is the covariance matrix of the error distribution; note C^{-1} is always positive definite and symmetric. We shall have much more to say about the Gaussian statistical model in chapter 3. In this case the components of $\mathbf{d}$ are the expected values of the observed quantities. Another example quite naturally expressed by this formalism is the one in which the true measurement is known to lie in a specified interval

$$d_j^- < d_j < d_j^+. \tag{10}$$

Let

$$e_j = \tfrac{1}{2}(d_j^+ - d_j^-) \tag{11}$$

$$\bar{d}_j = \tfrac{1}{2}(d_j^+ + d_j^-) \tag{12}$$

and define

$$\mathbf{d} = (\bar{d}_1, \bar{d}_2, \bar{d}_3, \cdots \bar{d}_N). \tag{13}$$

The appropriate norm for E^N is a version of the infinity norm:

$$\|x\| = \max\{|x_1|/e_1, |x_2|/e_2, |x_3|/e_3, \cdots |x_N|/e_N\} \tag{14}$$

and the value of T is unity.

The question of whether there is a satisfactory solution $m \in H$ for approximate data is seen to be trivial when the elements g_j are linearly independent: we know that the misfit

$$\|\mathbf{d} - \mathbf{g}[m]\| \tag{15}$$

can always be made zero, which is of course less than T. With linearly dependent elements we must find how small the misfit can be made by choosing an appropriate $m \in H$. If the smallest possible value (actually, the greatest lower bound) exceeds or equals T, no solution $m \in H$ is satisfactory and the mathematical model is incompatible with the observations; conversely, when the least misfit is less than T, there is a solution. We need consider only elements m_0 in the subspace G spanned by the representers because the misfit is quite independent of those parts of the solution that lie in $G^{\perp}$ (recall the identical argument in 1.12); therefore let

$$m = m_0 = \sum_{j=1}^{N} a_j g_j \qquad (16)$$

and then the minimization problem becomes one in the finite-dimensional space E^N:

$$\min_{\mathbf{a} \in E^N} \| \mathbf{d} - \Gamma \, \mathbf{a} \| . \qquad (17)$$

Here the Gram matrix Γ is singular. Techniques for solving this problem vary depending upon the norm in use for E^N.

The foregoing paragraphs should not obscure a most important fact: in almost all practical linear inverse problems the representers are linearly independent and then solutions satisfying the data always exist no matter what the actual values of d_j are. This fortunate state of affairs is in complete contrast to the situation with nonlinear problems or with artificial linear problems based upon infinite data sets. The existence of solutions does not mean they are always easy to find; requiring an exact fit to a finite data set often demands numerical computation of extreme accuracy; see section 2.08 for further discussion.

2.04 Uniqueness and Stability

Let us assume that we have established the existence of a satisfactory solution to a standard linear inverse problem; is there just one such solution? Unless the representers span the whole of H the answer is no. There are infinitely many different models in the form

$$m = m_0 + q \qquad (1)$$

where m_0 is a satisfactory solution and $q \in G^{\perp}$, the orthogonal complement to the subspace G spanned by the representers g_j. We can show this directly without appealing to the Decomposition Theorem or the completeness of H. We assume the representers are linearly independent, or if they are not, we choose a linearly independent subset of them which spans G, that is, a basis for G.

We explicitly construct an element $q_1 \neq 0$ satisfying the equations

$$(g_j, q_1) = 0, \quad j = 1, 2, 3, \cdots N . \tag{2}$$

Assume that the dimension of H is greater than N; then there must be an element $\tilde{g}_1 \in$ H such that the elements $\tilde{g}_1, g_1, g_2, \cdots g_N$ form a linearly independent set. To the N constraints on q_1 we add

$$(\tilde{g}_1, q_1) = \tilde{d}_1 \tag{3}$$

where $\tilde{d}_1 \neq 0$. The combined system (2) and (3) consists of $N + 1$ equations for the element $q_1 \in$ H; the representers of the system are linearly independent and therefore we can form the minimum norm solution for any value of $\tilde{d}_1$. This solution is not the zero element if $\tilde{d}_1 \neq 0$, because by Schwarz's inequality

$$0 < |\tilde{d}_1| = |(\tilde{g}_1, q_1)| \tag{4}$$

$$\leq \|\tilde{g}_1\| \|q_1\| . \tag{5}$$

The family of all elements q satisfying

$$(g_j, q) = 0, \quad j = 1, 2, 3, \cdots N \tag{6}$$

is a subspace of H which we have denoted by $G^\perp$; it is known as the *null space* of the problem or the *annihilator*. If H is infinite dimensional we may repeat the above process, adding further linearly independent elements $\tilde{g}_2, \tilde{g}_3, \cdots$; in this way we can construct arbitrarily many elements $q_1, q_2, \cdots \in G^\perp$, and thus show that the null space is also infinite dimensional. When the dimension of H is $M > N$, that of $G^\perp$ is $M - N$. Finally, when H has dimension N, the null space is just the element 0; if exact observations were specified in the problem, the solution m_0 is then unique.

We can understand the strange magnetization of figure 2.01a a little better now: it is just one of the infinite collection of mathematically correct functions that reproduce the data. The reader must not make the mistake of thinking that the solution $m = m_0$ is *the* solution and that the others obtained by adding components in $G^\perp$ are somehow irrelevant. There is no reason to believe that the true magnetization happens to be the one that exactly minimizes the norm of m, although, as we argue in the next section, with a reasonable choice of norm, the correct answer should not lie far away from it. Of course, we may choose any norm from among an infinite variety of inner-product norms; this in itself shows that ambiguity in the solution is inescapable.

It may have occurred to the reader that the difficulties of nonuniqueness are largely self-imposed: perhaps we are setting the problem up in quite the wrong way by seeking solutions which are functions but permitting only finite sets of observations. An apparently promising alternative is a more classical approach: assume the availability of ideal observations (for example a complete anomaly curve in 2.01, not just the finite set of measurements) and an algorithm that constructs the unique model associated with them; if imperfect measurements are to be used, they must be completed (by interpolation and extrapolation in our first example) and the exact theory is then applied to the reformed data. If this procedure is to work, the details of the completion process should not matter: any reasonable filling-in process should yield approximately the same answer. A precise formulation of this idea leads to the definition of a *stable* or *well-posed* problem. Unfortunately most geophysical inverse problems are *ill-posed* in this sense.

To study this question we must depart from our custom of working in finite-dimensional spaces for the observations. The observations are represented by an element d of D, a normed linear vector space, while the solution is as usual $m \in V$, where V is another normed space (not necessarily an inner-product space). In some suitable subset $D' \subseteq D$ every data element is associated with a unique solution element m; symbolically we write $m = \phi(d)$, $d \in D' \subseteq D$, $m \in V$ where ϕ is a function mapping D' onto V. The solution is *stable* at some fixed $d_0 \in D'$ if $\phi(d) \to \phi(d_0)$, whenever $d \to d_0$. In the "strong topology" (which means we define convergence in the sense of the norm) this is equivalent to

$$\lim_{\|d - d_0\| \to 0} \|\phi(d) - \phi(d_0)\| = 0 . \tag{7}$$

Some readers will recognize this as a statement that ϕ depends continuously on d at d_0. It may come as a surprise that continuity is not automatic when the solution is unique; after all, we have $\|\phi(d) - \phi(d_0)\| = 0$ whenever $\|d - d_0\| = 0$. Nonetheless, as mentioned earlier, the majority of geophysical inverse problems are not stable. Lack of continuity means that no matter how small the difference is between two distinct data elements, the corresponding pair of solutions can still be significantly separated from each other. When this situation prevails, different methods of perfecting incomplete data sets may yield quite different models and the diversity of solutions we had hoped to avoid by using an analytic solution remains.

As an aside it should be mentioned that until fairly recently it was widely believed that stability always attended the solution to the forward problem where the behavior of a physical system with known parameters must be predicted from known physical laws. With flourishing of the study of nonlinear dynamical systems, we now appreciate that this is often untrue: examples of extreme sensitivity to initial conditions that make prediction practically impossible can be found in celestial mechanics, chemical kinetics, and fluid dynamics, to name only a few subject areas.

A geophysical illustration is called for at this point; we turn again to the one-dimensional magnetic anomaly of 2.01. We now assume that the anomaly a is known exactly everywhere on the real line. The forward problem is solved with the convolution

$$a(x) = \int_{-\infty}^{\infty} g(x-v)m(v)dv \tag{8}$$

$$= g * m \tag{9}$$

where from 2.02(15)

$$g(s) = -\frac{\mu_0}{2\pi} \frac{s^2-h^2}{(s^2+h^2)^2} . \tag{10}$$

In keeping with the policy of using inner-product spaces where convenient, the magnetization m is an element of $L_2(-\infty, \infty)$. We also need to define a space for the data elements a; it will be seen in a moment that $\|a\|_2$ always exists; therefore let us choose $L_2(-\infty, \infty)$ as a home for a as well as for m. First we show that the inverse problem for m is unstable, that is, two elements m_1, $m_2 \in L_2$ may not be close, yet the corresponding data elements a_1, a_2 can be separated by an arbitrarily small positive distance in L_2. This would not be surprising but for the fact that if $a_1=a_2$ then $m_1=m_2$, in other words the solution is unique.

To demonstrate the instability, we examine a special magnetization function which exhibits lack of continuity under the norms we have chosen. A simple and indeed widely effective test function is the "boxcar": consider two magnetizations that differ by an amount Δ given by

$$\Delta(x) = m_2(x) - m_1(x) \tag{11}$$

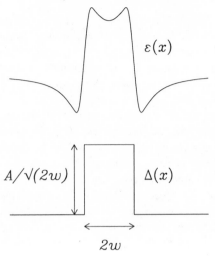

Figure 2.04a: The crustal magnetization $\Delta(x)$, a boxcar function, generates a magnetic anomaly $\varepsilon(x)$ at the ocean surface in the idealized system under consideration.

$$= \begin{cases} A/(2w)^{\frac{1}{2}}, & |x| \leq w \\ \phantom{A/(2w)^{\frac{1}{2}},} 0, & |x| > w \end{cases} \qquad (12)$$

where A, $w > 0$. It is easily seen that $\|\Delta\|_2 = A$ independently of the width parameter w. We substitute (12) into (8) to obtain the corresponding anomaly difference

$$\varepsilon(x) = a_2(x) - a_1(x) \qquad (13)$$

$$= \frac{\mu_0 A}{\pi(8w)^{\frac{1}{2}}} \left[\frac{x+w}{(x+w)^2+h^2} - \frac{x-w}{(x-w)^2+h^2} \right]. \qquad (14)$$

The shapes of Δ and ε are shown in figure 2.04a. The norm of ε can be found most elegantly by Parseval's Theorem (see next paragraph):

$$\|\varepsilon\|_2 = \frac{\mu_0 A}{(8\pi h)^{\frac{1}{2}}} \left[\frac{w}{w^2+h^2} \right]^{\frac{1}{2}}. \qquad (15)$$

Let us fix the value of A, but let w vary; we see that $\|\varepsilon\|_2 \to 0$ when $w \to \infty$, but $\|\Delta\|_2$ remains fixed; similarly as $w \to 0$, we have $\|\varepsilon\|_2 \to 0$, while $\|\Delta\|_2$ maintains its constant value. These two limits express the unstable nature of the inverse problem: there exist virtually identical anomaly functions (because $\|a_2 - a_1\|_2$ can be made as small as one pleases), whose corresponding source magnetizations differ by an arbitrary amount (since $\|m_2 - m_1\|_2 = A$ can be any positive number). Stability is obviously defined relative to a pair of norms, and it is always possible to choose norms that will make any linear problem appear stable; a traditional choice in stability studies is the uniform norm. The boxcar test function shows that our example is unstable in this norm as well.

To prove the uniqueness of solutions in L_2 we appeal to the theory of Fourier transforms. The reader is expected to be familiar with Fourier theory at least to the level of Bracewell's book (1978); for a rigorous yet readable treatment in a Hilbert space setting see Dym and McKean (1972). We first sketch the development of some of the necessary results. For a suitable subspace S of well-behaved functions in complex $L_2(-\infty, \infty)$, one may define a linear transformation F of $f \in S$ by $\hat{f} = Ff$ where

$$\hat{f}(\lambda) = \int_{-\infty}^{\infty} f(x) e^{-2\pi i \lambda x} \, dx \tag{16}$$

which is of course the *Fourier transform*. Then it can be shown that $\|\hat{f}\|_2 = \|f\|_2$ which is known as *Plancherel's Theorem*. Norm preservation in S allows F to be extended from that subspace to all of L_2, maintaining the norm preserving property (see the references for details). A consequence of the norm preservation is *Parseval's Theorem*, which is proved as follows (we drop the subscript on the norm): for $f, g \in L_2(-\infty, \infty)$

$$\|f + g\|^2 = \|\hat{f} + \hat{g}\|^2 \tag{17}$$

$$\|f\|^2 + \|g\|^2 + 2\mathrm{Re}(f, g) = \|\hat{f}\|^2 + \|\hat{g}\|^2 + 2\mathrm{Re}(\hat{f}, \hat{g}) \tag{18}$$

$$\mathrm{Re}(f, g) = \mathrm{Re}(\hat{f}, \hat{g}) \ . \tag{19}$$

Similarly, when $\|f + ig\|$ is considered, we find that the imaginary parts of the inner products are identical as well. Therefore

$$(f, g) = (\hat{f}, \hat{g}) \tag{20}$$

or

$$\int_{-\infty}^{\infty} f(x) g(x)^* \, dx = \int_{-\infty}^{\infty} \hat{f}(\lambda) \hat{g}(\lambda)^* d\lambda \ . \tag{21}$$

This is Parseval's Theorem. Another result is found by a simple change of variable in (16):

$$\int_{-\infty}^{\infty} g(x - v) e^{-2\pi i \lambda v} \, dv = \hat{g}(\lambda)^* e^{2\pi i \lambda x} \ . \tag{22}$$

Combining (21) and (22) we have the *Convolution Theorem*: for f, $g \in L_2(-\infty, \infty)$

$$\int_{-\infty}^{\infty} g(x-v)f(v)^* dv = \int_{-\infty}^{\infty} \hat{g}(\lambda)\hat{f}(\lambda) e^{2\pi i x \lambda} d\lambda \qquad (23)$$

where the left side is guaranteed to exist for f, $g \in L_2$ by Schwarz's inequality. Notice that the function given by the product $\hat{g}(\lambda)\hat{f}(\lambda)$ is not necessarily in L_2. If, however, $\hat{g}(\lambda)$ is bounded the product is in L_2 since, when $|\hat{g}(\lambda)| \le \alpha$,

$$|\hat{g}(\lambda)\hat{f}(\lambda)| \le \alpha|\hat{f}(\lambda)| \qquad (24)$$

and

$$\|\hat{g}\,\hat{f}\,\| \le \alpha\|\hat{f}\,\| . \qquad (25)$$

A final, almost obvious result which we need is this: for $f \in L_2(-\infty, \infty)$, if $Ff = 0$, $f = 0$; this follows at once from norm preservation of F.

We return to the discussion of uniqueness. Suppose there are two magnetizations m_1, $m_2 \in L_2(-\infty, \infty)$ both reproducing a given anomaly function a. Then by (8)

$$\int_{-\infty}^{\infty} g(x-v)[m_2(v)-m_1(v)]\,dv = 0 . \qquad (26)$$

And if $m_2 - m_1 = \Delta$ as before

$$\int_{-\infty}^{\infty} g(x-v)\Delta(v)\,dv = 0 . \qquad (27)$$

Using the Convolution Theorem we have

$$0 = \int_{-\infty}^{\infty} \hat{g}(\lambda)^* \hat{\Delta}(\lambda) e^{2\pi i x \lambda} d\lambda \qquad (28)$$

since $\Delta^* = \Delta$ and g, $\Delta \in L_2$. Moreover a short calculation shows that for the convolution kernel (10)

$$\hat{g}(\lambda) = \hat{g}(\lambda)^* = \mu_0 \pi |\lambda| e^{-2\pi h |\lambda|} \qquad (29)$$

and it is easily verified that $|\hat{g}(\lambda)| \le \mu_0/2h\,e$. Thus the integral (28) above can be written symbolically

$$(F\hat{k})^* = F\hat{k} = 0 \qquad (30)$$

where $\hat{k} \in L_2(-\infty, \infty)$ and $\hat{k}(\lambda) = \hat{g}(\lambda)\hat{\Delta}(\lambda)$. It follows that $\hat{k} = 0$, in other words, that $\hat{g}(\lambda)\hat{\Delta}(\lambda) = 0$ almost everywhere on the real line.

We see from (29) that $\hat{g}(\lambda) > 0$ except at $\lambda = 0$, and therefore $\hat{\Delta}(\lambda)$ vanishes almost everywhere. Equivalently $\hat{\Delta} = 0$ and finally then $\Delta = 0$. We have shown that, when the anomaly functions a_1 and a_2 are identical and the magnetizations m_1, m_2 are in L_2, these two elements are the same, and so the solution to the inverse problem is unique.

It is instructive to see that the space in which solutions are defined can affect their uniqueness. Since

$$\int_{-\infty}^{\infty} g(x)\,dx = 0 \qquad\qquad (31)$$

two magnetizations m and $m + c$, where c is a nonzero constant function, give rise to identical anomaly values. Had we chosen a model space that could contain both these magnetizations, the solution would not be unique even with ideal data. Obviously, if $m \in L_2(-\infty, \infty)$ then $m + c$ is not in L_2 since c does not possess a bounded norm.

Instability usually arises, roughly speaking, when the observations are a smoothed version of the model. This happens in many geophysical problems besides the one we have just discussed. Naturally nonlinear problems can be unstable too. For example, in the magnetotelluric problem of section 5.02, the conductivity model may be an arbitrary element in L_2, while the measured magnetotelluric response is an analytic function infinitely differentiable at all frequencies. This is precisely the same situation as we find with the magnetic anomaly problem. The smoothing action of the (nonlinear) forward operator makes the inverse magnetotelluric problem unstable.

Exercise

2.04(i) Retaining the space L_2 for the magnetic anomalies, devise a norm for the space of models so that relative to the pair of norms, the inverse problem is stable.

2.05 Some Special Solutions

In the standard linear inverse problem, the observations are expressible as bounded linear functionals of an element in a linear vector space. It has been established that if there are any solutions, there are infinitely many; on the basis of the data alone there is no way to choose between them. Two strategies will be

developed for dealing with this dilemma. The subject of the
present section is the construction of special solutions that max-
imize some desirable property through a norm (or normlike functional). The other approach, dealt with in a later chapter, aban-
dons the attempt to delineate a particular model; instead inferences about the whole class of acceptable solutions are made
using properties that are common to all members of the class.

One of the unsatisfactory aspects of the model in figure
2.01b is its very large amplitude. In a realistic geophysical problem it is usually possible to estimate a rough upper limit on plausible values of the unknown parameter. The estimates may
derive from samples of material or from physical arguments; it is
always possible to identify some value as ridiculously large. To
discriminate against very large solutions we can find the model
with the smallest norm. For a given inner-product space, the
norm-minimizing solution is easily found as described in 1.12; we
have already made great use of this particular solution, and
therefore we shall not repeat the details here. Remember, however, that all our considerations have been for models that fit the
data precisely. The question of how to handle inexact observation
will be postponed until chapter 3.

Another way of producing plausible solutions is to find the
one that is as close as possible to some preconceived model. For
example, in the lower mantle, a particularly simple model for density takes self-compression as the sole cause of density increase;
one would be suspicious of a solution that deviated too far from
this ideal, the Adams-Williamson profile (see Stacey, 1977). Let
$n \in H$ be the preferred model; we can easily find the element
$m \in H$ nearest to n in the sense of the norm, which satisfies

$$(g_j, m) = d_j, \quad j = 1, 2, 3, \cdots N . \tag{1}$$

Write $\Delta = m - n$, then clearly

$$(g_j, \Delta) = d_j - (g_j, n) = \tilde{d}_j . \tag{2}$$

The numbers $\tilde{d}_j$ can be calculated since n is known; we need only
solve the new system for the element Δ_0 with the smallest norm in
the usual fashion to obtain $n + \Delta_0$, the element nearest n.

A second illustration is provided by modeling the geomagnetic field through time. The simplest model of *secular variation*
(the changes in the field) would be that of least change. If one
constructs a simple model from today's data and another simple

one from observations of fifty years ago, it might appear that the
field is increasing in complexity because today's data are so much
more dense and complete and could therefore support much more
detail. One suggestion is to seek the model based on the old data
that resembles today's as much as possible, thus keeping the secular variation to a minimum. For an extensive discussion see
Constable et al. (1993).

The case when n can be specified precisely is relatively rare.
More commonly we can specify a particular *type* of solution as
more desirable than any other. For instance in marine paleomagnetism, a uniformly magnetized seamount would be considered ideal, although the direction and intensity of magnetization could not be specified ahead of time. Another important class
of problems is the one in which smooth models are required; then
we might use a norm that is sensitive to gradients in the model
and again a uniform model would be the preferred candidate.
Within the context of a linear theory, these types of problem can
all be presented as follows. We are given a collection of fixed elements $h_1, h_2, \cdots h_K \in H$. It will be natural to assume that the
elements h_k are linearly independent so that they form a basis for
a subspace K. The preferred solution would be in the form

$$h = \sum_{k=1}^{K} \beta_k \, h_k \, . \tag{3}$$

Our task is to discover the element $m_* \in H$, satisfying the data so
that departure from this ideal is as small as possible; that is, if

$$m = h + r \tag{4}$$

and

$$(g_j, m) = d_j, \quad j = 1, 2, 3, \cdots N \tag{5}$$

then m_* is the element obeying these equations with $\| r \|$ as small
as possible. We shall see this is equivalent to seeking the data-
fitting model with the smallest component in $K^\perp$, the orthogonal
complement of K: in the language of projection operators, the
model minimizing $\| P_{K^\perp} m \|$. The value of $\| r \| = \| P_{K^\perp} m \|$ is a semi-
norm of m—it obeys all the conditions required of a norm except
that $\| r \|$ may vanish without m necessarily being 0 (see section
1.05); we call m_* a *seminorm-minimizing* solution. When uncertainty is permitted in the data, finding models that minimize
$\| P_{K^\perp} m \|$ is called *regularization*; this is a natural extension of the

procedure about to be developed for noise-free observations (see 3.02).

We assume K, the number of free parameters in the ideal model, is less than N; also it is evident that for a unique solution none of the elements h_k must lie in $G^\perp$, the subspace orthogonal to G which is spanned by the representers g_j. If for some i, $h_i \in G^\perp$, the contribution to m from this element cannot be deduced from the data because $(g_j, h_i) = 0$ for all j. Actually, as we shall shortly see, a somewhat stronger version of this condition is necessary and sufficient.

To calculate m_* observe that if we knew the values of β_k we could find the smallest $\|r\|$ as in the case when the ideal element was known precisely. Then r is a norm-minimizing element and so $r \in G$ or equivalently

$$r = \sum_{j=1}^{N} \alpha_j \, g_j \, . \tag{6}$$

Another property of r follows from the Decomposition Theorem; let K be the subspace spanned by the elements $h_1, h_2, \cdots h_K$ and write $r = p + q$ where $p \in K$ and $q \in K^\perp$. Clearly $\|r\|^2 = \|p\|^2 + \|q\|^2$; for the smallest possible $\|r\|$ we must have $p = 0$, since any part of $m_* \in K$ can be supplied in h without increasing $\|r\|$ unnecessarily. Thus we find the property that $r \in K^\perp$, which is something we might even choose to view as part of the definition of r; in any case the model that minimizes $\|r\|$ is the one that minimizes $\|P_{K^\perp} m\|$. Because h_k are basis elements of K, the fact that $r \in K^\perp$ is the same as

$$(h_k, r) = 0, \quad k = 1, 2, 3, \cdots K \tag{7}$$

or from (6)

$$\sum_{j=1}^{N} (h_k, g_j)\alpha_j = 0, \quad k = 1, 2, 3, \cdots K \, . \tag{8}$$

Now we indulge in a little algebra; take the expansions for h and r, that is (3) and (6), and substitute them into (4) and then apply (5), the equations stating that m satisfies the data:

$$\sum_{k=1}^{K} (g_j, h_k)\beta_k + \sum_{i=1}^{N} (g_j, g_i)\alpha_i = d_j, \quad j = 1, 2, 3, \cdots N \, . \tag{9}$$

The picture is clarified by some matrix notation: let $\alpha, \mathbf{d} \in \mathbb{R}^N$, $\beta \in \mathbb{R}^K$, $A \in M(N \times K)$ and $\Gamma \in M(N \times N)$; then (9) and (8) become

$$A\beta + \Gamma\alpha = \mathbf{d} \qquad\qquad (10)$$

$$A^T\alpha = 0 \qquad\qquad (11)$$

where

$$A_{jk} = (g_j, h_k), \quad j = 1, 2, 3, \cdots N, \quad k = 1, 2, 3, \cdots K$$
$$\Gamma_{ij} = (g_i, g_j), \quad i = 1, 2, 3, \cdots N, \quad j = 1, 2, 3, \cdots N \qquad (12)$$

and the meaning of the vectors α, $\mathbf{d}$, and β is obvious. The equations above can be merged into one elegant package if so desired:

$$\begin{bmatrix} \Gamma & A \\ A^T & O \end{bmatrix} \begin{bmatrix} \alpha \\ \beta \end{bmatrix} = \begin{bmatrix} \mathbf{d} \\ 0 \end{bmatrix}. \qquad\qquad (13)$$

Here O and 0 are a matrix and a vector of zero elements with the appropriate sizes. The above composite linear system can be solved as it stands or we can perform some of the operations ourselves and arrive at explicit expressions for α and β separately as follows, which has the advantage that we can derive the condition for the solvability of the system. Under the usual assumption that the representers g_j are linearly independent, Γ is nonsingular. Therefore we may premultiply the first matrix equation (10) by $A^T\Gamma^{-1}$

$$A^T\Gamma^{-1}A\beta + A^T\alpha = A^T\Gamma^{-1}\mathbf{d} \qquad\qquad (14)$$

and applying (11) we see

$$A^T\Gamma^{-1}A\beta = A^T\Gamma^{-1}\mathbf{d}. \qquad\qquad (15)$$

Let us assume that $A^T\Gamma^{-1}A$ is nonsingular; then we can solve this equation for β and put the result in (10) to obtain

$$\beta = (A^T\Gamma^{-1}A)^{-1}A^T\Gamma^{-1}\mathbf{d} \qquad\qquad (16)$$

$$\alpha = \Gamma^{-1}(\mathbf{d} - A\beta). \qquad\qquad (17)$$

These equations give us the coefficients for the desired solution

$$m_* = \sum_{k=1}^{K} \beta_k h_k + \sum_{j=1}^{N} \alpha_j g_j. \qquad\qquad (18)$$

The condition that $A^T\Gamma^{-1}A$ may be inverted reduces to the requirement that

$$A\gamma = 0 \qquad\qquad (19)$$

only if $\gamma = 0 \in \mathbb{R}^K$, because Γ^{-1} is nonsingular. Equivalently

$$(g_j, \sum_{k=1}^{K} \gamma_k h_k) = 0, \quad j = 1, 2, 3, \cdots N \quad (20)$$

only if $\gamma_1 = \gamma_2 = \cdots = \gamma_K = 0$. This may be stated as the condition that the orthogonal projections of the elements h_k onto G be linearly independent. This is the stronger version of the condition that none of the elements of K should lie in $G^{\perp}$.

A different derivation gives an interpretation of β and is useful for certain numerical calculations. Consider $m - m_0$ where m is any solution satisfying the data and m_0 is the norm-minimizing solution. It may readily be verified (see 1.12) that

$$m - m_0 \in G^{\perp} \quad (21)$$

and this will be used as the condition that guarantees m satisfies the data constraints. Let us expand m in this expression with (4)

$$h + r - m_0 \in G^{\perp} . \quad (22)$$

The orthogonal projection operators of Hilbert space described in section 1.11 prove very useful here; apply the projection operator P_G to the element $h + r - m_0$. From our earlier discussion $r \in G$, and m_0 also belongs to this subspace; thus they are unaffected by the projection operator. Therefore

$$P_G h + r - m_0 = 0 . \quad (23)$$

Thus

$$r = m_0 - P_G h . \quad (24)$$

Our objective is to minimize $\|r\|$. While m_0 is fixed we are free to choose the element h provided it remains within the subspace K. This is the standard minimization problem solved by the Projection Theorem; to make use of the answer given in 1.12 we need a basis for the subspace of approximating elements. Since a basis for K is composed of the elements h_k, a basis for the space of projected elements is just

$$l_k = P_G h_k, \quad k = 1, 2, 3, \cdots K . \quad (25)$$

provided of course that these are linearly independent, which we shall assume. Hence the smallest $\|r\|$ is found by constructing the least-squares approximation to m_0 from the linear combination

$$l = \sum_k \beta_k \ l_k \ . \tag{26}$$

A little algebra will verify that the best approximating coefficients β_k here are identical to the numbers defined by (16). Substituting this solution for r into (4) and (18) we find

$$m_* = m_0 + \sum_k \beta_k (h_k - l_k) \ . \tag{27}$$

Despite our use of the Decomposition Theorem, which requires H to be Hilbert (that is complete) the result remains true in any inner-product space. An uninteresting proof can be supplied by writing the smallest value of $\|r\|^2$ consistent with the data as an explicit functional of the coefficients β_k; this functional is a quadratic form which must be minimized over β_k. No use of completeness appears in this demonstration.

A limiting case of the seminorm minimization is a technique called *collocation*. Instead of a few elements with special status, there is an infinitely large family of favored elements, arranged in a hierarchy of decreasing quality, usually ordered according to decreasing smoothness. Typical examples are, in one dimension, the powers of x, or on the surface of a sphere, the spherical harmonics, arranged in order of increasing degree. With N data constraints one chooses the first N elements of the family. If the orthogonal projections $P_G h_k$ are linearly independent, the observations can be satisfied; in other words, the solution is obtained by expanding in a basis for the subspace K, which has been chosen with precisely the right dimension to yield a unique result. If, as is usually the case, the objective is to produce smoothly varying models, collocation is not a very satisfactory approach. As illustrated in section 2.07, the results can be astonishing erratic even when the elements of K are all very nicely behaved. Collocation is still widely used in geophysical modeling problems, particularly in situations where the observations are contaminated with noise; then a model is constructed in a subspace of ideal elements with dimension less than N. Since the norm- or seminorm-minimizing solutions we advocate are also just expansions in a particular basis, the moral of the story is that it is not easy to guess what a good basis might be in any given problem; it is simpler to start from a principle like minimum roughness and let that idea generate the basis for the expansion.

We now have the tools to solve a number of interesting problems. Some examples will be expanded in detail before we tackle the next major question—how to incorporate observational uncertainty into the theory.

2.06 The Magnetic Anomaly Profile

The present section and several subsequent ones describe applications of the foregoing theory for exact data to a variety of simple geophysical problems. The intention is to illustrate the flexibility of our approach by solving problems in different ways making use of a number of Hilbert or pre-Hilbert spaces.

In the magnetic anomaly problem, it will be recalled from sections 2.02 and 2.04 that the observations $d_1, d_2, d_3, \cdots d_N$ are vertical components of anomaly measured at $x_1, x_2, x_3 \cdots x_N$ and that

$$d_j = \int_{-\infty}^{\infty} g_j(x)\, m(x)\, dx \tag{1}$$

where m is the density of vertical dipoles in the source layer and

$$g_j(x) = \frac{-\mu_0}{2\pi} \frac{(x - x_j)^2 - h^2}{[(x - x_j)^2 + h^2]^2}\,. \tag{2}$$

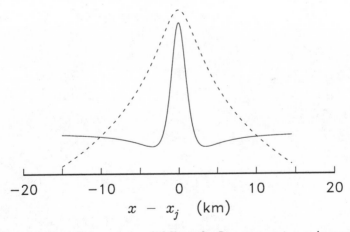

$$x - x_j \quad (\text{km})$$

Figure 2.06a: Shown as a solid line, the L_2 representer g_j in equation (1) for the magnetic anomaly data. Dashed curve, the representer $\tilde{g}_j$ of (29). Since the units of the representers differ, the functions have been scaled and shifted arbitrarily for ease of comparison.

For the numerical example I chose $h = 2$ km and an even spacing between observation points of 1.0 km; a typical function $g_j(x)$ is illustrated in figure 2.06a. Since $g_j \in L_2(-\infty, \infty)$ let us initially assume that $m \in L_2(-\infty, \infty)$ also. With the usual inner product we have the familiar expression

$$d_j = (g_j, m), \quad j = 1, 2, 3, \cdots N . \tag{3}$$

First we find the solution m_0 which has the smallest norm of all those satisfying these data.

We have already relied upon the norm-minimizing solution to help us solve a number of problems. Recall from 1.12 that we must first calculate the Gram matrix:

$$\Gamma_{jk} = (g_j, g_k) \tag{4}$$

$$= \int_{-\infty}^{\infty} \left[\frac{\mu_0}{2\pi} \right]^2 \frac{(x-x_j)^2 - h^2}{[(x-x_j)^2 + h^2]^2} \cdot \frac{(x-x_k)^2 - h^2}{[(x-x_k)^2 + h^2]^2} \, dx . \tag{5}$$

We can avoid an orgy of algebraic manipulation if we make use of Parseval's Theorem (see 2.04):

$$\Gamma_{jk} = (\hat{g}_j, \hat{g}_k) \tag{6}$$

where

$$\hat{g}_j(\lambda) = \int_{-\infty}^{\infty} g_j(x) e^{-2\pi i \lambda x} \, dx \tag{7}$$

$$= \mu_0 \pi |\lambda| e^{-2\pi h |\lambda|} e^{-2\pi i x_j \lambda} . \tag{8}$$

Thus

$$\Gamma_{jk} = \int_{-\infty}^{\infty} \mu_0^2 \pi^2 \lambda^2 e^{-4\pi h |\lambda|} e^{-2\pi i (x_j - x_k)\lambda} \, d\lambda \tag{9}$$

$$= 2\mu_0^2 \pi^2 \, \text{Re} \int_0^{\infty} \lambda^2 e^{-2\pi [2h - i(x_j - x_k)]\lambda} \, d\lambda \tag{10}$$

$$= 2\mu_0^2 \pi^2 \, \text{Re} \, \frac{1}{(2\pi [2h - i(x_j - x_k)])^3} \int_0^{\infty} s^2 e^{-s} \, ds \tag{11}$$

$$= \frac{\mu_0^2 h}{\pi} \frac{4h^2 - 3(x_j - x_k)^2}{[4h^2 + (x_j - x_k)^2]^3} . \tag{12}$$

If $x_{j+1} - x_j = $ constant, this Gram matrix is a *Toeplitz matrix*, that is, one in which the value of Γ_{jk} depends only upon $j - k$. Toeplitz matrices can be very efficiently inverted with special algorithms (Golub and Van Loan, 1983); see section 3.07.

The next step of course is to solve for $\alpha \in \mathbb{R}^N$ the linear system

$$\Gamma\alpha = \mathbf{d} . \tag{13}$$

Solutions can be guaranteed if Γ is not singular, which is equivalent to the requirement that the representers g_j should be linearly independent. We show this to be the case, provided that the observation coordinates are distinct.

Our proof depends again on the theory of the Fourier transform. It follows from the fact that the Fourier transform maps $L_2(-\infty, \infty)$ linearly and uniquely onto itself and that if the transforms $\hat{g}_j$ form a linearly independent set, then the representers are linearly independent too. Linear dependence of the transforms would mean by definition that

$$\sum_{j=1}^{N} a_j \hat{g}_j = 0 \tag{14}$$

where the a_j are complex constants not all zero. Then

$$0 = \sum_{j=1}^{N} a_j \mu_0 |\lambda| \, e^{-2h\,|\lambda|} \, e^{-2\pi i x_j \lambda} \tag{15}$$

$$= (\mu_0 \pi |\lambda| \, e^{-2\pi h\,|\lambda|}) \sum_{j=1}^{N} a_j \, e^{-2\pi i x_j \lambda} \tag{16}$$

where this equation holds for almost all λ (that is the exceptional set has measure zero). The term in parentheses is clearly positive except at $\lambda = 0$, so that for linear dependence we must have

$$\sum_{j=1}^{N} a_j \, e^{-2\pi i x_j \lambda} = 0 . \tag{17}$$

Consider the real number

$$I = \int_{-L}^{L} \left| \sum_{j=1}^{N} a_j \, e^{-2\pi i x_j \lambda} \right|^2 d\lambda \tag{18}$$

$$= 2L \sum_{j=1}^{N} |a_j|^2 + \sum_{j \neq k} a_j a_k^* \, \frac{\sin 2\pi (x_j - x_k)L}{\pi(x_j - x_k)} \tag{19}$$

$$\geq 2L \sum_{j=1}^{N} |a_j|^2 - \sum_{j \neq k} |a_j a_k^*|/|\pi(x_j - x_k)| . \tag{20}$$

Table 2.06A: Artificial Magnetic Anomalies

x_j (km)	d_j (nT)	x_j (km)	d_j (nT)	x_j (km)	d_j (nT)
−15	−168.72	−4	192.36	7	−28.557
−14	−192.57	−3	177.66	8	−137.87
−13	−197.78	−2	165.33	9	−104.91
−12	−199.79	−1	157.87	10	−135.62
−11	−188.49	0	155.41	11	−188.49
−10	−135.62	1	157.87	12	−199.79
−9	−104.91	2	165.33	13	−197.78
−8	−137.87	3	177.66	14	−192.57
−7	−28.557	4	192.36	15	−168.72
−6	138.00	5	195.60		
−5	195.60	6	138.00		

For fixed values of the constants a_j it is clearly possible to choose L large enough to ensure that $I > 0$, unless all the a_j are zero. We see therefore that $\Sigma\, a_j e^{-2\pi i x_j \lambda}$ cannot vanish for almost all λ; it follows then that $\hat{g}_j$ and hence g_j form a linearly independent set of elements. Notice how the proof breaks down if $x_j = x_k$ for some $k \neq j$.

Let us proceed to the numerical solution for the example shown in figure 2.01a and tabulated in table 2.06A: the explicit evaluation of m_0 is shown in figure 2.06b. The results are quite gratifying: the magnetization is of reasonable intensity with the expected symmetry and polarity reversals. The magnetization decays away to zero outside the central 30 km, which is the interval in which anomalies are available. An unsatisfactory aspect of this solution is perhaps the number of unsightly undulations in the central anomaly region (within ±6 km of the ridge axis) and at about ±12 km from the origin. Are these a necessary part of a solution? This could be an important matter because in paleomagnetism there is a great deal of uncertainty in the behavior of the strength of the main geomagnetic field and the possibility exists that there may have been short periods of high or low intensity. One way of deciding this question is by examining the resolution of the data; this is the topic of a later chapter. Another approach is to construct as smooth a model as possible by

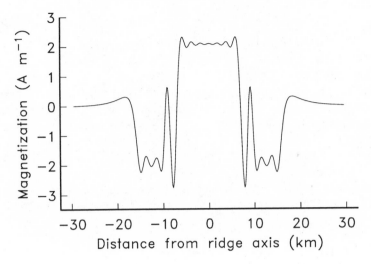

Figure 2.06b: The model satisfying the field values shown in figure 2.01a exactly and possessing the smallest L_2 norm. The solution shown is the equivalent magnetization in a 1-km-thick layer.

minimizing a functional that discriminates against oscillations. We shall follow this course.

Now we choose a different Hilbert space in which to work. We want the norm of the space to contain the first derivative of m and so the inner product must involve that derivative instead of plain $m(x)$. This is the idea mentioned in section 2.02 where the inner product was modified but the changes were made in such a way as to preserve the relationship between the observations and the model. We know that merely changing the norm cannot affect the fact that the data are *linear* functionals of the model, but that does not necessarily mean that the relation can be expressed as an inner product, because that requires the functional to be bounded. To get a derivative of m into the functional we integrate the linear relation (1) for d_j by parts:

$$d_j = - \int_{-\infty}^{\infty} g_j^{(-1)}(x) m^{(1)}(x) \, dx \qquad (21)$$

where $m^{(1)}(x) = dm/dx$ and

$$g_j^{(-1)}(x) = \int_{-\infty}^{x} g_j(y) \, dy \qquad (22)$$

$$= \frac{\mu_0}{2\pi} \frac{x - x_j}{(x - x_j)^2 + h^2} . \tag{23}$$

Note we have used the fact that $g^{(-1)}(\infty) = 0$. We see that integration by parts has expressed d_j as a linear functional of the derivative of m. In an appropriate space where the norm also depends on the derivative, a small model would be one with small gradients, that is, a smooth model. It would be convenient to view

$$(u, v)' = \int_{-\infty}^{\infty} u^{(1)}(x) v^{(1)}(x) \, dx \tag{24}$$

as an inner product on the space of continuously differentiable functions. There is at first sight an objection: when $u = \text{constant} \neq 0$,

$$(u, u)' = 0 \tag{25}$$

contradicting $1.10(I\,4)$, one of the properties of a proper inner product. We can escape this difficulty by declaring that all differentiable functions differing from each other by a constant amount are equivalent to a single element in our space, which we call $W_2^1(-\infty, \infty)$. Thus $u = \text{constant}$ is the zero element 0 in $W_2^1(-\infty, \infty)$. This is exactly the same as the construction of an equivalence class for elements of L_2 in which functions differing in value only on a set of zero measure are considered to be one element (see section 1.11).

With this definition we have

$$d_j = (\tilde{g}_j, m)' \tag{26}$$

where $\tilde{g}_j \in W_2^1(-\infty, \infty)$ and is explicitly given by

$$\tilde{g}_j(x) = g_j^{(-2)}(x) = -\int^x g_j^{(-1)}(y) \, dy \tag{27}$$

$$= -\frac{\mu_0}{4\pi} \ln\{h^2 + (x - x_j)^2\} + c . \tag{28}$$

Proceeding in the usual way

$$\Gamma_{jk} = (\tilde{g}_j, \tilde{g}_k)' \tag{29}$$

$$= \frac{\mu_0^2}{2\pi} \frac{h}{4h^2 + (x_j - x_k)^2} . \tag{30}$$

Linear independence of the representers $\tilde{g}_j$ can be deduced from the linear independence of the functions g_j. The norm-

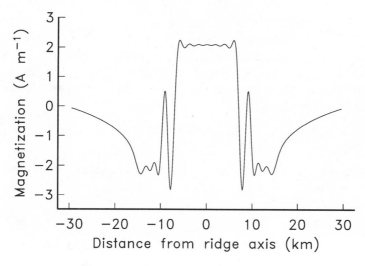

Figure 2.06c: The magnetization model with the smallest norm in the space W_2^1 generating the given magnetic anomaly values. Notice the somewhat smaller oscillations of the magnetization compared with those of figure 2.06b.

minimizing solution in W_2^1 is of course given by

$$m_0 = \sum_{j=1}^{N} \alpha_j \tilde{g}_j \; . \tag{31}$$

Notice m_0 is defined only up to an arbitrary constant magnetization. A particular solution is depicted in figure 2.06c, where the constant has been selected to yield a magnetization agreeing with the minimum L_2 norm value at $x = 0$. The new solution is slightly smoother in a subjective sense. There is less overshoot near $x = 8$ km and the oscillations around $x = 12$ km are greatly reduced; the behavior on the flanks is less abrupt than in our first model. The improvement in smoothness is admittedly modest.

Perhaps an even smoother solution will be found if we use the second derivative of m in the norm; presumably all we need to do is to integrate by parts again. A nontrivial obstacle presents itself. Formally we find

$$d_j = -g_j^{(-2)}(\infty) m^{(1)}(\infty) + \int_{-\infty}^{\infty} g_j^{(-2)}(x) m^{(2)}(x) \, dx \; . \tag{32}$$

Now $g_j^{(-2)}(\infty)$ is not defined and the integral

$$\int_{-\infty}^{\infty} g_j^{(-2)}(x)^2 dx \qquad\qquad (33)$$

is divergent. In the context of an inner-product space, d_j cannot be written as a *bounded* linear functional of $m^{(2)}$, the second derivative of m. It is actually possible to discover magnetization functions m, for which

$$\int_{-\infty}^{\infty} m^{(2)}(x)^2 dx \qquad\qquad (34)$$

can be made arbitrarily close to zero while satisfying the constraint equations

$$d_j = \int_{-\infty}^{\infty} g_j(x)m(x)\, dx \ . \qquad\qquad (35)$$

The use of second derivative information in this problem is feasible, but involves the introduction of more complicated normed spaces, for example, a Sobolev space where

$$\|m\| = \{\int_{-\infty}^{\infty} [m(x)^2 + \kappa^4 m^{(2)}(x)^2]\, dx \}^{1/2}. \qquad\qquad (36)$$

The constant κ represents a length scale; oscillations with wavelengths smaller than this are suppressed in the norm-minimizing solution. The investigation is best carried out entirely in the Fourier domain but it is not of sufficient general interest to merit further discussion here. We remark that the troubles of nonexistent integrals arising in this example when higher smoothing is desired can be traced to the unbounded interval on which the model is defined; when the model is confined to a finite region, which is after all a more realistic assumption, these difficulties evaporate.

The incorporation of second derivatives into a norm is of paramount importance in the construction of certain smooth models. To study these ideas we need a different example, one that may be quite familiar to many readers.

Exercises

2.06(i) Dorman (1975) gives the following example of a linear inverse problem, which he solves by an analytic technique. A vertical fault separates two regions in which density varies only with depth z (positive downwards); gravity measurements on a level profile are to be used to infer the density contrast across the fault. If $\Delta\rho(z)$ is the density difference across the fault, the observed horizontal gravity gradient on a profile across the fault is given by

$$d_j = \int_0^\infty \frac{2Gz}{x_j^2 + z^2} \Delta\rho(z) \, dz$$

where G is Newton's gravitational constant, and x_j is the horizontal coordinate of the jth observation measured from the fault plane. Treat this relation as a bounded linear functional of $\Delta\rho$ in the Hilbert space $L_2(0, \infty)$. Calculate the Gram matrix of the representers. Show the representers are linearly independent if the values of x_j^2 are distinct.

2.06(ii) If you have access to a computer find the smallest model in the sense of the norm of L_2 that exactly satisfies the data in the table below. Plot the model. Is it a reasonable profile for contrast across a transform fault in the oceanic lithosphere? Construct one element of the subspace $G^\perp$, the orthogonal complement to the subspace of representers and use it to find a totally outrageous model fitting the data exactly.

x_j (km)	d_j (Ms^{-2})
2	12056
5	4779.6
10	1760.5
20	518.49
30	239.21
50	87.88

2.06(iii) Shen and Beck (1983) consider the problem of determining the history of surface temperature from measurements of temperature in a borehole. If $\Theta(\tau)$ is the temperature at time τ before the present on the surface, and the borehole is drilled in rock of uniform thermal properties (diffusivity η), it can be shown that the diffusion of heat into the ground leads to the following equation for $T(z)$, the present-day temperature profile in the hole:

$$T(z) = \int_0^\infty \frac{z \, e^{-z^2/4\eta\tau}}{(4\pi\eta\tau^3)^{\frac{1}{2}}} \Theta(\tau) \, d\tau.$$

By making the change of variables $z^2 = \zeta$ and $1/4\eta\tau = u$, convert the above integral into a Laplace transform and hence obtain an exact solution to the inverse problem of deducing Θ from observations of T. Why is this solution worthless for practical purposes? Find the Gram matrix for the problem when Θ is chosen to be in L_2.

2.06(iv) Demonstrate the linear independence of the representers in exercise (i) and for the simplified magnetic anomaly problem of this section by means of the theory of analytic functions of a complex variable. Examine the consequences of linear dependence in the complex coordinate plane (x or z), specifically near the singularities of the representers.

2.07 Interpolation

A very common situation in every experimental science is the one in which measurements are made of a quantity for a number of values of a variable parameter; from this collection of measurements a curve is generated by interpolation. The calibration of a seismograph, for example, is often performed at a few frequencies and a complete curve is then constructed by interpolation based upon the assumption that the response is a smooth function of frequency. Interpolation can be seen as a linear inverse problem; this perspective will give us insight into other problems. Suppose we wish to find a curve on the unit interval $[0, 1]$; values of the function are known at the points $x_1, x_2, x_3, \cdots x_N$, where $0 \le x_j < x_{j+1} \le 1$. We may write formally

$$d_j = \int_0^1 \delta(x - x_j) f(x) \, dx, \quad j = 1, 2, 3, \cdots N. \tag{1}$$

Here δ is the Dirac delta function. This expression brings out the fact that d_j can be expressed as a linear functional of f, but it is not a proper inner product in L_2. It often comes as a surprise for a student to learn that the value of the function f at the point x cannot be expressed as a valid inner product of L_2. There are several ways of seeing this. First recall the elements of L_2 are not really functions but sets of equivalent sequences; the value at a particular point is in fact not even defined for such elements. Less abstractly perhaps, we may understand the question through the property that an inner product (g, f) is a *bounded* linear functional of f (Schwarz's inequality), which means that its magnitude must be limited by some constant times the norm of f. It is easy to see from the example $x^{-1/4}$, a valid element of $L_2(0, 1)$ with norm $\sqrt{2}$, that the magnitude of the function evaluated at a particular point (in this case near $x = 0$) is not bounded. We need a different inner-product space on which the act of evaluation of a function can be written as a proper inner product. Furthermore, to discover a smooth curve we prefer a space in which the norm is sensitive to oscillations in f. Whatever norm we choose to impose on the space of functions, a linear functional of an element in that space remains a linear functional; if it is a bounded functional then by Riesz's Theorem, we must be able to write it as an inner product.

Our objectives are achieved if we integrate (1) by parts:

$$d_j = f(1) - \int_0^1 H(x-x_j)f^{(1)}(x)\,dx \qquad (2)$$

$$= f(0) + \int_0^1 H(x_j-x)f^{(1)}(x)\,dx \qquad (3)$$

where H is the Heaviside unit step function and $f^{(n)}$ denotes $d^n f/dx^n$. This is the same device used in the previous section where we preserved the functional relation but altered the form of the inner product. Custom in interpolation theory suggests we should repeat the process:

$$d_j = f(0) + x_j f^{(1)}(0) + \int_0^1 (x_j-x)H(x_j-x)f^{(2)}(x)\,dx \ . \qquad (4)$$

For functions $f \in C^2[0, 1]$ this is just a strange way of writing Taylor's Theorem with an integral remainder. Suppose we equip $C^2[0, 1]$ with the inner product

$$(f,g) = f(0)g(0) + f^{(1)}(0)g^{(1)}(0) + \int_0^1 f^{(2)}(x)g^{(2)}(x)\,dx \ . \qquad (5)$$

The inner-product space $C^2 L_2$ created from $C^2[0, 1]$ in this way is not complete under the associated norm; although this has no serious consequences for our purposes we shall always work with the completion of the space called W_2^2, an example of a Sobolev space (see 1.07). In the interpolation problem we now have

$$d_j = (g_j, f), \quad j = 1, 2, 3, \cdots N \ . \qquad (6)$$

Evaluation at a particular point x is a bounded functional in this space because all the elements of $C^2[0, 1]$ are so smooth. Notice that in the inner product both elements must be treated equally (this is property $(I1)$ of section 1.10) so that if d_j is to be written with (5) the function $(x_j-x)H(x_j-x)$ appearing under the integral in (4) is not g_j but its second derivative. A certain amount of integration must be done to show that

$$g_j(x) = \begin{cases} 1 + x\,x_j + \dfrac{x^2 x_j}{2} - \dfrac{x^3}{6} , & x \le x_j \\[3mm] 1 + x_j x + \dfrac{x_j^2 x}{2} - \dfrac{x_j^3}{6} , & x > x_j \ . \end{cases} \qquad (7)$$

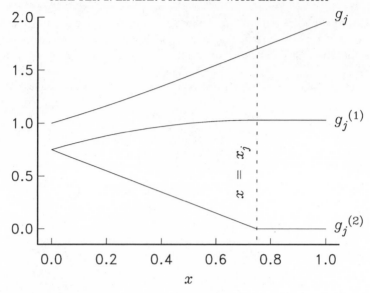

Figure 2.07a: Representer for evaluation in the space W_2^2 under the inner product (5). The evaluation point is $x = x_j$, about 0.75. Also shown are the first and second derivatives of g_j which are continuous on the entire interval.

The function g_j is in $C^2[\,0,\,1\,]$: it consists of a cubic polynomial in x when $x \le x_j$ and a straight line when $x > x_j$, joining at $x = x_j$ in such a way that g_i, $g_j^{(1)}$ and $g_j^{(2)}$ are continuous (see figure 2.07a).

We could find the function f_0 that minimizes the norm $\|f\|$, where

$$\|f\|^2 = f(0)^2 + f^{(1)}(0)^2 + \int_0^1 f^{(2)}(x)^2 \, dx \ . \tag{8}$$

This would certainly yield a fairly smooth solution because those functions having large second derivatives will tend to be eliminated. We can improve on this solution if we remove the contributions from $f(0)$ and $f^{(1)}(0)$. Let us write

$$f(x) = f(0) + x\,f^{(1)}(0) + r(x) \ . \tag{9}$$

Then obviously $r(0) = 0$ and $r^{(1)}(0) = 0$ and so

$$\|r\|^2 = \int_0^1 f^{(2)}(x)^2 \, dx \ . \tag{10}$$

In 2.05 we saw how to minimize $\|r\|$ when

$$f = \sum_k \beta_k h_k + r \tag{11}$$

where $h_1, h_2, \cdots h_K$ are fixed elements. Here we choose $h_1=1$ and $h_2=x$, so that $\beta_1=f(0)$ and $\beta_2=f^{(1)}(0)$; in this way we avoid penalizing the offset at $x=0$ and the slope of the curve there. With this choice the techniques of 2.05 give us f_*, the function possessing the smallest 2-norm of the second derivative of any function that passes through the given values $d_1, d_2, \cdots d_N$. In this sense it is the smoothest interpolating curve.

To be certain the linear systems of 2.05 always have solutions we need to demonstrate that the representers g_j are linearly independent and that $(\sum \gamma_k h_k, g_j) \neq 0$ unless $\gamma_k =0$. The second condition is easily verified. Linear independence of the representers is shown as follows. First form the linear combination

$$f(x) = \sum_{j=1}^{N} \alpha_j g_j(x) \tag{12}$$

and suppose that $f(x)=0$ for all x in $[0, 1]$. Now consider the third derivative $f^{(3)}(x)$; each component $g_i^{(3)}$ in the sum is a step function equal to -1 on the left of x_j and zero on the right ($g^{(3)}(x_j)$ is not defined). We see that

$$
\begin{aligned}
f^{(3)}(x) = 0 = & -\alpha_N , & x_{N-1} < x < x_N \\
f^{(3)}(x) = 0 = & -\alpha_{N-1} - \alpha_N , & x_{N-2} < x < x_{N-1} \\
f^{(3)}(x) = 0 = & -\alpha_{N-2} - \alpha_{N-1} - \alpha_N , & x_{N-3} < x < x_{N-2}
\end{aligned} \tag{13}
$$

$$\cdots .$$

Obviously from these relations $\alpha_N = \alpha_{N-1} = \alpha_{N-2} = \cdots \alpha_1 = 0$, and so, by definition, the functions g_j are linearly independent. This proof runs into trouble if $x_1=0$, but can be repaired by consideration of $f(0)$ instead of $f^{(3)}(0)$.

The smooth curve f_* is the *cubic spline* interpolator. A number of important properties of f_* can be inferred without engaging in numerical evaluations. The spline function consists of a set of polynomial pieces

$$A_j x^3 + B_j x^2 + C_j x + D_j, \quad x_j \leq x \leq x_{j+1} . \tag{14}$$

These are cubic polynomials, unless, of course, $A_j =0$. The polynomial in each interval joins smoothly onto those in the neighboring

Table 2.07A: Seismometer response

Period (s)	Gain relative to 1s	Period (s)	Gain relative to 1s
10.0	0.00207	0.833	1.29
5.0	0.0160	0.667	1.61
2.5	0.120	0.5	1.91
2.0	0.221	0.333	1.93
1.67	0.350	0.25	1.66
1.25	0.671	0.2	1.34
1.0	1.0	0.143	0.837

intervals preserving continuity of $f_*^{(2)}$ (obviously, since $f_* \in W_2^2[0, 1]$). From 2.05(18)

$$f_* = \sum \beta_k h_k + \sum \alpha_j g_j \tag{15}$$

and notice that $h_k^{(2)}(1) = g_j^{(2)}(1) = 0$; thus $f_*^{(2)}(1) = 0$. The second derivative of f_* vanishes at $x = 0$ also; this follows from the fact that the problem is invariant under the coordinate transformation $x' = 1 - x$. If $x_N < 1$ then $f^{(2)}(x) = 0$ for all x with $x_N \leq x \leq 1$ and similarly at the other end. These few properties are in fact sufficient to define the cubic spline interpolating curve. Numerically, it is far simpler to find the polynomial coefficients by matching values and derivatives at the interfaces $x = x_j$ than by solving the systems in 2.05. This is the way cubic splines are found in the standard computer programs for interpolation. Indeed, the whole process can be simplified even more by use of different basis for the space spanned by the representers, the so-called B-spline basis; for a thorough treatment of this and many related matters, the reader should consult de Boor (1978).

Instead of integrating by parts twice, suppose we had stopped with

$$d_j = f(0) + \int_0^1 H(x_j - x) f^{(1)}(x) \, dx \ . \tag{16}$$

The reader may find it amusing to work through this case, finding the interpolating function that minimizes the 2-norm of $f^{(1)}$. Note that it is now necessary to complete the normed vector space derived from $C^1[0, 1]$ because the representers g_j are not themselves continuously differentiable.

A geophysical example mentioned earlier is the calibration of a seismometer. Shown in table 2.07A are the results of an

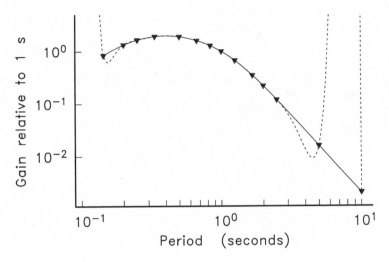

Figure 2.07b: The response of a seismometer to sinusoidal vertical ground motion of various periods. The measured values are shown by the triangles. The solid curve is the spline interpolation, while the dashed one is a high-degree polynomial interpolation.

experiment measuring the gain of an electronically fed-back instrument at fourteen frequencies as reported by Peterson et al. (1980); these values are the triangles in figure 2.07b. The solid curve in the figure is the cubic spline interpolation of the response, using the logarithm of the gain as the dependent variable, and the logarithm of the period as the independent one. Also shown for comparison is the interpolation of the same data by Lagrange's method, a procedure that generates the polynomial fitting the given values exactly. This rather old-fashioned method is an example of collocation, the expansion of the model completely in terms of fixed elements h_k (here successive powers of x) which have no special relation to the observations:

$$f = \sum_{k=1}^{N} \beta_k h_k \ . \tag{17}$$

As figure 2.07b shows, even when each of the elements h_k is very smooth the results can be far from satisfactory.

The idea of seminorm minimization provides a rational basis for interpolation of functions in more than one independent variable. We treat interpolation on the unit sphere, a problem typical of many in geophysics where smooth functions over the Earth's surface are required. Let S^2 be the surface of the unit sphere and we define the collection of unit vectors $\hat{\mathbf{r}}_1, \hat{\mathbf{r}}_2, \cdots \hat{\mathbf{r}}_N$ to be points at which a function has been measured; we ask for a smooth interpolating function on S^2. The traditional method for providing an interpolator is the one that worked so poorly in the last illustration, expansion in an arbitrary set of basis functions. We shall discover the function f satisfying

$$d_j = f(\hat{\mathbf{r}}_j), \quad j = 1, 2, 3, \cdots N \tag{18}$$

which possesses the smallest value of

$$\int_{S^2} (\nabla_s^2 f)^2 \, d^2\hat{\mathbf{r}} \tag{19}$$

where ∇_s^2 is the surface Laplacian operator; like d^2/dx^2, this operator is second order and independent of position and orientation on S^2. Although it is not necessary to solve the problem, it is extremely convenient for our purposes to expand f in spherical harmonics. We therefore take this opportunity for a digression: we mention without proof some of the properties of these ubiquitous functions.

We need a spherical polar coordinate system, with θ the colatitude and ϕ the longitude of a point in S^2. The complex valued function $Y_l^m(\theta, \phi)$, called a *spherical harmonic* of *degree l* and *order m* is defined by

$$Y_l^m(\theta, \phi) = (-1)^m \left[\frac{2l+1}{4\pi}\right]^{1/2} \left[\frac{(l-m)!}{(l+m)!}\right]^{1/2} P_l^m(\cos\theta) \, e^{im\phi} \tag{20}$$

where $-l \le m \le l$ and $P_l^m(\mu)$ is the *associated Legendre function*:

$$P_l^m(\mu) = \frac{(1-\mu^2)^{m/2}}{2^l l!} \left[\frac{d}{d\mu}\right]^{l+m} (\mu^2-1)^l . \tag{21}$$

The special case $m=0$ of P_l^m is written simply $P_l(\mu)$; it is a *Legendre polynomial* and, as the name suggests, is a polynomial in μ of degree l. It is easily seen from the definition that $P_0(\mu)=1$; less obvious is the fact that $|P_l(\mu)| < 1$ and $|P_l(\mu)| < (1-\mu^2)^{-1/2} l^{-1/2}$ when $|\mu| < 1$. Furthermore, $P_l(1)=1$. The associated Legendre function $P_l^m(\mu)$ vanishes at $l - |m|$ points in the open μ interval $(-1, 1)$; hence $Y_l^m(\theta, \phi)$ vanishes on $l - |m|$ lines of constant latitude. There are many symmetry relations between the spherical harmonics; the simplest is

$$Y_l^m(\theta, \phi)^* = (-1)^m \, Y_l^{-m}(\theta, \phi) . \tag{22}$$

More profound is the *Addition Theorem* for spherical harmonics; if θ_1, ϕ_1, and θ_2, ϕ_2 are polar coordinates for two points $\hat{\mathbf{r}}_1$, $\hat{\mathbf{r}}_2 \in S^2$, then

$$P_l(\hat{\mathbf{r}}_1 \cdot \hat{\mathbf{r}}_2) = \frac{4\pi}{2l+1} \sum_{m=-l}^{l} Y_l^m(\theta_1, \phi_1)\, Y_l^m(\theta_2, \phi_2)^* \, . \tag{23}$$

Notice that if we set $\hat{\mathbf{r}}_1 = \hat{\mathbf{r}}_2$ we may deduce at once that $|Y_l^m(\theta, \phi)| \leq \pi^{-\frac{1}{2}}/2 < 1$. Many of the properties of the spherical harmonics follow directly from the fact that they are eigenfunctions of the operator ∇_s^2 under the "boundary condition" that solutions shall be continuous at every point of S^2:

$$\nabla_s^2 Y_l^m = -l(l+1) Y_l^m \, . \tag{24}$$

Referred to the polar coordinate system

$$\nabla_s^2 = \frac{\partial^2}{\partial\theta^2} + \cot\theta \, \frac{\partial}{\partial\theta} + \frac{1}{\sin^2\theta} \, \frac{\partial^2}{\partial\phi^2} \, , \quad 0 < \theta < \pi \, . \tag{25}$$

Let us introduce the complex inner-product space of complex functions on S^2 under the inner product

$$(f, g) = \int_0^{2\pi} d\phi \int_0^{\pi} \sin\theta \, d\theta \, f(\theta,\phi) \, g(\theta, \phi)^* \, . \tag{26}$$

The space can be completed in the usual way to the complex Hilbert space $L_2(S^2)$. Each spherical harmonic is an element of $L_2(S^2)$ and indeed

$$\|Y_l^m\| = 1 \, . \tag{27}$$

Furthermore they are orthogonal to each other

$$(Y_l^m, Y_j^k) = 0 \tag{28}$$

unless $l = j$, and $m = k$. The spherical harmonics are a complete orthonormal basis for elements in $L_2(S^2)$; this means that if $f \in L_2(S^2)$

$$f = \sum_{l=0}^{\infty} \sum_{m=-l}^{l} b_l^m \, Y_l^m \tag{29}$$

where

$$b_l^m = (f, Y_l^m) \tag{30}$$

and the convergence of the infinite sum is in the sense of the norm (see 1.08). It is easily seen that

$$\|f\| = \left[\sum_{l=0}^{\infty} \sum_{m=-l}^{l} |b_l^m|^2 \right]^{1/2}. \tag{31}$$

Returning to interpolation on S^2, let us suppose that f is so smooth that

$$\int |\nabla_s^2 f|^2 \, d^2\hat{\mathbf{r}} \tag{32}$$

is finite. It follows that $\nabla_s^2 f$ possesses a convergent spherical harmonic series expansion, which, from the eigenfunction property (24), can be written

$$\nabla_s^2 f = \sum_{l=1}^{\infty} -l(l+1) \sum_{m=-l}^{l} b_l^m Y_l^m(\theta, \phi) \tag{33}$$

where the coefficients b_l^m are those for a spherical harmonic expansion of f itself. Also we see

$$\int |\nabla_s^2 f|^2 \, d^2\hat{\mathbf{r}} = \sum_{l=1}^{\infty} l^2(l+1)^2 \sum_{m=-l}^{l} |b_l^m|^2. \tag{34}$$

These expressions suggest a new normed space in which to perform further calculations. Let b be the infinite sequence of complex numbers

$$b = \{b_0^0, b_1^{-1}, b_1^0, b_1^1, b_2^{-2}, b_2^{-1}, b_2^0, b_2^1, b_2^2, \cdots \}. \tag{35}$$

Such sequences are elements of a linear vector space with complex scalars under the obvious laws of addition and scalar multiplication. We may make this into an inner-product space B_2 by introducing an inner product for $a, b \in B_2$:

$$(a, b)_B = a_0^0 (b_0^0)^* + \sum_{l=1}^{\infty} l^2(l+1)^2 \sum_{m=-l}^{l} a_l^m (b_l^m)^* \tag{36}$$

with associated norm

$$\|b\|_B = [\, |b_0^0|^2 + \sum_{l=1}^{\infty} l^2(l+1)^2 \sum_{m=-l}^{l} |b_l^m|^2 \,]^{1/2} \tag{37}$$

$$= [\, |b_0^0|^2 + \int |\nabla_s^2 f|^2 \, d^2\hat{\mathbf{r}} \,]^{1/2}. \tag{38}$$

The introduction of the factor $l^2(l+1)^2$ into the inner product (36) has the same effect as the double integration by parts had on the inner product (5); only very smooth functions belong to the associated space. In fact B_2 can be shown to be complete under the corresponding norm, that is, B_2 is Hilbert.

When a function $f \in L_2(S^2)$ admits expansion in spherical harmonics with coefficient element $b \in B_2$ we can write the value d_j, which is a sample of f at the point with coordinates θ_j, ϕ_j, as

$$d_j = \sum_{l=0}^{\infty} \sum_{m=-l}^{l} b_l^m \, Y_l^m(\theta_j, \phi_j) \tag{39}$$

$$= b_0^0 \, Y_0^0(\theta_j, \phi_j) + \sum_{l=1}^{\infty} l^2(l+1)^2 \sum_{m=-l}^{l} b_l^m \, \frac{Y_l^m(\theta_j, \phi_j)}{l^2(l+1)^2} \tag{40}$$

$$= (b, y_j)_B \tag{41}$$

where

$$y_j = \{ Y_0^0(\theta_j, \phi_j)^*, \frac{Y_1^{-1}(\theta_j, \phi_j)^*}{4}, \frac{Y_1^0(\theta_j, \phi_j)^*}{4}, \frac{Y_1^1(\theta_j, \phi_j)^*}{4}$$

$$\ldots \frac{Y_l^m(\theta_j, \phi_j)^*}{l^2(l+1)^2}, \ldots \}. \tag{42}$$

The sequence y_j is an element of B_2 because $\|y_j\|_B$ exists; the series for $\|y_j\|_B^2$ converges as we may see:

$$\|y_j\|_B^2 = |Y_0^0(\theta_j, \phi_j)|^2 + \sum_{l=1}^{\infty} l^2(l+1)^2 \sum_{m=-l}^{l} \left| \frac{Y_l^m(\theta_j, \phi_j)}{l^2(l+1)^2} \right|^2. \tag{43}$$

Recall the property that for any l, m, θ_j, ϕ_j $|Y_l^m(\theta_j, \phi_j)| < 1$; so

$$\|y_j\|_B^2 < 1 + \sum_{l=1}^{\infty} \sum_{m=-l}^{l} \frac{1}{l^2(l+1)^2} \tag{44}$$

$$= 1 + \sum_{l=1}^{\infty} \frac{2l+1}{l^2(l+1)^2} = 2. \tag{45}$$

Thus the samples of a sufficiently smooth function f can be written as bounded linear functionals on B_2:

$$d_j = (b, y_j)_B \tag{46}$$

$$= (y_j, b)_B^* \tag{47}$$

$$= (y_j, b)_B \tag{48}$$

since the data d_j are real numbers.

For interpolation we want the function fitting the data that possesses the smallest value of $\int (\nabla_s^2 f)^2 d^2\hat{r}$. Write the element $b \in B_2$ corresponding to f as the sum of two parts

$$b = \beta_1 h_1 + r \qquad (49)$$

where $\beta_1 = b_0^0$ and

$$h_1 = \{\ 1,\ 0,\quad 0,\quad 0,\quad \cdots\ \}$$
$$r = \{\ 0,\ b_1^{-1},\ b_1^0,\ b_1^1,\quad \cdots\ \}. \qquad (50)$$

Then clearly, since r has no b_0^0 term

$$\|r\|_B^2 = \int\limits_{S^2} (\nabla_s^2 f)^2\, d^2\hat{\mathbf{r}} \qquad (51)$$

and the seminorm minimization of 2.05 can be brought into play. The matrix A is a single column consisting of constant components:

$$A_{jk} = (y_j,\ h_k)_B, \quad j = 1, 2, 3,\ \cdots\ N;\ k = 1 \qquad (52)$$

$$= Y_0^0 (\theta_j,\ \phi_j)^* \qquad (53)$$

$$= \pi^{-1/2}/2\ . \qquad (54)$$

For the Gram matrix we find

$$\Gamma_{jk} = (y_j,\ y_k)_B \qquad (55)$$

$$= Y_0^0 (\theta_j,\ \phi_j)\ Y_0^0 (\theta_k,\ \phi_k)^*$$
$$+ \sum_{l=1}^{\infty} l^2 (l+1)^2 \sum_{m=-l}^{l} \frac{Y_l^m(\theta_j,\phi_j)}{l^2(l+1)^2}\ \frac{Y_l^m(\theta_k,\phi_k)^*}{l^2(l+1)^2} \qquad (56)$$

$$= Y_0^0 (\theta_j,\ \phi_j)\ Y_0^0 (\theta_k,\ \phi_k)^*$$
$$+ \sum_{l=1}^{\infty} \frac{1}{l^2(l+1)^2} \sum_{m=-l}^{l} Y_l^m(\theta_j,\ \phi_j)\ Y_l^m(\theta_k,\ \phi_k)^*. \qquad (57)$$

We recognize the sum over m as the one occurring in (23), the Addition Theorem:

$$\Gamma_{jk} = \frac{1}{4\pi} + \frac{1}{4\pi} \sum_{l=1}^{\infty} \frac{2l+1}{l^2(l+1)^2}\ P_l(\hat{\mathbf{r}}_j \cdot \hat{\mathbf{r}}_k)\ . \qquad (58)$$

This sum over l can be reduced to an integral. The work is heavy and will not be set out here but a simpler result of this kind is given a full derivation in section 4.04. The interested reader will find many calculations in Whaler and Gubbins (1981). The Gram matrix elements can be written

$$\Gamma_{jk} = \frac{1}{4\pi} [2 - \frac{1}{6}\pi^2 + \text{dilog}\,(\frac{1}{4}|\hat{\mathbf{r}}_j - \hat{\mathbf{r}}_k|^2)] \qquad (59)$$

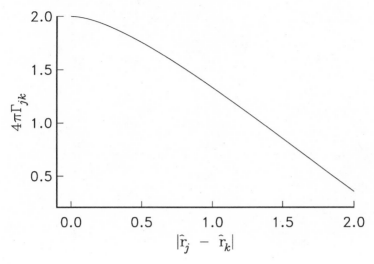

Figure 2.07c: The inner product Γ_{jk} in the normed vector space B_2 as a function of observer separation. Notice that such functions also form a basis for the interpolation.

where $\mathrm{dilog}(x)$ is the *dilogarithm*, a special function defined by

$$\mathrm{dilog}(1+x) = - \int_0^x \frac{\ln(1+t)}{t}\, dt \;, \quad x \geq -1 \,. \tag{60}$$

For a numerical algorithm to evaluate this function, see appendix A. Notice that writing $\hat{\mathbf{r}}_j \cdot \hat{\mathbf{r}}_k$ as $\cos \Delta$, we have $|\hat{\mathbf{r}}_j - \hat{\mathbf{r}}_k| = 2\sin \Delta/2$. See figure 2.07c.

To be sure the equations in 2.05 always have a unique solution we must verify the conditions discussed there. First, it is trivially seen that the orthogonal projection of h_1 onto the subspace spanned by the elements y_j does not vanish. Linear independence of the representers y_j can be shown by a procedure just like the one given for the functions $e^{2\pi i x_j \lambda}$ in 2.06. We form $\Sigma\, \alpha_j\, y_j$ and consider the sum

$$I = \sum_{l=0}^{L} \sum_{m=-l}^{l} \frac{1}{l^2} \left| \sum_{j=1}^{N} \alpha_j\, Y_l^m(\theta_j, \phi_j) \right|^2 . \tag{61}$$

After some algebra we find that, for L large enough, $I > 0$ if any α_j is nonzero; to show this we need the inequality

$$|P_l(\cos \theta)| \leq (l\, \sin \theta)^{-\frac{1}{2}} . \tag{62}$$

The details are left to the interested reader.

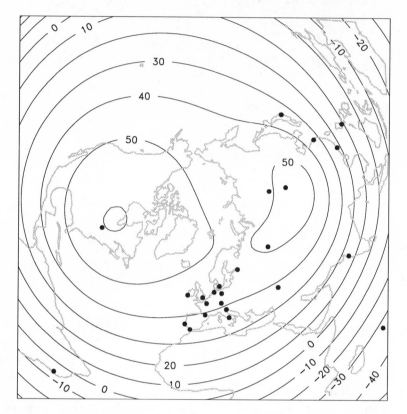

Figure 2.07d: Contours of the spherical spline interpolation of the vertical magnetic field in the year 1900, superposed on a map in Lambert equal area projection. The magnetic observatories are shown as solid dots.

Having obtained the element $b_* \in B_2$

$$b_* = \beta_1 h_1 + \sum_{j=1}^{N} \alpha_j y_j \tag{63}$$

we are still left with the task of evaluating the interpolating function on S^2. Equation (41) expresses the function value at a point $\hat{\mathbf{r}} \in S^2$ as an inner product in B_2

$$f_*(\hat{\mathbf{r}}) = (b_*, y(\hat{\mathbf{r}}))_B \tag{64}$$

$$= (\beta_1 h_1, y(\hat{\mathbf{r}}))_B + \sum_{j=1}^{N} \alpha_j (y(\hat{\mathbf{r}}), y_j)_B . \tag{65}$$

The inner products in the sum are exactly in the form of those comprising the components of Γ; thus

$$f_*(\hat{\mathbf{r}}) = \frac{\beta_1}{2\pi^{1/2}} + \sum_{j=1}^{N} \frac{\alpha_j}{4\pi} \text{ dilog } (\frac{1}{4}|\hat{\mathbf{r}} - \hat{\mathbf{r}}_j|^2) . \tag{66}$$

Notice we can eliminate the constant terms in the representers because 2.05(8) assures us that in the sum the constants make no contribution, the constant being supplied solely by the leading term. For a more complete and mathematically dense treatment of splines on the sphere and the plane (called *thin-plate* splines) see Wahba (1990, chapter 2) and the references therein. We observe in passing that the thin-plate splines corresponding to minimization of the 2-norm of the surface Laplacian on the plane give rise to representers of the form

$$g_j(\mathbf{x}) = |\mathbf{x} - \mathbf{x}_j|^2 \ln |\mathbf{x} - \mathbf{x}_j| \tag{67}$$

a result we might obtain from the dilogarithm basis by letting the radius of the sphere grow very large.

Magnetic fields measured at 25 magnetic observatories provide a numerical application of this theory. In 1900 thirty-three stations were in operation; the vertical field values (positive downward) at twenty-five of them were used as a basis for an interpolated field. The observations are tabulated in table 2.07B; figure 2.07d shows the location of the stations and contours of the spherical spline model. Note that despite the uneven distribution of the sampling points, the model is smooth and well behaved even in regions devoid of observations such as the Pacific Ocean. Eight observatories were not used in the model; they provide a test for the interpolation. These additional stations are all in Europe, where one would expect interpolation to be a valid exercise in view of the density of sample points. The results are shown in table 2.07C. The difference between the spherical spline prediction and the measured value is the column labeled SS; the agreement is good with a maximum error of less than 1 percent. For comparison with the collocation method of interpolation I also fitted a fourth degree spherical harmonic model to the twenty-five data; there are exactly twenty-five unknown coefficients in such an expansion and so the fit can be performed exactly. The spherical harmonic model is very badly behaved: outside the immediate area of the stations, the model grows rapidly, reaching values of over 6×10^8 nT. In table 2.07C we see that the values interpolated by spherical harmonic fitting (SH) for the eight test stations

Table 2.07B: Magnetic Observatory Data in 1900

Latitude	Longitude	Observatory	Field (nT)
43.79	−79.27	Agincourt	59691.5
43.62	1.47	Toulouse	39408.0
41.72	44.80	Tiflis	37784.5
40.87	14.25	Capodimonte	36311.0
38.72	−9.15	Lisbon	37522.0
36.47	−6.20	San Fernando	35432.5
35.69	139.75	Tokyo	34410.5
31.22	121.43	Zi-ka-wei	33744.0
22.30	114.18	Hongkong	22447.5
18.90	72.82	Bombay	14605.5
14.59	120.98	Manila	11112.0
−19.43	57.55	Mauritius	−32985.0
−22.23	−43.17	Rio de Janeiro	−5905.0
59.92	104.50	Oslo	47016.0
59.69	30.48	Slutsk	47064.0
56.84	60.63	Sverdlovsk	50825.0
55.69	12.58	Copenhagen	44793.0
53.54	8.15	Wilhelmshaven	44183.0
52.39	13.07	Potsdam	43013.5
52.27	104.27	Irkutsk	57722.0
51.94	−10.25	Valentia	45116.5
51.49	0.00	Greenwich	43798.5
48.82	2.50	Paris	42119.5
48.15	11.62	Munich	41011.0
44.87	13.85	Pola	38899.5

are generally grossly inferior to those of the spherical spline interpolation.

We have treated the vertical magnetic field as if it were some arbitrary smooth function of position, and not as the radial component of a vector field governed by the differential equations of physics. Those equations impose constraints of their own that ought to be acknowledged: for further discussion of this issue see section 4.04 and Shure et al. (1982).

Finally, we mention how the spherical problem might have been solved without the use of spherical harmonics. The essential idea is to use a space whose norm contains the L_2 norm of the operator ∇_s^2; this space is really the same as B_2 (that is, they are isometric). Evaluation of the function at a point involves Green's function for the operator, which must appear in the representer

Table 2.07C: Comparison of Interpolated Fields

Lat.	Long.	Observatory	Field (nT)	Misfit SS	Misfit SH
51.47	−0.32	Kew	43827.0	−8.7	−61.9
50.80	4.37	Uccle	42937.5	109.5	−115.4
50.15	−5.08	Falmouth	43538.0	−13.6	−7060.0
52.10	5.18	de Bilt	43620.0	67.3	53.7
54.37	12.75	Neufahrwasser	44510.0	−405.8	−737.6
53.85	−2.47	Stonyhurst	44742.5	438.0	12196.6
40.22	−8.42	Coimbra	38527.5	−109.3	−3501.7
42.70	2.88	Perpingnan	38828.5	−147.5	−4033.3

for evaluation. In fact, the reader can see how this works from the simple one-dimensional case because it is precisely Green's function for the operator d^2/dx^2, under appropriate boundary conditions, that appears in (7). Also see exercise 2.07(ii) below.

Exercises

2.07(i) By using the complex Fourier basis solve the problem of interpolation for functions $f(\theta)$ that are periodic on $[0, 2\pi)$. Use the 2-norm of the second derivative as the measure of roughness to be minimized. Show that the interpolating function apparently consists of quartic polynomials in θ between the sample points, not cubics as in the ordinary spline problem. But show this is only an illusion!

2.07(ii) Solve the spline interpolation in a simply connected domain D in the plane $\mathbb{R}^2$ using calculus of variations and Lagrange multipliers. Minimize the functional

$$\int_D (\nabla^2 f)^2 \, d^2\mathbf{x}$$

subject to the formal integral constraints

$$d_j = \int_D f(\mathbf{x}) \delta(\mathbf{x}_j - \mathbf{x}) \, d^2\mathbf{x}, \quad j = 1, 2, 3, \cdots N .$$

Show that the interpolating basis functions are Green's functions for $\nabla^4 f$ on D subject to suitable boundary conditions.

2.08 Seismic Dissipation in the Mantle

We come now to an example that illuminates an important aspect of the geophysical inverse problem which has so far been ignored. The issue is the proper treatment of the numerical calculations in order to avoid, or reduce as far as possible, the effects of inaccuracies in practical computations. The problem of accumulation of numerical errors is particularly acute in systems where the functionals cannot be found exactly, either because no closed-form expression exists for the inner products or because the functionals themselves are the result of numerical calculations. To illustrate this matter we consider the determination of the dissipation of seismic energy in the mantle from observations of the Earth's free oscillations. First we need a little background material.

After a large earthquake it is found that the Earth continues to vibrate for a long time; in some cases vibrations have been detected six weeks after the original event. The vibrations can be considered to be composed of a sum of simple oscillations, each with a single frequency and characteristic pattern of movement called a *normal mode* of oscillation. In a perfectly spherical system the modes may be divided into two species, the *toroidal* and *spheroidal* types; the displacements of the two kinds are characterized as follows:

$$\mathbf{u}_T = \nabla \times (\mathbf{r}\tau) \tag{1}$$

$$\mathbf{u}_S = \nabla \chi + \nabla \times \nabla \times (\mathbf{r}\psi) \tag{2}$$

where τ, χ, and ψ are scalar functions of position within the Earth and $\mathbf{r}$ is the radius vector from the center. On surfaces of constant radius the modes have shapes that can be described by the spherical harmonic functions $Y_l^m(\theta,\phi)$, but in a system with spherical symmetry the $2l+1$ modal patterns of degree l (one for each m) all have the same frequency. For every spherical harmonic shape there is a sequence of modes differing from each other in their radial behavior; to identify them, each mode is assigned an *overtone number*. The lowest frequency mode for a fixed l is given overtone number $n = 0$, the next $n = 1$, and so on; generally speaking, the complexity of the internal motion increases with increasing n. The symbols $_nT_l$ and $_nS_l$ describe toroidal and spheroidal modes of degree l and overtone n. Over a thousand modes in the frequency range 0.309 mHz to 12.5 mHz have been identified in seismic records and their frequencies measured. The classic paper on this endeavor is by Gilbert and Dziewonski

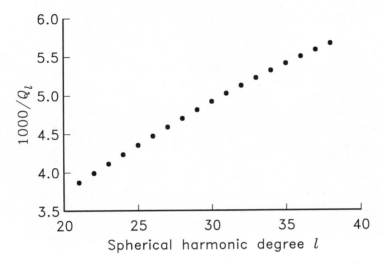

Figure 2.08a: Decay parameters calculated for a series of 18 fundamental spheroidal modes of free oscillation $_0S_{21}$, $_0S_{22}$, $\cdots$ $_0S_{38}$. These numbers are derived from a simple Earth model and will be used as input data for inversion to obtain a loss function varying with radius.

(1975). Since the variations with radius of the elastic parameters and density control the frequencies of oscillation, the observations of the modal frequencies can be used to infer the internal mechanical structure of the Earth. This famous inverse problem is extremely complex; it is moreover nonlinear, a difficulty which we cannot yet handle. Instead, we shall look at the simpler question of the decay of the oscillations.

To be specific, let us concentrate on a set of modes designated by $_0S_l$, modes with $n = 0$, called fundamental spheroidal modes. The oscillations in these modes may be considered to be independent of each other: after initial excitation by the earthquake, each one loses its energy by friction in the mantle and its amplitude decays eventually to zero. At a point on the surface of the Earth at times $t > 0$ the vertical motion obeys an equation of the form

$$A_l(t) = A_l(0)\, e^{-\omega_l t/2Q_l} \cos(\omega_l t + \Phi) \tag{3}$$

where Q_l is the *qualify factor* of the mode. Obviously large values of Q_l correspond to very persistent modes; in the sequence $_0S_l$, values of Q_l range from about 150 for $_0S_{40}$ to about 500 for $_0S_2$

(the mode $_0S_0$ is found to have an anomalously high value of over 6000). For our numerical illustration we examine eighteen modes: $_0S_{21}$, $_0S_{22}$, $\cdots$ $_0S_{38}$; values of their decay parameters calculated from a simple Earth model appear in figure 2.08a. These numerical values are not the ones actually observed because, as we shall see, our inability to handle the uncertainty in the true observations would cause us great difficulties. The genuine observations, drawn from a set of over seventy values presented by Masters and Gilbert (1983) will be treated later in section 3.06. The decay of a mode is caused mostly by frictional losses associated with the shearing components of its motion. The size of Q_l is governed by two factors: the variation with radius of the intrinsic loss factor of the material and the different patterns of motions of the modes. The attenuation of the Earth is greatest in the upper mantle and so the modes with motion concentrated there (those with higher values of l) decay most rapidly. The very large value of Q_l measured for $_0S_0$ is explained by the small amount of shearing in its motion. We wish to determine the distribution of the intrinsic attenuation within the mantle. For small amplitudes, the reciprocal of Q is linearly related to the loss function of the material. Let us define $1000/Q$ to be the observed quantity and renumber the data for convenience; then

$$d_j = 1000/Q_{j+20} = \int_c^a g_j(r)q(r)\,dr/a \qquad (4)$$

where a is the radius of the Earth, c that of the core, q is the unknown function describing shear dissipation, and the functions g_j are known. (The normalization of the integral by $1/a$ is to avoid annoying units of length in the representers and norms.)

None of the modes in question is associated with appreciable motion in the core, so that there is no need to extend the range of integration into that region. To define q we consider a small volume of material inside the Earth which as a whole is vibrating in one of the normal modes. During a single oscillation of the mode the mean kinetic energy in shear is E; also during one cycle an amount of energy δE is dissipated. The loss function used here is defined by

$$q = 1000\, \delta E/E . \qquad (5)$$

This number is assumed to depend only on distance from the center of the Earth and to be independent of frequency. The

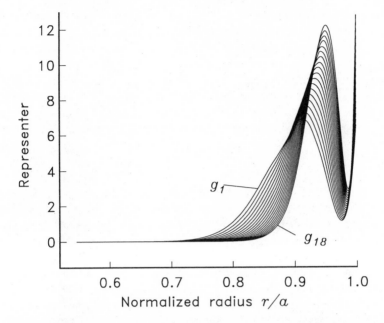

Figure 2.08b: The 18 representers in L_2 for dissipation associated with the fundamental modes $l = 21, 22, \cdots 38$ whose decay parameters appear in figure 2.08a.

factor of 1000 in this equation and the previous one are simply to put the values into a convenient numerical range. The functions g_j, which give the shear kinetic energy density averaged over a spherical shell, are calculated by solving coupled systems of ordinary differential equations in which the elastic parameters of the Earth appear as coefficients; the set corresponding to the modes $_0S_{21}$ to $_0S_{38}$ are shown in figure 2.08b. Details of those calculations appear in the references given by Masters and Gilbert (1983).

Regarding the numbers d_j as observations and knowing the functions g_j, we have an ordinary linear inverse problem for q. Suppose we interpret the functional for d_j as the inner product on $L_2(c, a)$ and we set out to find q_0, the function satisfying the data that minimizes the squared norm

$$\int_c^a q(r)^2 dr / a . \tag{6}$$

Then proceeding in the usual way we must evaluate the elements of the Gram matrix

$$\Gamma_{jk} = \int_c^a g_j(r) g_k(r) \, dr / a \ . \tag{7}$$

Since the functions g_j are not simple elementary expressions these integrals must be performed numerically. Usually, this involves evaluation of $g_j(r)$ at a number of sample points, $r = r_1, r_2, \cdots r_L$, and the approximation of the integral by a sum

$$\int_c^a f(r) \, dr / a \approx \sum_{i=1}^L w_i f(r_i) \tag{8}$$

where the numbers w_i are weights chosen to optimize the approximation. Well-known examples include the trapezoidal rule in which the sampling points are equally spaced and the more precise Gaussian quadrature formulas; consult any book on numerical analysis, for example, Ralston (1965). An alternative technique is to assume that the functions g_i are interpolated by known elementary functions between sample points, for example, straight lines or cubic splines, and then the integrals are performed exactly for the interpolating functions. While the obvious numerical realization of our standard methodology is in principle satisfactory, in practice there are circumstances in which it breaks down dramatically. The trouble arises from the poor conditioning (that is, large condition number) of the matrix Γ, which is intimately related to what may loosely be called "approximate linear dependence" of the representers g_j. With the functions of figure 2.08b we find that coefficients α_j exist so that

$$g_1 = \sum_{j=2}^{18} \alpha_j g_j + \varepsilon \tag{9}$$

where $\|\varepsilon\| / \|g_1\| = 1.41 \times 10^{-6}$. In other words, although the functions are strictly speaking linearly independent, a very good approximation to g_1 can be built from linear combinations of the other representers. This is reflected directly in κ, the condition number of Γ; see exercise 2.08(i). Here, in the matrix 2-norm, κ is 1.44×10^{16}. Recall from section 1.06 that with such a large condition number it is impossible to guarantee a correct numerical solution to the linear equations for α_j unless the relative accuracy of the computer storage and arithmetic is considerably better than 1 part in 10^{16}; the same sort of accuracy seems to be required of the numerical integration. Since the computer used by the author carries only about fifteen significant figures in "double precision," its most accurate mode of operation, it is quite unable to solve the problem in its present form.

Let us take a more thoroughly numerical viewpoint of the problem. We examine an approximation to the original problem in which the numerical integration formula used represents exactly the linear functional connecting an observation to the model. The same quadrature formula is used with each representer. We can imagine this particular problem to be one in a sequence of approximations in which L, the number of sampling points in r, increases; provided the quadrature formula is valid for the functions g_j, the elements in the corresponding sequence of matrices Γ converge to their true values, and the sequence of solutions obtained also converges to the correct one (putting aside the question of inadequacy of finite-precision arithmetic). For the approximate problem we have

$$d_j = \sum_{i=1}^{L} g_j(r_i) w_i q(r_i) \tag{10}$$

$$= \sum_{i=1}^{L} G_{ji} w_i q_i \tag{11}$$

where $G_{ji} = g_j(r_i)$ and $q_i = q(r_i)$. In matrix notation the constraint equations are

$$\mathbf{d} = G\, W\, \mathbf{q} \tag{12}$$

with $\mathbf{d} \in E^N$, $\mathbf{q} \in E^L$, $G \in M(N \times L)$, and $W \in M(L \times L)$; here W is the diagonal matrix $\text{diag}(w_1, w_2, \cdots w_L)$. We discussed in section 1.13 how to find the shortest vector satisfying such an underdetermined system; that vector would obviously be the one with the smallest Euclidean norm $\|\mathbf{q}\|_E$. The solution we want has the smallest value of

$$\left[\sum_{i=1}^{L} w_i q_i^2 \right]^{\frac{1}{2}} = \|W^{\frac{1}{2}} \mathbf{q}\|_E \tag{13}$$

where $W^{\frac{1}{2}} = \text{diag}(w_1^{\frac{1}{2}}, w_2^{\frac{1}{2}}, \cdots w_L^{\frac{1}{2}})$ since this minimizes the approximation to $\|q\|$. (We assume $w_i > 0$, something true of most quadrature formulas.) To find this vector write

$$\mathbf{n} = W^{\frac{1}{2}} \mathbf{q} \tag{14}$$

and rewrite (12) in terms of $\mathbf{n}$ using (14):

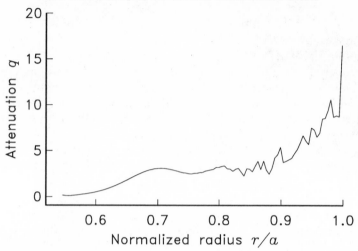

Figure 2.08c: An approximation to the attenuation model with smallest L_2 norm exactly satisfying the decay parameters shown in figure 2.08a. Note how rough the model is.

$$\mathbf{d} = G\,W^{\frac{1}{2}}\mathbf{n}\,. \tag{15}$$

The vector $\mathbf{n}_0$ satisfying this system and possessing the smallest value of $\|\mathbf{n}\|_E$ is obtained in the standard way and then the required answer is

$$\mathbf{q}_0 = W^{-\frac{1}{2}}\mathbf{n}_0 = (W^{\frac{1}{2}})^{-1}\mathbf{n}_0\,. \tag{16}$$

Expressed in matrix notation the standard solution forms an approximation to the Gram matrix

$$\Gamma \approx G\,W\,G^T = \Gamma_1\,. \tag{17}$$

A solution vector $\alpha \in E^N$ is found for the system

$$\mathbf{d} = \Gamma_1\alpha \tag{18}$$

and a model constructed via linear combinations of the representers

$$\mathbf{q}_0 = G^T\,\alpha\,. \tag{19}$$

With exact arithmetic the two formulations are identical. If, however, the smallest solution $\mathbf{n}_0$ is discovered via QR factorization as described in 1.13, the condition number of the system is given by the square root of that of Γ_1. Therefore the problem is practically solvable in a larger number of cases. The improvement for the dissipation problem is critical: with a value of 1.2×10^8 for κ, sensible calculations can be done with 14-decimal computer. (Actually, the condition number of Γ_1 was found by squaring that of the more stable scheme; it is impossible to calculate the value unless it is small enough to permit the inverse to be estimated.)

An approximate solution for the L_2 minimizing dissipation function is shown in figure 2.08c. For simplicity let us use 101 equally spaced samples in the interval and adopt Simpson's rule as the integration procedure. Thus

$$W = \mathrm{diag}(\tfrac{1}{3}h, \tfrac{4}{3}h, \tfrac{2}{3}h, \tfrac{4}{3}h, \cdots \tfrac{1}{3}h). \tag{20}$$

Since the ratio $c/a = 0.547$ in the Earth, $h/a = (1-0.547)/100 = 0.00453$. The "observations" used in the calculations are not the actual values found by measurement but are artificial numbers derived from a smooth model. Attempting to satisfy the slightly noisy measured values exactly leads to extraordinary models, again a result of the poor conditioning of the system. To overcome this tendency we must use some additional techniques to be discussed in the next chapter. The model of figure 2.08c is rather unsatisfactory—it is very rough, particularly in the upper mantle. This irregularity is not a result of arithmetic imprecision, but is a real feature of the smallest solution in L_2. To test the adequacy of the numerical quadrature scheme let us create a series of solutions with increasingly many sample points. For $L = 201$ the solution was even more irregular than the one shown in the figure and norm was a few percent higher. Clearly the sequence has not converged properly to the true norm minimizer: either still more sampling points must be taken or a more sophisticated quadrature rule is needed. In fact there seems to be no good reason to pursue this course: these models are evidently exceedingly rough. We must regard the smallest model in L_2 as implausible, particularly as we can discover much smoother solutions satisfying the data.

As in earlier examples we turn to solutions that are maximally smooth. Again we minimize the 2-norm of a linear roughening operator like the second derivative. If we follow the by now familiar steps of integration by parts, we arrive at

$$d_j = g_j^{(-1)}(a)q(a)/a - g_j^{(-2)}(a)q^{(1)}(a)/a + \int_c^a g_j^{(-2)}(r)q^{(2)}(r)\,dr/a. \tag{21}$$

A straightforward numerical implementation of these equations is very cumbersome, however. To find $g_j^{(-2)}$ we must integrate g_j twice; of course, this must be done numerically and a value produced at every sample point r_i. Similarly, to find q, $q^{(2)}$ must be integrated twice numerically. This seems like a lot of work, particularly if reasonable precision is to be maintained. A much simpler formulation presents itself if, once more, a purely numerical perspective is adopted.

Let us retain the same numerical integration method as in the L_2 study and therefore we also keep equally spaced sample points r_i. (There is no difficulty in taking unequally spaced samples as the reader may easily verify.) Instead of exact differentiation we introduce a differencing operation defined on E^L by a matrix:

$$\mathbf{f}' = \partial \mathbf{f} \tag{22}$$

where $\partial \in M(L \times L)$ is the matrix

$$\partial = \frac{1}{h} \begin{bmatrix} 1 & & & & & \\ -1 & 1 & & & \mathbf{0} & \\ & -1 & 1 & & & \\ & & -1 & 1 & & \\ & & & \cdots & & \\ \mathbf{0} & & & & -1 & 1 \end{bmatrix}. \tag{23}$$

The first component of $\mathbf{f}'$ is just f_1/h; the rest are the scaled differences $(f_i - f_{i-1})/h$. The matrix has a simple inverse:

$$\partial^{-1} = h \begin{bmatrix} 1 & & & & \\ 1 & 1 & & \mathbf{0} & \\ 1 & 1 & 1 & & \\ 1 & 1 & 1 & 1 & \\ & & \cdots & & \end{bmatrix}. \tag{24}$$

The action of ∂ is a first approximation to differentiation (apart from the first row) and ∂^{-1} is a primitive integrator. Consider now the norm on E^L defined by

$$\| \mathbf{f} \|_{\partial^2} = \| W_\gamma^{\frac{1}{2}} \partial^2 \mathbf{f} \|_E \tag{25}$$

where $W_\gamma^{\frac{1}{2}}$ is a diagonal matrix:

$$W_\gamma^{\frac{1}{2}} = \text{diag}(\gamma, \gamma, w_2^{\frac{1}{2}}, w_3^{\frac{1}{2}}, \cdots w_{L-1}^{\frac{1}{2}}) \tag{26}$$

and γ is any positive constant. A short calculation shows that

$$\| \mathbf{f} \|_{\partial^2} = \frac{\gamma^2}{h^4} f_1^2 + \frac{\gamma^2}{h^4} (f_2 - 2f_1)^2 + \sum_{i=2}^{L-1} \frac{w_i}{a} \left[\frac{f_{i-1} - 2f_i + f_{i+1}}{h^2} \right]^2 . \quad (27)$$

The sum at the end of this expression can be recognized as an approximation for the squared seminorm of a function $f \in W_2^2$:

$$\int_c^a f^{(2)}(r)^2 \, dr / a . \quad (28)$$

It is easily seen from Taylor's theorem that as h becomes small

$$f^{(2)}(r_i) = \left. \frac{d^2 f}{dr^2} \right|_{r=r_i} = \frac{f_{i-1} - 2f_i + f_{i+1}}{h^2} + O(h) \quad (29)$$

for functions $f \in C^3[c,a]$. From our earlier analysis the norm-minimizing solution is certainly in C^3 because it is composed of linear combinations of $g_j^{(-4)}$. Let us minimize $\|\mathbf{q}\|_{\partial^2}$ while satisfying the data, that is,

$$G \, W \, \mathbf{q} = \mathbf{d} . \quad (30)$$

Define the vector $\mathbf{n} \in E^L$ by

$$\mathbf{n} = W_\gamma^{\frac{1}{2}} \partial^2 \mathbf{q} \quad (31)$$

and then by (25) the Euclidean norm of $\mathbf{n}$ is the same as the ∂^2-norm of $\mathbf{q}$. From (12) the correct prediction of the data values may be written in terms of $\mathbf{n}$ thus

$$G \, W \, \partial^{-2} W_\gamma^{-\frac{1}{2}} \mathbf{n} = \mathbf{d} \quad (32)$$

which is of course an underdetermined system because $\mathbf{n} \in E^L$, $\mathbf{d} \in E^N$, and $N < L$. We may find the ordinary Euclidean norm minimizer using QR factorization of $G \, W \, \partial^{-2} W_\gamma^{-\frac{1}{2}}$; call the solution vector $\mathbf{n}_0$. Given $\mathbf{n}_0$ we calculate

$$\mathbf{q}_0 = \partial^{-2} W_\gamma^{-\frac{1}{2}} \mathbf{n}_0 \quad (33)$$

which is the solution we want: it minimizes $\|\mathbf{q}\|_{\partial^2}$ over the set of solutions to the data equations.

In contrast to a careful numerical realization of the integration-by-parts functional, this treatment has not attempted to maintain high accuracy of approximation everywhere. Specifically, the errors in approximating $d^2 f / dr^2$ are only $O(h)$ and the end contributions are absent in the Simpson rule integration; $\|\mathbf{q}\|_{\partial^2}^2$ is a relatively poor approximation to its parent integral. Nevertheless, our approach is quite satisfactory. The actual value of the norm of q in W_2^2 is not of the slightest importance, nor is its precise definition—it is merely one way of penalizing the roughness of functions. Therefore, our approximation need not be very faithful provided essential features are retained, for example, that constant functions or linear trends affect only the two leading γ^2 terms. We could, without real loss, replace $W_\gamma^{\frac{1}{2}}$ by the unit matrix. Where accuracy is important, in matching the data vector **d** with linear functionals of **q**, it is preserved.

We are not quite finished. The two terms in γ^2 in the expression for $\|\mathbf{f}\|_{\partial^2}^2$ serve no useful purpose in discriminating against rough solutions. We should minimize a seminorm in which these two terms are absent. The reader will correctly surmise that the seminorm minimizing solution 2.05(13) involving the matrices Γ and A is quite unsuitable for reasons of numerical inaccuracy. The alternative form 2.05(27) matches the matrix treatment very nicely. We set

$$\mathbf{q} = \beta_1\mathbf{h}_1 + \beta_2\mathbf{h}_2 + \mathbf{r} \tag{34}$$

where $\mathbf{h}_1 = (1, 1, 1, \cdots 1)$ and $\mathbf{h}_2 = (0, 1, 2, \cdots L-1)$; and we seek q_*, the solution that makes $\|\mathbf{r}\|_{\partial^2}$ as small as possible. The steps described in 2.05 can be carried out verbatim provided that it is remembered that the orthogonal projections are carried out in the Hilbert space E^L with inner product

$$(\mathbf{f}, \mathbf{g}) = (W_\gamma^{\frac{1}{2}}\partial^2\mathbf{f}) \cdot (W_\gamma^{\frac{1}{2}}\partial^2\mathbf{g}) . \tag{35}$$

All the calculations can be written in ways that use Householder rotations (see 1.13 for a stable projection algorithm) so that no unnecessary loss in precision is incurred.

To secure the seminorm minimizer $\mathbf{q}_*$ rather than the norm minimizer $\mathbf{q}_0$ does not in fact require much additional computational effort, yet the amount of programming effort needed is fairly substantial. There is a remarkably simple alternative which is completely satisfactory for practical use. The definition of $\|\mathbf{f}\|_{\partial^2}$ includes the positive constant γ; since the terms in γ are eliminated when $\mathbf{q}_*$ is found, the precise value is unimportant. Notice, however, that when γ is very small, the value of the norm is dominated by the terms in the sum corresponding to the integral, that is, the seminorm contribution. One must not set γ to zero because $W_\gamma^{-\frac{1}{2}}$ appears in the calculations; nevertheless, we might suspect that simply by finding $\mathbf{q}_0$, the norm minimizer, when γ is small we can obtain an adequate approximation for $\mathbf{q}_*$. A major concern is

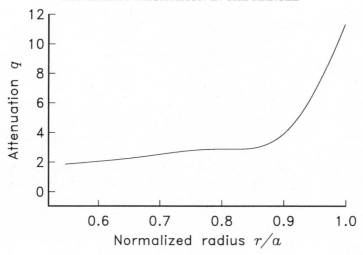

Figure 2.08d: A close approximation to the seminorm-minimizing attenuation model in W_2^2, which is the smoothest model in the sense of smallest 2-norm of the second derivative. The model fits the decay parameter data of figure 2.08a.

that this procedure may introduce numerical instability leading to unacceptable loss in precision, particularly in a delicate system like the ill-conditioned problem of seismic dissipation. Surprisingly, the fear is groundless: when a computer code with proper column interchanges is used, no instability is introduced. This can be shown by an argument similar to one given by Lawson and Hanson (1974) when they treat another weighted least squares problem. Indeed, if we choose γ to be about $(h\eta)^{1/2}$, where η is the relative computer precision, the numerical solution for $\mathbf{q}_0$ ought to be indistinguishable from $\mathbf{q}_*$ within the accuracy limitations imposed by the computer.

Both approaches to seminorm minimization were used to find smooth models for the dissipation program; the results were gratifyingly similar and appear in figure 2.08d. A series of solutions was obtained with γ varying; in this series L was fixed at 101 with $h/a = 0.00453$. The parameters of the computer are such that $(\eta h/a)^{1/2} = 5.6 \times 10^{-9}$; in fact significant differences between the models vanished when $\gamma < 10^{-5}$. Comparison between these limiting solutions and the one obtained by proper calculation of $\mathbf{q}_*$ revealed differences of at most three parts in 10^3 at the base of the mantle; with the exception of a few values near the deeper endpoint, agreement between $\mathbf{q}_0$ and $\mathbf{q}_*$ was generally better than one part in 10^4. This must be regarded as completely satisfactory agreement in view of the large condition number of the system. The convergence of the quadrature approximation is also satisfactory to this level of accuracy; this is to be expected because the solutions are so smooth.

The very smooth model of figure 2.08d might be viewed with suspicion by seismologists. Perhaps a discontinuity in q should be allowed in the "transition zone" of the mantle. The reader should have no difficulty in seeing how this demand can be accommodated within our scheme. We could construct the model that most closely resembles a two-layer system with constant values in the upper and lower mantles; or we might want to insist on smoothness in the two separate regions. For those interested in an up-to-date treatment of this geophysical problem, the paper by Widmer et al. (1991) is recommended.

The requirement that the predictions of the model match the observations exactly had no serious consequences in our earlier examples because they were relatively well conditioned. On the other hand, the seismic dissipation problem is very unstable and even minute amounts of noise would be amplified to produce wild oscillations in the solution. Since the uncertainty of the measured values of the data in this problem is never less than 1 percent, the question of observational uncertainty must be dealt with. That is the topic of the next chapter.

Exercise

2.08(i) We study the relationship between poor conditioning of the Gram matrix and "approximate linear dependence" of the representers. Find the coefficients γ_j that minimize

$$S = \left\| \sum_{j=1}^{N} \gamma_j g_j \right\| \quad \text{subject to} \quad \sum_{j=1}^{N} \gamma_j^2 = 1 .$$

Obviously, when $S_{\min}$ is zero, the representers are linearly independent and we could guess the smaller $S_{\min}$, the more ill-conditioned is Γ, the Gram matrix. We can see this more precisely in the following way; define

$$\rho = \frac{S_{\min}}{\left(\sum \|g_j\|^2 \right)^{1/2}} .$$

Show that

$$\frac{1}{N\kappa} \le \rho^2 \le \frac{1}{\kappa}$$

where $\kappa = \text{cond}(\Gamma)$, the condition number of the Gram matrix measured in the 2-norm. See exercise 1.06(iii).

CHAPTER 3

LINEAR PROBLEMS WITH UNCERTAIN DATA

3.01 Tolerance and Misfit

Actual measurements of geophysical quantities are never exact: in fact, a datum measured with an accuracy of two or three significant figures is usually considered to be rather precisely known. It is therefore essential for any practical scheme that solves inverse problems to take this into account. We continue to develop methods for constructing solutions to linear inverse problems in a way that recognizes the nonuniqueness associated with incomplete observations, and the almost inevitable instability that attends most geophysical realizations of the theory. The emphasis will continue to be on regularization, the selection of a well-behaved model from the infinite set of them that satisfy the data. Generally the solution sought will be the least complicated or least unusual one as defined by its size or distance from an archetype. When the data are exact, there is no ambiguity about what is meant by "fitting the data": each number must be matched precisely by the predictions of the model. This demanding requirement must be relaxed when the measurements are themselves uncertain. Two things are now required: a misfit measure, that is, a way of quantifying the disagreement between the observations and their theoretical counterparts; and a tolerance, which is a level of misfit that is considered acceptable so that when a model achieves that value, it may be said that the model adequately honors the demands of the measurements. In data-fitting problems solved by familiar least-squares regression analysis (see, for example, Seber, 1977) it is not necessary to say how accurate we think the measurements are ahead of time because the residuals—the misfit of the model—tells us this after the fit. The situation in inverse theory is different because, for linear problems at least, we can always fit the data exactly however much noise there is: our models are very accommodating in this respect. It should come as no surprise to the reader that we shall always assume the misfit can be represented as a norm in an N-dimensional linear vector space of measured numbers. It should also be obvious, at least in principle, how the theory will unfold: we seek the smallest model (or one with least semi-norm, or the one nearest a cherished structure) in the normed space of models, subject to the condition that the distance in the data space is acceptably small according to the norm of that space.

The task of this section is to set out the kinds of misfit functionals we shall employ and their associated tolerances. Superficially, the most natural kind of assessment of uncertainty in a measurement is the one asserting that the value lies between two definite limits:

$$d_j^- \le d_j \le d_j^+ . \tag{1}$$

Then when the model satisfies the data, each prediction of the model lies within the appropriate interval; viewing the collection of predictions as a vector in $\mathbb{R}^N$, we see that this is just the requirement that the element lie inside, or on the surface of, a certain N-dimensional rectangular box. As we saw in 2.03 we can supply a norm for the space so that this condition is stated by

$$\|\mathbf{d} - \mathbf{p}\| \le 1 \tag{2}$$

where $\mathbf{p}$ is the vector of theoretical values, the norm is

$$\|\mathbf{x}\| = \max_j 2|x_j|/(d_j^+ - d_j^-) \tag{3}$$

and the vector representing the data $\mathbf{d} \in E^N$ is

$$\mathbf{d} = (\tfrac{1}{2}[d_1^+ + d_1^-], \tfrac{1}{2}[d_2^+ + d_2^-], \cdots \tfrac{1}{2}[d_N^+ + d_N^-]) . \tag{4}$$

Thus uncertainty of the kind where every measurement has a definite known range can be handled by a normed data space. The only practical situation in which measurement uncertainty takes this form is the case of digitally recorded signals, when the discretization of the digitization is the sole source of error. This is in fact rare, because it is easy to design the digitization and recording electronics with more than adequate dynamic range; the limitations on the accuracy of a measurement are usually imposed by instrumental noise (like Brownian motion of the seismometer beam) or environmental noise (such as the pounding of the surf on the nearby shore). In these circumstances, the mathematical description of the uncertainty is much better handled by a statistical model.

Let us briefly set out some elementary statistics all of which will be familiar to most readers. A real *random variable* (conventionally represented by an upper-case letter) is a quantity whose value cannot be specified precisely; instead one imagines it to be the potential outcome of a measurement, or a series of measurements. The only things known precisely about a random variable are probabilities or averages of infinitely many repeated observations. We make the obvious generalization and treat a collection of N random variables as a random vector $\mathbf{X}$

$$\mathbf{X} = (X_1, X_2, X_3, \cdots X_N) \tag{5}$$

in an N-dimensional vector space which we will call RV^N. All the information that we have about the random vector is contained in its *probability density function* (pdf). When X is distributed according to a pdf f, this is written $X \sim f$. The pdf permits any statistical property of the random vector to be calculated. For example, the probability that the random vector X lies in the set of points $S \subset \mathrm{IR}^N$ is given by

$$\Pr(\mathbf{X} \in S) = \int_S f(\mathbf{x}) \, d^N \mathbf{x}. \tag{6}$$

Suppose a long series of repeated realizations of the random variable is obtained and for each one a particular functional $F[\mathbf{X}]$ is computed; an important statistic is the average or *expected* value of the functional denoted by $E[F[\mathbf{X}]]$ and calculated thus:

$$E[F] = \int_{\mathrm{IR}^N} F[\mathbf{x}] \, f(\mathbf{x}) \, d^N \mathbf{x}. \tag{7}$$

If a general vector-valued function is taken instead of a functional, the pdf can be used to find its expectation in just the same way.

The most frequently assumed form for the pdf of a random noise process is the *normal* or *Gaussian* distribution; its simplest form is the one in which each of the components of the random vector is identical and has zero *mean* (that is, expected value). The pdf of any single component X_n of the random vector is

$$f(x_n) = \frac{e^{-x_n^2/2\sigma^2}}{(2\pi)^{1/2}\sigma} \tag{8}$$

where σ, the parameter appearing in the definition of the pdf, is called the *standard deviation* of the distribution. The value of σ specifies the width of the distribution and, for a zero-mean normal random variable, the chance that X_n lies in the interval $(-\sigma, \sigma)$ is $0.6816894 \cdots$. The complete pdf of the random vector $\mathbf{X} \in \mathrm{RV}^N$ composed of these components is:

$$f(\mathbf{x}) = \frac{e^{-\mathbf{x} \cdot \mathbf{x}/2\sigma^2}}{(2\pi)^{N/2}\sigma^N} \tag{9}$$

which is just the product of the individual pdfs. When such a factorization is possible, the components are said to be *statistically independent*. By putting (8) into (7) it is easily seen that statistically independent variables are always *uncorrelated*, which means that

$$E[(X_i - \bar{X}_i)(X_j - \bar{X}_j)] = 0 \tag{10}$$

where $\bar{X}_i$ is the mean of X_i. Positive correlation implies that two zero-mean correlated random variables have a tendency to have the same sign. Conversely of course, uncorrelated random variables in a Gaussian process have no such association with each other. When the true value of an observation is disturbed in an additive way by a random Gaussian process of the kind we have just described, the measured values will be distributed randomly around the proper value according to the Gaussian law. If an experimentalist believes the noise in his or her observations obey this statistical model, it is conventional to describe the result of a single measurement in the form

$$d_j \pm \sigma. \tag{11}$$

That particular interval is often plotted as an "error bar" with central point at the best estimated value of the datum in graphs where uncertainty is to be indicated.

Our objective is to develop a criterion for deciding when the predictions of the theory are reasonably similar to the observations. The statistical model allows us to calculate how far the random disturbance is likely to take the measured value from the correct vector $\mathbf{d}$, and we may perform the calculation for any norm of our choice. A cursory look at the Gaussian distribution function suggests that the ordinary Euclidean norm would be the one leading to the easiest calculations and, for the moment, we shall be guided by that (far from trivial) consideration. First we calculate the average distance of the random vector from the true one:

$$E[\|(\mathbf{d}+\mathbf{X}) - \mathbf{d}\|] = E[\|\mathbf{X}\|] \tag{12}$$

$$= \int_{E^N} \|\mathbf{x}\| f(\mathbf{x}) \, d^N\mathbf{x} \tag{13}$$

$$= \int_{-\infty}^{\infty} \cdots \int_{-\infty}^{\infty} (\sum_i x_i^2)^{\frac{1}{2}} \frac{\exp\sum_i -x_i^2/2\sigma^2}{(2\pi)^{N/2}\sigma^N} \, dx_1 \cdots dx_N \, . \qquad (14)$$

To evaluate the integral we introduce N-dimensional spherical coordinates with $\rho = \|\mathbf{x}\|$ and Ω_N the surface of the unit ball, that is, the set of vectors with $\|\mathbf{x}\| = 1$; then $d^N\mathbf{x} = \rho^{N-1}d\rho \, d\Omega_N$ and we find

$$E[\|\mathbf{X}\|] = \int_{\Omega_N} \int_0^{\infty} \rho \, \frac{e^{-\rho^2/2\sigma^2}}{(2\pi)^{N/2}\sigma^N} \, \rho^{n-1}d\rho \, d\Omega_N \, . \qquad (15)$$

There is no difficulty in showing the integral may be treated as a repeated integral. Recall the definition of the Γ function:

$$\Gamma(z) = \int_0^{\infty} t^{z-1}e^{-t} \, dt \, . \qquad (16)$$

To perform the ρ integral in (15) we make the substitution $t = \rho^2/2\sigma^2$ and use the Γ function:

$$E[\|\mathbf{X}\|] = 2^{-\frac{1}{2}}\pi^{-N/2}\sigma \, \Gamma(\tfrac{1}{2}N+\tfrac{1}{2}) \int_{\Omega_N} d\Omega_N \, . \qquad (17)$$

The remaining integral is the measure of the surface (surface "area") of the unit ball in N dimensions. Apart from 2π for the circle and 4π for the sphere these are numbers with which most of us are unfamiliar. To complete the calculation we use the basic fact that the probability that the random vector $\mathbf{X}$ lies somewhere in the space must be unity; this implies

$$1 = \int_{E^N} f(\mathbf{x}) \, d^N\mathbf{x} \, . \qquad (18)$$

After the same change of variables and reduction to a Γ function this produces

$$\int_{\Omega_N} d\Omega_N = \frac{2\pi^{N/2}}{\Gamma(\tfrac{1}{2}N)} \, . \qquad (19)$$

Putting this into (17) we obtain the expected distance in the 2-norm of the randomly perturbed vector from the true one:

$$E[\|\mathbf{X}\|] = \frac{\sqrt{2} \, \sigma \, \Gamma(\tfrac{1}{2}N+\tfrac{1}{2})}{\Gamma(\tfrac{1}{2}N)} \, . \qquad (20)$$

Application of the Stirling's asymptotic formula for the Γ function for large argument (Abramowitz and Stegun, 1968, chapter 6)

$$\Gamma(z) = e^{-z} \, z^{z-\frac{1}{2}} (2\pi)^{\frac{1}{2}} \left[1 + \frac{1}{12z} + \frac{1}{288z^2} + O(z^{-3}) \right] \qquad (21)$$

provides a more informative expression:

$$E[\, \|\mathbf{X}\| \,] = \sigma N^{\frac{1}{2}} \left[1 - \frac{1}{4N} + \frac{1}{32N^2} + O(N^{-3}) \right]. \qquad (22)$$

Since the approximation in (22) of neglecting the terms of order $1/N^3$ incurs an error of less than 0.05 percent when $N \geq 3$, there will be little occasion to employ the exact expression. It seems perfectly reasonable to use the expected value as T, the tolerance, to define a satisfactory value of the misfit between theory and noisy observation.

Naturally, there are other possible choices, even if we retain the 2-norm as our measure of misfit. For example, one might decide that the proper choice of tolerance is the value below which the norm falls exactly 50 percent of the time in repeated, independent realizations of the N-observation experiment, the *median value*. Then we must solve for $T_{0.5}$ the equation

$$0.5 = \Pr(\|\mathbf{X}\| \leq T_{0.5}) \qquad (23)$$

$$= \Pr(\|\mathbf{X}\|^2 \leq T_{0.5}^2). \qquad (24)$$

Scaling with σ we find

$$\|\mathbf{X}\|^2/\sigma^2 = \sum_{n=1}^{N} Y_n^2. \qquad (25)$$

The components Y_n are said to be *standardized* independent Gaussian random variables because they have zero mean and unit standard deviation. A random variable formed in this way is said to be distributed as χ^2 with N degrees of freedom; after the normal distribution, this is the best-known distribution in all of statistics. Tables of the value exceeded 50 percent of the time as a function of N (and many other standard probability values, for example, 90, 95, and 99 percent critical values) are widely available (for example, in Abramowitz and Stegun, 1968); they must be computed numerically because no closed-form elementary expression exists. Alternatively, large-N asymptotic theory yields

$$T_{0.5} = \sigma N^{\frac{1}{2}} \left[1 - \frac{1}{3N} - \frac{13}{810N^2} + O(N^{-3}) \right] \qquad (26)$$

which is an approximation with an error of less than 1 part in 10^3

if $N \geq 3$. Comparing (22) and (26) we see that numerically the median tolerance is not substantially different from the one based upon the mean. Probability values other than 0.5 might be selected as being sensible, but it is hard to justify a tolerance that would be very rarely met; this would demand too good a fit to the data and would yield an unnecessarily exciting model. More defensibly, one could argue for a large tolerance, one likely to be satisfied very often by chance, on the grounds that the computed model would then be very unlikely to contain false undulations. We shall investigate this choice in the illustrations.

Let us now generalize the distribution law of the random errors: we allow the different observations to possess different amounts of uncertainty, although at this point they remain uncorrelated. The pdf for the collection of random variables X_1, X_2, $\cdots$ X_N, with associated standard deviations σ_1, σ_2, $\cdots$ σ_N, is

$$f(x_1, x_2, \cdots x_N) = \frac{\exp[-\tfrac{1}{2}(x_1^2/\sigma_1^2 + x_2^2/\sigma_2^2 + \cdots x_N^2/\sigma_N^2)]}{(2\pi)^{N/2}\sigma_1\sigma_2 \cdots \sigma_N}. \tag{27}$$

In vector notation this is

$$f(\mathbf{x}) = \frac{\exp[-\tfrac{1}{2}\mathbf{x} \cdot \Sigma^{-2}\mathbf{x}]}{(2\pi)^{N/2}\sigma_1\sigma_2 \cdots \sigma_N} \tag{28}$$

where the diagonal matrix $\Sigma = \mathrm{diag}(\sigma_1, \sigma_2, \cdots \sigma_N)$. Suppose we define a new, standardized set of random variables by scaling each component by its inverse standard deviation:

$$X'_n = X_n/\sigma_n . \tag{29}$$

From their pdfs we see that the standardized variables are identically distributed with standard deviation unity and zero mean. Naturally, in the subsequent inversion process, the same scaling must be applied to the measurements themselves (and to their representers in the linear theory) but this is quite trivial. Standardization has reduced the more general problem to the simple one we treated initially. We may proceed to use the ordinary Euclidean norm in the space of scaled measurements and calculate the appropriate tolerance as before. In the original space, this amounts to using a weighted 2-norm to measure discrepancy between theory and observation. Is there a good reason not to employ the apparently more natural Euclidean norm in the original variables? As we mentioned earlier, one reason is that the calculation of the tolerances is straightforward if standardization is adopted but the task becomes quite complex if some other 2-norm is chosen. The additional complication might be tolerable if there were good statistical reasons for another choice but, in a certain sense, the scaled random variables are optimal from a statistical viewpoint as well.

We have tried to avoid being dogmatic about the criterion for choosing a tolerance level once the norm has been selected: it ought not to matter very much whether we use the expected value of $\|\mathbf{X}\|$ or choose a critical probability because, ideally, the actual value of T ought to be quite similar in either case. Statistically, this is equivalent to saying that the pdf for $\|\mathbf{X}\|$ is highly concentrated near its expected value; we shall see in a moment that this is the case for the standard norm. The usual measure of broadness of the pdf of a single random variable Y is the *variance* which is the second moment about the mean value:

$$\mathrm{var}[\,Y\,] = \mathrm{E}[\,(Y - \bar{Y})^2\,] \tag{30}$$

$$= \mathrm{E}[\,Y^2\,] - \bar{Y}^2 \tag{31}$$

where the mean is $\bar{Y} = \mathrm{E}[\,Y\,]$. Roughly speaking, the variance is the square of the width of the peak of the pdf, on the assumption that the pdf is a function with one maximum. The reader surely knows that, if X is normally distributed with standard deviation σ, $\mathrm{var}[\,X\,] = \sigma^2$. When $\mathbf{X} \in \mathrm{RV}^N$ is distributed according to (9), the simplest multivariate Gaussian model, $\mathrm{var}[\,\|\mathbf{X}\|\,] = \sigma^2 N - \mathrm{E}[\,\|\mathbf{X}\|\,]^2$; as N becomes large

$$\mathrm{var}[\,\|\mathbf{X}\|\,] = \tfrac{1}{2}\sigma^2\left[1 - \frac{1}{2N} + O\,(N^{-2})\right]. \tag{32}$$

Retaining the same distribution for the random vector, suppose we ask, of all the norms on RV^N in the form

$$\|\mathbf{X}\| = (w_1^2 X_1^2 + w_2^2 X_2^2 + \cdots + w_N^2 X_N^2)^{\frac{1}{2}} \tag{33}$$

subject to a normalization of the weights, $\Sigma\, w_n^2 = N$, which one possesses the smallest variance? It is not hard to show by elementary calculus that $w_n = 1$ for all n for the least-variance norm. Thus the ordinary Euclidean norm is optimal in having the smallest variance of 2-norms in this class. The conclusion remains the same even when we expand the class of 2-norms on RV^N to the fullest extent by permitting any positive definite quadratic form under the radical; this is shown most easily by spectral factorization of the weight matrix, a topic discussed in section 3.03. Returning to the random vector of uncorrelated components with differing standard deviations, we see that the standardization of the random vector also yields the minimum variance norm. In simple terms, the scaling merely insures that the influence of a particular observation on the misfit is inversely proportional to its uncertainty. Thus the case for the standard scaling is quite strong.

The final step to a completely general multivariate Gaussian pdf is to allow the components of the random vector to be correlated. Then for $\mathbf{X} \in \mathrm{RV}^N$

$$f(\mathbf{x}) = \frac{\exp[-\tfrac{1}{2}\mathbf{x} \cdot C^{-1}\mathbf{x}]}{(2\pi)^{N/2}(\det C)^{\frac{1}{2}}} \tag{34}$$

where $C \in \mathrm{M}(N \times N)$ is a positive definite, symmetric matrix, called the *covariance matrix* of the random vector and det stands for the determinant. From this definition it can be shown, upon application of the theory of determinants, that

$$\mathrm{E}[X_i X_j] = C_{ij} \tag{35}$$

and so the element C_{ij} is called the *covariance* between components X_i and X_j. It is scarcely an exaggeration to say that covariance estimates are never available for experimental geophysical data. One reason is that in a properly designed experiment, the errors in one measurement should be independent of those in any other, which would preclude correlation between the noise contributions; geophysicists are generally optimistic about the quality of their experimental procedures. Another factor is that the number of quantities to be estimated in a full covariance matrix that describes N measurements is $\tfrac{1}{2}N(N+1)$, an enormous amount of work if N is moderately large. Usually, one assumes a diagonal covariance matrix in absence of information about the off-diagonal terms which corresponds of course to the case of uncorrelated data. It is sometimes possible to calculate covariances from a theoretical model of the noise process, for example, when crustal signals are regarded as noise in the measurement of the magnetic field made by a satellite; see, for example, Jackson (1990). For completeness we set out the theory for dealing with the most general case. Recall from section 1.13 the Cholesky factorization of a positive definite matrix; we write

$$C = L L^T \tag{36}$$

where L is a lower triangular matrix. Then if a new random vector $\mathbf{X}'$ is formed by a linear transformation of the original vector according to

$$\mathbf{X}' = L^{-1}\mathbf{X} \tag{37}$$

the pdf for the primed vector can be found as follows. First we note that, because $\det L = \det L^T$ and the determinant of the product of two square matrices is just the product of their determinants,

$$\det L = \det L^T = (\det C)^{\frac{1}{2}}. \tag{38}$$

(From its definition as a kind of square root of C, either L or $-L$ could serve; traditionally the sign is chosen to make the determinant positive.) But the Jacobian of the transformation from $\mathbf{X}$ to $\mathbf{X}'$ is just $\det L$ and so the transformation to primed variables gives a pdf corresponding to (9), that is, identical, uncorrelated components with unit standard deviation.

This is not the only linear transformation that can take the original, correlated vector into an uncorrelated one. For example, any rotation of L, corresponding to premultiplication with an N-by-N orthogonal matrix, preserves the desired property. Furthermore, it is not necessary to perform the factorization of C to produce explicit primed variables; it is sufficient to measure misfit in the data space with the 2-norm

$$\|\mathbf{x}\| = (\mathbf{x} \cdot C^{-1}\mathbf{x})^{\frac{1}{2}} \tag{39}$$

and to compute the tolerance according to our rules, setting $\sigma = 1$.

We have suggested that the Euclidean norm is the most natural one for measuring the size of Gaussian random perturbation. Certainly it is the most compatible with the Hilbert space calculations that we have performed so far. In chapter 4 we shall develop the machinery for minimizing or constraining the 1-norm and the uniform norm by means of linear programs. Therefore we shall quickly sketch their statistical properties. Recall that the uniform norm of a vector $\mathbf{X} \in \mathrm{RV}^N$ is given by

$$\|\mathbf{X}\|_\infty = \max_{1 \le n \le N} |X_n|. \tag{40}$$

Under the assumption of independent Gaussian components, with zero mean and standard deviation σ, the distribution function for this quantity is easily found as follows:

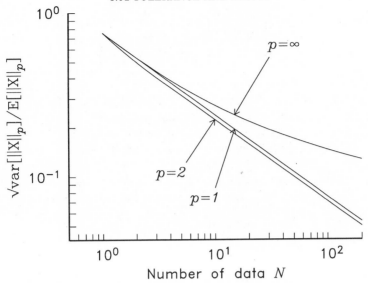

Figure 3.01a: Normalized square-root variance of three p-norms as functions of N. Observe how the 2-norm has the best performance, but by only a tiny margin over the 1-norm.

$$\Pr(\|\mathbf{X}\|_\infty \le U) = \Pr(|X_1| \le U \text{ and } |X_2| \le U$$
$$\cdots \text{ and } |X_N| \le U) \tag{41}$$

$$= [\Pr(|X_1| \le U)]^N \tag{42}$$

$$= [\operatorname{erf}(U/\sqrt{2})]^N \tag{43}$$

where erf is the *error function*; see Abramowitz and Stegun (1968). The mean and variance are easy to compute numerically and satisfy the following asymptotics:

$$E[\|\mathbf{X}\|_\infty]/(\ln N)^{1/2} \to \sigma\sqrt{2} \tag{44}$$

$$\operatorname{var}[\|\mathbf{X}\|_\infty] \cdot (\ln N) \to \sigma^2. \tag{45}$$

In comparing this norm with the Euclidean we examine the width of the distribution after normalization by the mean value. The behavior is shown in figure 3.01a; notice how slowly the ratio diminishes for the uniform norm. According to our earlier discussion this would already be reason enough for dismissal. But there is another, more powerful argument against the norm. In practice random disturbances are never exactly Gaussian in their distribution; a common imperfection takes the form of occasional very large values, called *outliers*. As (44) shows, large values are very rare under a true Gaussian distribution; their more frequent appearance totally wrecks inferences based upon extreme values, such as $\|\mathbf{X}_\infty\|$. For those interested in such matters there is a statistical literature on the treatment of outliers (see, for example Barnett and Lewis, 1984), but we shall be content simply to avoid the use of the uniform norm on the observations.

The Euclidean norm is also regarded by some as unnecessarily sensitive to outliers since, through the square, large deviations tend to influence the 2-norm disproportionately. Among the p -norms the 1-norm is the least susceptible to the problem. As we noted in 1.06 this norm is given by

$$\|\mathbf{X}\|_1 = \sum_{n=1}^{N} |X_n| \, . \tag{46}$$

The computations of statistical values for the 1-norm is somewhat involved and so we shall merely quote some results from Parker and McNutt (1980): when the components of $\mathbf{X}$ are the familiar independent Gaussian variables

$$E[\,\|\mathbf{X}\|_1\,] = \sigma N (2/\pi)^{\frac{1}{2}} \tag{47}$$

$$\mathrm{var}[\,\|\mathbf{X}\|_1\,] = \sigma^2 N (1{-}2/\pi) \, . \tag{48}$$

The performance ratio is plotted for this norm in figure 3.01a; we see that while the 2-norm is superior for all $N > 1$ the advantage is trivial. Tables of the distribution function for the 1-norm statistic are difficult to find, and therefore a short one is provided in appendix B. For the rest of this chapter we shall continue to employ the 2-norm of misfit chiefly because the calculations associated with it are the simplest.

We shall not dwell upon methods for making estimates of the statistical parameters of the noise because they properly belong outside the scope of this book, in the realm of statistical theory and signal processing. For our purposes it is sufficient to say that the Gaussian model for the random element in a data set is often a moderately good approximation and that, by analysis of repeated measurements for example, the experimentalist can provide, at the very least, a rough estimate of the uncertainty in the observations. Only relatively rarely in geophysics can a precise statistical characterization of the uncertainty be obtained. It is therefore sensible not to attempt fanatical exactitude using

sophisticated statistical techniques. Another reason for not plac-
ing too much emphasis on elaborate statistical models is the inevitable presence of errors arising from approximations of the
mathematical model that embodies the solution to the forward
problem. Because of computational expense or even lack of an
adequate mathematical theory, it is often necessary to simplify
the model in a radical and perhaps unverifiable way. The most
common kind of approximation is the assumption of some kind of
symmetry in the model: geophysicists frequently assume that the
relevant properties of the Earth vary only with one coordinate, for
example, depth below the surface, or distance from the Earth's
center. A typical example is the one of the previous chapter in
which the magnetization of the oceanic crust was assumed to be
concentrated into a thin layer and to be quite constant in one
direction. With a model of this kind it would be unwarranted to
demand an exact fit to the data because the model itself is not a
very good rendering of reality. If one can make a rough calculation of the kinds of errors that are introduced by the approximation (which is possible in this case) it is a good idea to include in
the tolerance an allowance for such effects. The situation
becomes more difficult in the study of nonlinear problems, when it
may be impossible to assess the magnitude of the perturbations
created by ignoring departures from the assumed ideal conditions.
The geophysicist should always be alert to the possibility that the
underlying model may be deficient and that random noise may
not be the major source of disagreement between the theory and
the observation.

3.02 Fitting within the Tolerance

In this section we develop the necessary techniques for constructing a model in agreement with the uncertain data. The material
is rather dry because the process is a good deal more complex
than it was in the case of exact fit. We shall have to wait until
the next section to get some geometrical insight into what is going
on. It will be assumed from now on that a weighted 2-norm is in
use to assess the misfit of the model's predictions to the observations and that a tolerance defining the acceptable distance
between the two has been chosen. For simplicity we consider just
the case of uncorrelated noise, but we allow for possibly different
standard deviations of the noise components, although it should
be clear from the methods of the previous section how to

generalize to the case with covariance; the more general result will be indicated without proof. As in chapter 2, the predictions of the model are given by bounded linear functionals on a Hilbert space. Thus the statement that there is a satisfactory match between theory and observation takes the form

$$\left[\sum_{j=1}^{N} [d_j - (g_j, m)]^2 / \sigma_j^2 \right]^{\frac{1}{2}} \leq T \tag{1}$$

where $g_j \in H$ are linearly independent representers in the solution to the forward problem and $m \in H$ is the unknown model. To keep the manipulations to a minimum we treat the question of finding the element with smallest norm in H satisfying the data; we shall give the parallel result for seminorm minimization without derivation.

As in section 1.12 we separate the model m into parts in the two subspaces G and $G^{\perp}$ by the Decomposition Theorem; it is obvious that the component in $G^{\perp}$ contributes nothing to the misfit functional above because it is orthogonal to every one of the representers, but it does increase the norm. Therefore, exactly as before, we know that the norm-minimizing element satisfying the data in this weaker sense lies entirely in G; in other words, it can be expressed as a linear combination of representers:

$$m_0 = \sum_{j=1}^{N} \alpha_j g_j . \tag{2}$$

The search for a model has again been reduced to the problem of finding the N real expansion coefficients α_j. It should be noticed that the same considerations apply no matter what data misfit norm is in use here. In terms of these coefficients we may write the model norm squared thus:

$$\|m_0\|^2 = (\sum_{j=1}^{N} \alpha_j g_j, \sum_{k=1}^{N} \alpha_k g_k) \tag{3}$$

$$= \sum_{j=1}^{N} \sum_{k=1}^{N} \alpha_j \Gamma_{jk} \alpha_k \tag{4}$$

$$= \boldsymbol{\alpha} \cdot \Gamma \boldsymbol{\alpha} \tag{5}$$

Γ is of course the Gram matrix. Similarly, the model acceptance criterion (1) can be written in matrix and vector language:

$$\| \Sigma^{-1}(\mathbf{d} - \Gamma \boldsymbol{\alpha}) \| \leq T \tag{6}$$

where $\Sigma \in M(N \times N)$ is the diagonal matrix of σ_j as in 3.01(28) and the norm is the Euclidean length in E^N. Because the noise is effectively standardized (that is normalized by the standard errors) in (1), the appropriate value of T should be found from 3.01(22), for example, by taking $\sigma = 1$.

There are three different norms with which we might equip our N-dimensional space: for numerical purposes the ordinary Euclidean norm is most convenient and will be denoted in the usual way. We can make E^N a Hilbert space under the Γ-norm given by

$$\|\alpha\|_\Gamma = (\alpha \cdot \Gamma\alpha)^{1/2} . \tag{7}$$

This norm has the nice property that the norm of the model in H is the same as the Γ-norm of α in E^N. Of course this is only true of models that are confined to the subspace G, but these are the only ones that concern us here. The subspace G and E^N are said to be *isometric*. Finally, we might want to define a norm associated with misfit to the data, say the T-norm:

$$\|\alpha\|_T = \|\Sigma^{-1}\alpha\| . \tag{8}$$

On the whole it is simplest to stick with the Euclidean norm for the algebraic work, but it is useful occasionally to refer to the Γ-norm by name rather than to write out "the norm of model associated with the coefficient vector α."

Our task is to minimize the quadratic form $\alpha \cdot \Gamma\alpha = \|\alpha\|_\Gamma^2$ over all vectors α obeying the data misfit inequality (6). The situation is illustrated graphically in figure 3.02a, showing the zone of acceptable models and the contours of the model norm as they appear in a two-dimensional space when $N = 2$. The diagram strongly suggests that the solution vector lies on the boundary of the allowable misfit region, in other words, the smallest norm model is associated with the largest allowable misfit and *equality* holds in the fitting criterion (6), not inequality. This is easily proved as follows. Assume the contrary—that the smallest value of $\|m_0\|^2$ occurs for a model misfit $T_1 < T$ with a coefficient vector α_1. Now consider the vector $\alpha_2 = \mu\alpha_1$ where μ is a real number and $0 < \mu < 1$. If the second vector is substituted into (6) we find after a little algebra that

$$\|\Sigma^{-1}(\mathbf{d} - \Gamma\alpha_2)\|^2 = T_1^2 + (1-\mu)c_1 + (1-\mu)^2 c_2 \tag{9}$$

where c_1 and c_2 are constants depending upon α_1 and the other given vectors and matrices. If μ is sufficiently close to one, we

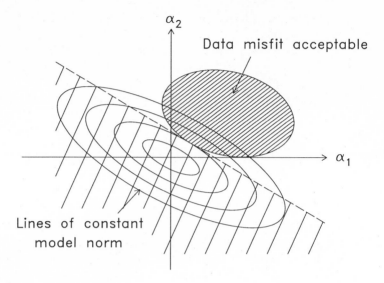

Figure 3.02a: Contours of model norm and, densely shaded, the region of acceptable misfit in E^2, the two-dimensional normed space of expansion coefficients α_j. Shown lightly shaded, the region LL.

can keep the misfit measure under T because $T_1 < T$; then the vector α_2 lies inside the acceptance zone. But $\|\alpha_2\|_\Gamma = \mu\|\alpha_1\|_\Gamma$ and, because μ is less than one, this is less than the supposed minimum norm. Therefore α_1 cannot give rise to the smallest norm model and so the best model cannot be obtained unless equality holds in the misfit criterion.

There is one situation in which this proof fails: if $\alpha_1 = 0$; then the norm does not decrease when it is multiplied by $\mu < 1$. The smallest norm model is $m_0 = 0$. So if the zero element fits the data, the model predictions for the data (all zeros) may lie strictly inside the acceptable region E^N, not just on its boundary. The following analysis applies to the case in which the zero element is not an acceptable model (something easily checked!).

We must now find the smallest value of the quadratic form (5) subject to an equality constraint. This is a problem most readily solved with a Lagrange multiplier (see section 1.14). We form an unconstrained functional $U[\alpha, \nu]$ that must be stationary with respect to all its arguments; it is simpler if the fitting criterion is squared here:

$$U[\alpha, \nu] = \alpha \cdot \Gamma\alpha - \nu(T^2 - \|\Sigma^{-1}(\mathbf{d} - \Gamma\alpha)\|^2) . \tag{10}$$

An argument pair where U is stationary will be called α_0, v_0. To find these points we first differentiate with respect to α_j, or equivalently take the gradient operator with respect to α, and set this vector to zero:

$$\nabla_{\alpha_0} U = 0 \tag{11}$$

$$= 2\Gamma\alpha_0 - 2v_0(\Sigma^{-1}\Gamma)^T\Sigma^{-1}(\mathbf{d} - \Gamma\alpha_0) . \tag{12}$$

After a little rearrangement this becomes

$$(v_0^{-1}\Sigma^2 + \Gamma)\alpha_0 = \mathbf{d} . \tag{13}$$

Naturally differentiating U with respect to the Lagrange multiplier just gives us the condition that the misfit is T:

$$\|\Sigma^{-1}(\mathbf{d} - \Gamma\alpha_0)\|^2 = T^2 . \tag{14}$$

The pair of equations (13) and (14) must be solved together for α_0 and v_0.

One difficulty is that the system is nonlinear, something obvious from (14). If, however, v_0 is known, (13), the equations for α_0 are linear. Provided we are not concerned with efficiency, the solution is straightforward: we simply sweep through all values of v_0 solving for α_0 and substituting this vector into (14) to see if it is satisfied; if it is we have a stationary solution. There are numerically better ways to do this than stepping along the real line and we shall discuss one in a moment; an important property is that the value of v_0 we are interested in is positive. There are other stationary solutions but we shall see that the one corresponding to the norm minimizer is unique and positive. This brings us to a weakness of the elementary calculus approach to minimizing $\|m\|$. We must demonstrate that our answer really is the global minimizer of the norm; all we have done so far is write equations that identify stationary points of the functional. We shall prove that if a positive v_0 can be found, the corresponding α_0 generates a model with the least norm.

The following proof is suggested by the figure: the dashed line, tangent to the contours of $\|\alpha\|_\Gamma$ and the boundary of the misfit constraint region, separates the plane into two halfspaces. Every point strictly within the lower left halfspace has a misfit larger than the required one; every point within the upper right halfspace has a norm larger than the one at the tangent point. Only one point has the best of both worlds—the point of contact. Indeed, our elementary proof is really an example of a more

profound theorem concerning optimization over convex sets, the
Separating Hyperplane Theorem (see Luenberger, 1969, p. 133).
Remember we assume a positive value of v_0 is known with the
property that the corresponding vector α_0 found by solving (13)
satisfies the misfit equation (14). We define the set $LL \subset E^N$ as
the set of points β obeying

$$\beta \cdot (\Gamma\alpha_0) \leq \alpha_0 \cdot \Gamma\alpha_0 = \|m_0\|^2 . \tag{15}$$

The set LL is the generalization of the shaded half plane below
and to the left of dashed line in the figure; this is because $2\Gamma\alpha_0$ is
the gradient of $\alpha \cdot \Gamma\alpha$ (the squared norm) at α_0 and the gradient
vector points in a direction normal to the surface of constant norm
through that point. First we show that every α with Γ-norm less
than $\|\alpha_0\|_\Gamma$ lies in LL, something intuitively clear from the picture.
If $\alpha \neq \alpha_0$

$$0 < \|(\alpha - \alpha_0)\|_\Gamma^2 \tag{16}$$

$$= (\alpha - \alpha_0) \cdot \Gamma(\alpha - \alpha_0) \tag{17}$$

$$= \alpha \cdot \Gamma\alpha + \alpha_0 \cdot \Gamma\alpha_0 - 2\alpha \cdot \Gamma\alpha_0 . \tag{18}$$

Thus, rearranging

$$\alpha \cdot \Gamma\alpha_0 < \tfrac{1}{2}(\alpha \cdot \Gamma\alpha + \alpha_0 \cdot \Gamma\alpha_0) . \tag{19}$$

But we have asserted that the Γ-norm of α is less than that of α_0
and so

$$0 < \tfrac{1}{2}(-\alpha \cdot \Gamma\alpha + \alpha_0 \cdot \Gamma\alpha_0) . \tag{20}$$

Adding (19) to (20) we obtain

$$\alpha \cdot \Gamma \alpha_0 < \alpha_0 \cdot \Gamma\alpha_0 \tag{21}$$

which by (15) tells us that $\alpha \in LL$.

Next we prove that every point of LL not equal to α_0 has a
misfit greater than T, and so it will follow that vectors smaller in
the Γ-norm than α_0 must generate larger misfits, which is the
result we want. We write the difference of the associated data
misfits squared

$$\Delta = \|\Sigma^{-1}(\mathbf{d} - \Gamma\alpha)\|^2 - \|\Sigma^{-1}(\mathbf{d} - \Gamma\alpha_0)\|^2 . \tag{22}$$

We show that Δ is positive. Expanding and canceling we obtain
what is essentially equivalent to $x^2 - y^2 = (x + y)(x - y)$:

$$\Delta = (\Sigma^{-1}(2\mathbf{d} - \Gamma\alpha_0 - \Gamma\alpha)) \cdot (\Sigma^{-1}(\Gamma\alpha_0 - \Gamma\alpha)) . \tag{23}$$

Up to this point we have not used the definition of α_0; at last we do—by means of (13) replace $\mathbf{d}$ in (23)

$$\Delta = (\Sigma^{-1}(\Gamma\alpha_0 - \Gamma\alpha) + 2v_0^{-1}\Sigma\alpha_0) \cdot (\Sigma^{-1}(\Gamma\alpha_0 - \Gamma\alpha)) \qquad (24)$$

$$= \|(\Sigma^{-1}(\Gamma\alpha_0 - \Gamma\alpha))\|^2 + 2v_0^{-1}[\alpha_0 \cdot \Gamma \alpha_0 - \alpha \cdot \Gamma\alpha_0] . \qquad (25)$$

Since $\alpha \neq \alpha_0$ the first term is positive; the second term is also positive by (21) and so we have established that all vectors with Γ-norms less than that of α_0 must have larger misfits, provided a solution to the equations can be found with positive v_0.

The above analysis also shows that if a stationary solution is found with $v_0 < 0$ it cannot be the minimum norm model. The proof is left as an exercise for the reader. (Hint: show the misfit can be made smaller than T.)

We pause here to point out a quite different way of arriving at our system, traditionally called *regularization*. There are two independent properties of a model, its size and its disagreement with observation. These are measured by the model norm and the misfit to the data (also a norm of course); we would like to keep both as small as possible. Since they usually cannot both be brought to zero together let us propose a penalty functional in H depending on a positive weight parameter v:

$$W[m, v] = \|m\|^2 + v\sum_{j}^{N} [d_j - (g_j, m)]^2/\sigma_j^2 . \qquad (26)$$

The idea is to minimize W over elements of H. When v is very small W effectively measures the squared norm of m; when v is very large W is proportional to the squared data misfit, while generally W is just an intermediate functional that penalizes both in some proportion set by v. We call v a *trade-off* parameter because it balances or trades off the two undesirable properties of the model. By making W stationary we arrive at exactly the same solution that we have been discussing already. An advantage of this approach is that it is obvious only positive v are relevant to the problem; a disadvantage is that we do not see immediately that for a fixed misfit we obtain the smallest possible associated norm. As a last remark in this digression, we note that an equation almost like (26) can be obtained with v as a Lagrange multiplier if we seek the minimum norm model with specified misfit but do not project m onto the finite-dimensional subspace G right from the start; the only difference is the explicit appearance of T^2 in the constraint term.

The next question is that of finding the ν_0 appropriate to the given value of T, having first shown that there is such a value and that it is unique. We adopt the view now that ν is an arbitrary positive real parameter that we substitute for ν_0 in (13); ν determines the vector $\boldsymbol{\alpha}_0(\nu)$ through the solution of the linear system. The linear equations are always soluble because Γ is positive definite and so is $\nu^{-1}\Sigma$; the sum of two positive definite matrices also has that property and so is nonsingular (check this; see exercise 3.02(ii)). The squared misfit associated with $\boldsymbol{\alpha}_0(\nu)$ will be called $X^2[\nu]$:

$$X^2[\nu] = \| \Sigma^{-1}(\mathbf{d} - \Gamma\boldsymbol{\alpha}_0(\nu)) \|^2 . \tag{27}$$

What we shall show is that there is a single positive value of ν, naturally called ν_0, such that

$$X^2[\nu_0] = T^2 \tag{28}$$

provided the given value of T is less than $\| \Sigma^{-1}\mathbf{d} \|$. A typical functional X^2 is shown in figure 3.02b. The simple monotone behavior for positive ν is a perfectly general property; the horrible antics for negative ν demonstrate the importance of establishing beforehand that the required solution lies on the right side of the origin.

The condition on the size of the given tolerance T is easy to deal with: if T exceeds or matches the critical value it is obvious from (1) that the model $m_0 = 0$ satisfies the data, and this has the smallest possible norm of all conceivable models and so the original problem is solved without more ado. We do not need to seek a positive value of ν_0 in this case and indeed this is fortunate because none exists.

We show next that for $\nu > 0$ the function X^2 decreases from an upper bound of $\| \Sigma^{-1}\mathbf{d} \|^2$, tending to zero as ν tends to infinity. Thus there must be one and only one value of ν achieving the given value of T^2. The two limiting values are easily established: first, as $\nu \to 0$ it is seen from (13) that $\boldsymbol{\alpha}_0 \to 0$ and then by (27) that $X^2 \to \| \Sigma^{-1}\mathbf{d} \|^2$. In the limit $\nu \to \infty$ (13) becomes the linear system for the expansion coefficients of the model of minimum norm *exactly* fitting the data that occupied so much of our attention in chapter 2. To prove that X^2 is a decreasing function of ν we merely differentiate X^2; the intervening details will be omitted except to note the usefulness of the following two nice results: combining (13) and (27)

$$X^2[\nu] = \nu^{-2}\|\Sigma\boldsymbol{\alpha}_0(\nu)\|^2 \tag{29}$$

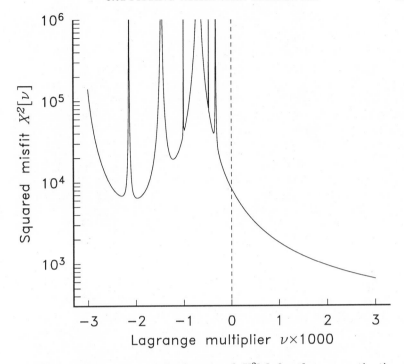

Figure 3.02b: The misfit functional $X^2[\nu]$ for the magnetization problem treated in section 3.04. The wild behavior for $\nu < 0$ is best understood in terms of the spectral expansion; see 3.03.

and for a general square nonsingular matrix A depending on the real parameter ν

$$\frac{dA^{-1}}{d\nu} = -A^{-1}\frac{dA}{d\nu}A^{-1}.\tag{30}$$

After straightforward but lengthy algebra then we find

$$\frac{dX^2}{d\nu} = -2\nu^{-3}\alpha_0(\nu) \cdot \Sigma^2(\nu^{-1}\Sigma^2 + \Gamma)^{-1}\Gamma\alpha_0(\nu)\tag{31}$$

$$= -2\nu^{-3}\alpha_0(\nu) \cdot (\Sigma^{-2} + \nu^{-1}\Gamma^{-1})^{-1}\alpha_0(\nu).\tag{32}$$

This equation says that $dX^2/d\nu < 0$ because the matrix is positive definite: it is the inverse of the sum of two positive definite matrices and this is easily seen also to be positive definite. See exercise 3.02(ii).

There is another reason for computing the derivative of X^2 besides its role in the proof of the uniqueness of v_0. To locate the value of v_0 we must solve the nonlinear equation (28). A familiar and quite efficient numerical method for such problems is Newton's iterative scheme: one rewrites the equation in the form

$$f(v_*) = 0 . \tag{33}$$

We assume f is continuously differentiable on an open interval containing the root v_* and we know an approximation to v_* (which we call v_1) contained in the interval. Then by Taylor's Theorem

$$0 = f(v_1 + [v_* - v_1]) \tag{34}$$

$$= f(v_1) + (v_* - v_1)\frac{df}{dv_1} + R \tag{35}$$

where R is a term that goes to zero more rapidly than $v_* - v_1$ as $v_1 \to v_*$. Newton's method is to obtain a better approximation than v_1 on the assumption that R can be neglected:

$$v_2 = v_1 - \frac{f(v_1)}{df/dv_1} . \tag{36}$$

The process is repeated for $n = 2, 3, \cdots$:

$$v_{n+1} = v_n - \frac{f(v_n)}{df/dv_n} \tag{37}$$

and if the sequence of v_n converges it must do so to the true root v_* if the derivative exists there. Furthermore, the asymptotic rate of convergence is second order, which means that if the error at iteration n is ε_n then $\varepsilon_{n+1} \to$ constant$\times\varepsilon_n^2$, a very rapid rate of progress. In our application we write $f(v) = X^2[v] - T^2$ and we have already found the derivative of X^2 which is in this case the same as df/dv. Convergence of Newton's method is not generally guaranteed but we know a lot about our special problem: for example, we have that $df/dv < 0$ and that $v_* > 0$. These conditions by themselves are not enough to insure convergence, but it can be shown that $d^2X^2/dv^2 > 0$. This requires a very determined application of elementary matrix algebra; alternatively, it can be demonstrated much more simply by the use of the spectral factorization of Γ (for which the reader must wait until it is introduced in the next section); see exercise 3.03(ii). With the property of positive second derivative, we may prove that the sequence of Newton's approximations converges monotonically to the correct answer from below if we start at $v = 0$, which is known to lie below the true root.

The proof is not hard, but the reader weary of proofs may skip this paragraph. When v_n is less than v_0, $f(v_n) > 0$ because f decreases monotonically, and conversely, if $f(v_n) > 0$ we know $v_n < v_0$. Thus if v_n is below the root v_0, when we use the fact that the derivative is negative in (37) we find $v_{n+1} > v_n$; hence the next approximation is larger than v_n. Setting $v_{n+1} - v_n = h$ we may write $f(v_{n+1})$ in terms of values and derivatives at v_n by means of equation 2.07(4), obtained during our discussion of interpolation:

$$f(v_n + h) = f(v_n) + h f^{(1)}(v_n) + \int_0^h (h - \mu) f^{(2)}(\mu) \, d\mu . \qquad (38)$$

Remember that the kth derivative is denoted by $f^{(k)}$; this is a form of Taylor's Theorem with a remainder. Substituting from (37) this becomes

$$f(v_{n+1}) = \int_0^h (h - \mu) f^{(2)}(\mu) \, d\mu . \qquad (39)$$

Again, assuming $v_n < v_0$ we know $h > 0$, and since $f^{(2)}(\mu) > 0$ we see from this equation that $f(v_{n+1}) > 0$. It follows that $f(v_{n+1}) < v_0$. Thus the sequence of approximants increases but is bounded above by v_0; such a sequence must converge from the fundamental completeness of the real numbers. Newton's method in its raw form is not the best way to find v_0; we return to this matter in section 3.04.

The final topic of this section is the generalization of these results in various ways. Suppose first that we have correlation in the uncertainties associated with the measurements and further that we know the covariance matrix C. It is not hard to show that one must replace Σ^2 by the matrix C in every equation, for example (13) or (32). None of the essential equations requires Σ alone, although a Cholesky factor of C could be used then.

Another most important generalization is to the minimization of a seminorm rather than a norm as we discussed in great detail for exact data. We state the results without proof. Recall from 2.05 that we choose a model in the form

$$m = h + r \qquad (40)$$

where $h \in K$, a finite-dimensional subspace of H with known basis elements $h_1, h_2, \cdots h_K$. The idea is to make r small but not to penalize h; since $r = P_{K^\perp} m$, that is the projection of m onto the orthogonal complement of K, $\|r\|$ is a seminorm of m. The seminorm minimizing model lies in the union of the two subspaces K and G; or

$$m_* = \sum_{j=1}^{N} \alpha_j g_j + \sum_{k=1}^{K} \beta_k h_k . \tag{41}$$

In place of (6) we have the misfit criterion

$$\|\Sigma^{-1}(\mathbf{d} - \Gamma\boldsymbol{\alpha} - A\boldsymbol{\beta})\| \leq T . \tag{42}$$

Here the submatrix $A \in M(N \times K)$ is the one we encountered in 2.05 with entries

$$A_{jk} = (g_j, h_k), \quad j = 1, 2, \cdots N; \quad k = 1, 2, \cdots K . \tag{43}$$

As in the norm minimization we must introduce a Lagrange multiplier v and we find stationary solutions at the coefficient vectors $\boldsymbol{\alpha}_0 \in \mathbb{R}^N$ and $\boldsymbol{\beta}_0 \in \mathbb{R}^K$ when the following system of equations is satisfied

$$\begin{bmatrix} v_0^{-1}\Sigma^2 + \Gamma & A \\ A^T & O \end{bmatrix} \begin{bmatrix} \boldsymbol{\alpha}_0 \\ \boldsymbol{\beta}_0 \end{bmatrix} = \begin{bmatrix} \mathbf{d} \\ 0 \end{bmatrix} \tag{44}$$

where $O \in M(K \times K)$ is a submatrix filled with zeros. Notice this is just a slightly modified form of 2.05(13), the equation for the solution with exactly fitting data; the upper left submatrix in the new system differs only by the addition of a diagonal matrix.

The critical tolerance below which solution of (44) is required does not correspond to $m = 0$ but to $r = 0$; in other words, if the ideal elements of K can fit the data to within T there is no need to bother with r. When the ideal elements can satisfy the observations, the model predictions can lie in the interior of the misfit zone, otherwise they must lie on the edge. The smallest value of misfit that can be obtained with the elements of K alone is found by solving the linear least squares problem

$$\min_{\boldsymbol{\beta} \in \mathbb{R}^K} \|\Sigma^{-1}(A\boldsymbol{\beta} - \mathbf{d})\| . \tag{45}$$

Assuming that the required tolerance T is bigger than this we must solve for v_0 numerically.

Like its counterpart (13), (44) is linear if v_0 is known and so we use the same tactics as before. A little algebra shows that (29) remains true in the present case; this allows us to borrow the differentiation formula (32) simply by reinterpreting (44) to be identical with (13) after a change of variables: we blow everything up into the vector space $\mathbb{R}^{N+K}$. Let $\boldsymbol{\alpha}'$ be the composite vector of unknown coefficients on the right of (44); and $\mathbf{d}'$ is the vector on the left with zeros below $\mathbf{d}$. Now make a Σ' diagonal matrix with zeros on the diagonal after the first N positive diagonal elements of Σ. It should be clear that we can define a new $\Gamma' \in M(N+K \times N+K)$ so that (13) in the primed variables is (44) in the original ones and (29) also applies for these new variables as well!

The important equations in this section are (13) and (44); they are obviously closely related to their counterparts 1.12(12) and 2.05(13) but it was much harder to derive them. Furthermore, these equations are more difficult to solve because of the presence of the unknown parameter ν_0. In the next section we shall consider a quite different way of looking at the problem that sheds light on our labors here.

Exercises

3.02(i) Calculate explicitly the coefficients c_1 and c_2 in the expansion of the data misfit functional

$$\| \Sigma^{-1}(\mathbf{d} - \mu\Gamma\boldsymbol{\alpha}_1)\|^2 = T_1^2 + (1-\mu)c_1 + (1-\mu)^2 c_2 .$$

Give a value of μ with $0 < \mu < 1$ such that the functional attains the value $(T^2 + T_1^2)/2$.

3.02(ii) A positive definite matrix $A \in M(N \times N)$ is one with the property that

$$\mathbf{x} \cdot A\mathbf{x} > 0$$

for every $\mathbf{x} \in \mathbb{R}^N$ except $x = 0$. Show the Gram matrix for linearly independent representers is positive definite. If A is positive definite show that A^{-1} exists and is also positive definite. Also show that if A and B are positive definite, so is $A + B$. These results are needed when we prove $dX^2/d\nu < 0$.

3.02(iii) It might be convenient to start Newton's iterations from this point because it certainly lies below ν_0 but, as it stands, (32) is not valid there. Calculate the derivative $dX^2/d\nu$ when $\nu = 0$.

3.02(iv) Suppose that all the measurement uncertainties are identical. Show then that the difference between the prediction of the norm-minimizing model for the jth datum and the actual observed value is proportional to the jth coefficient in the expansion of m_0 in the representers. What is the corresponding result for a seminorm-minimizing solution?

3.03 The Spectral Approach

The complexity of the solution to the problem of the last section seems somehow at odds with the simple goal of finding the smallest model with misfit less than or equal to a specified amount. We now describe a different approach which is only approximate and depends for its validity on special properties of the forward problem that cannot be guaranteed in general, but its simplicity and explanatory power easily make up for these shortcomings. Our point of departure is a set of elementary spectral properties of a linear transformation. In general, the *spectrum* of a linear transformation A is the set of values z for which the mapping $A - zI$ does not possess an inverse, where I is the identity operator; this definition is broad enough to cover operators in spaces of infinite dimension, although we shall be concerned only with matrices, the transformations on finite-dimensional spaces. The reader may wish to refer to one of the many excellent texts on this material, for example Strang (1980). The spectrum of every symmetric $A \in M(N \times N)$ consists of N *eigenvalues* $\lambda_1, \lambda_2, \lambda_3, \cdots \lambda_N$, real numbers that permit the solution of the equation

$$A \mathbf{u}_n = \lambda_n \mathbf{u}_n \tag{1}$$

for each $n = 1, 2, 3, \cdots N$ with a nonzero vector $\mathbf{u}_n \in \mathbb{R}^N$, called the *eigenvector* associated with λ_n. If in addition to being symmetric we demand that A be positive definite, the eigenvalues must all be positive. Let us order them to be nonincreasing:

$$\lambda_1 \geq \lambda_2 \geq \lambda_3 \geq \cdots \geq \lambda_N > 0 \quad . \tag{2}$$

As the possible equality suggests, there may be repeated eigenvalues, that is, a number may appear more than once in the list, and when repetition occurs the eigenvalue problem is said to be *degenerate*. An eigenvalue appearing r times is associated with an eigenvector that can be chosen arbitrarily from an r-dimensional subspace of $\mathbb{R}^N$. Nonetheless, it is always possible to choose a set of N mutually orthogonal eigenvectors

$$\mathbf{u}_m \cdot \mathbf{u}_n = 0, \ m \neq n \tag{3}$$

that are each of unit length in the traditional Euclidean norm:

$$\|\mathbf{u}_n\| = (\mathbf{u}_n \cdot \mathbf{u}_n)^{1/2} = 1 \ . \tag{4}$$

In the presence of degeneracy such a collection is obviously not unique; even without it there is ambiguity in the signs because when $\mathbf{u}_n$ is an eigenvector, so is $-\mathbf{u}_n$.

Let us write the matrix $U \in \mathrm{M}(N \times N)$ composed of columns made from the eigenvectors taken in order: our notation for this is

$$U = [\, \mathbf{u}_1, \mathbf{u}_2, \mathbf{u}_3, \ \cdots \ \mathbf{u}_N \,] \ . \tag{5}$$

The reader can easily verify that U is an orthogonal matrix, that is, one satisfying

$$U^T U = U\, U^T = I \tag{6}$$

where $I \in \mathrm{M}(N \times N)$ is the identity matrix; equivalently

$$U^{-1} = U^T \ . \tag{7}$$

We have already encountered orthogonal matrices in section 1.12. Recall that for every $\mathbf{x} \in E^N$ (with the 2-norm assumed henceforth)

$$\|U\mathbf{x}\| = \|\mathbf{x}\| \ . \tag{8}$$

Let us calculate the product $A\,U$: from (1) and (5)

$$A\,U = A\,[\, \mathbf{u}_1, \mathbf{u}_2, \mathbf{u}_3, \ \cdots \ \mathbf{u}_N \,] \tag{9}$$

$$= [\, \lambda_1 \mathbf{u}_1, \lambda_2 \mathbf{u}_2, \lambda_3 \mathbf{u}_3, \ \cdots \ \lambda_N \mathbf{u}_N \,] \tag{10}$$

$$= U\,\Lambda \tag{11}$$

where Λ is the diagonal matrix of eigenvalues $\mathrm{diag}(\lambda_1, \lambda_2, \lambda_3, \cdots \lambda_N)$. Suppose that we multiply (11) on the right with U^T; from property (6) we obtain the key result, called the *spectral factorization* of A

$$A = U\,\Lambda\,U^T \ . \tag{12}$$

Like the QR factorization and the Cholesky factorization, this is a means of expressing a given matrix as a product of special factors. Another way of writing (12) is

$$A = \sum_{n=1}^{N} \lambda_n \mathbf{u}_n \mathbf{u}_n \tag{13}$$

where you may remember from 1.13 that we introduced the notation $\mathbf{x}\mathbf{x}$ to be the dyad matrix $x\,x^T$ where x is the column matrix equivalent to the vector $\mathbf{x}$; thus $(\mathbf{x}\mathbf{x})_{ij} = x_i x_j$. Each of the dyads in the sum above is an orthogonal projection operator onto the one-dimensional subspace spanned by the associated eigenvector; when there is degeneracy, it is easy to see that the collection of elementary projections associated with the degenerate eigenvalue combines to become the orthogonal projection onto the appropriate r-dimensional subspace. Considered as a sum over projections, the representation is unique. The sum of projections is the only form that generalizes nicely to linear operators on Hilbert spaces of infinite dimension.

We return now to the solution of a linear inverse problem. We shall perform a series of variable changes that almost magically simplify the original equations and that suggest a very natural solution to the problem of keeping the norm small while fitting the data. As argued in the previous section there is no reason to consider models that lie outside the linear subspace $G \subset H$, the one spanned by the representers, and so we may work in the Euclidean space of expansion coefficients, which is isometric to G under the Γ norm; recall 3.02(5)

$$\| m \|^2 = \boldsymbol{\alpha} \cdot \Gamma \boldsymbol{\alpha} \ . \tag{14}$$

Also the misfit between the predictions of such a model and the observations has a simple expression in terms of $\boldsymbol{\alpha}$: if we call that misfit δ, then from 3.02(6)

$$\delta = \left[\sum_{j}^{N} [d_j - (g_j, m)]^2 / \sigma_j^2 \right]^{\frac{1}{2}} \tag{15}$$

$$= \| \Sigma^{-1}(\mathbf{d} - \Gamma\boldsymbol{\alpha}) \| \ . \tag{16}$$

Here we have assumed uncorrelated errors but allowed different standard deviations in each measurement. Suppose that we standardize the observations by weighting every d_j by $1/\sigma_j$; then each new "datum" has unit variance. In vector language the new data $\mathbf{e} \in E^N$ are given by

$$\mathbf{e} = \Sigma^{-1}\mathbf{d} \ . \tag{17}$$

The components of $\mathbf{e}$ are dimensionless, the magnitudes proportional to the accuracy of the measurement. We put this vector into (16) and perform some minor manipulations:

$$\delta = \| \mathbf{e} - \Sigma^{-1}\Gamma\boldsymbol{\alpha}\| \tag{18}$$

$$= \| \mathbf{e} - \Sigma^{-1}\Gamma\Sigma^{-1}\Sigma\boldsymbol{\alpha}\| \tag{19}$$

$$= \| \mathbf{e} - A\,\mathbf{x}\| \tag{20}$$

where we have transformed versions of the expansion coefficient vector and the Gram matrix:

$$A = \Sigma^{-1}\Gamma\Sigma^{-1} \tag{21}$$

$$\mathbf{x} = \Sigma\boldsymbol{\alpha} . \tag{22}$$

Equation (14) is unchanged in form by switching to the new variables:

$$\| m \|^2 = \mathbf{x} \cdot A\,\mathbf{x} . \tag{23}$$

At this point we introduce the spectral factorization of A since this matrix has the requisite properties of symmetry and positive definiteness. In (20) we have

$$\delta = \| \mathbf{e} - U\Lambda U^T\mathbf{x}\| \tag{24}$$

$$= \| U(U^T\mathbf{e} - \Lambda U^T\mathbf{x})\| \tag{25}$$

$$= \| (U^T\mathbf{e}) - \Lambda(U^T\mathbf{x})\| . \tag{26}$$

The second line follows from (6) and the last from (8). Another variable change suggests itself: rotation by the orthogonal matrix U^T:

$$\tilde{\mathbf{e}} = U^T\mathbf{e} \tag{27}$$

$$\tilde{\mathbf{x}} = U^T\mathbf{x} . \tag{28}$$

So that finally we have the simplest imaginable expression for misfit; written now in terms of the square, it is

$$\delta^2 = \| \tilde{\mathbf{e}} - \Lambda\tilde{\mathbf{x}}\|^2 \tag{29}$$

$$= \sum_{n=1}^{N} (\tilde{e}_n - \lambda_n\tilde{x}_n)^2 . \tag{30}$$

And the squared model norm is just

$$\| m \|^2 = \tilde{\mathbf{x}} \cdot \Lambda\tilde{\mathbf{x}} \tag{31}$$

$$= \sum_{n=1}^{N} \lambda_n\tilde{x}_n^2 . \tag{32}$$

Before describing how we may use these simplified forms to develop a model with a small norm and a small data misfit, let us see how the original problem has been transformed by the variable changes. In essence we have constructed new data from linear combinations of the original data and corresponding to these new "observations" there are appropriate representers. If the new data were exact the prediction problem would be exactly to satisfy

$$\tilde{e}_j = (\psi_j, m), \quad j = 1, 2, 3, \cdots N \tag{33}$$

where

$$\tilde{e}_j = \sum_{k=1}^{N} \kappa_{jk} d_k \tag{34}$$

$$\psi_j = \sum_{k=1}^{N} \kappa_{jk} g_k \tag{35}$$

and the transformation matrix κ has components

$$\kappa_{jk} = (U^T \Sigma^{-1})_{jk} . \tag{36}$$

It is simple to show that the new representers are orthogonal to each other:

$$\begin{aligned} (\psi_j, \psi_k) &= 0, \quad j \neq k \\ &= \lambda_j, \quad j = k \end{aligned} \tag{37}$$

so that the Gram matrix of the ψ_j is the diagonal matrix Λ. Considered as random variables, the transformed data $\tilde{e}_j$ are of unit variance and, like the original d_j, uncorrelated; this requires a little algebra to verify, which we omit. Because the ψ_j form an orthogonal basis for G there is an expansion in these functions for the solution and the coefficients are $\tilde{x}_j$:

$$m = \sum_{j=1}^{N} \tilde{x}_j \psi_j . \tag{38}$$

We call (38) the *spectral expansion* of the model; it is sometimes instructive to regard this as a natural kind of Fourier series for m.

Here is our recipe for using the spectral expansion: we retain only the first $J < N$ terms in (38) by making the coefficients $\tilde{x}_{J+1}, \tilde{x}_{J+2}, \cdots \tilde{x}_N$ vanish. We turn to (30) to choose the values of the nonzero coefficients: we exactly annihilate the first J terms of (30) by designating appropriate values of $\tilde{x}_n$.

Thus

$$\tilde{x}_n = \tilde{e}_n/\lambda_n, \quad n = 1, 2, \cdots J$$
$$= 0, \qquad n = J+1, J+2, \cdots N \, . \tag{39}$$

From (30) and (32) the squared misfit and model norm under this prescription become

$$\delta^2 = \sum_{n=J+1}^{N} \tilde{e}_n^{\,2} \tag{40}$$

$$\|m\|^2 = \sum_{n=1}^{J} \tilde{e}_n^{\,2}/\lambda_n \, . \tag{41}$$

How shall we select the proper value of J, the number of terms retained in the spectral expansion? As J grows from zero, the misfit diminishes and the norm increases. Because the eigenvalues have been arranged to decrease, (41) shows that a term early in (38) makes a smaller contribution to the norm than a later one of the same size, but they both reduce the squared misfit by the same amount. It seems reasonable to continue to accept terms until δ in (40) just falls below the specified tolerance T of section 3.01 or, alternatively, comes as close to the critical value as possible. Then the model has acceptable misfit but we have built it with components that do not needlessly inflate the norm.

This prescription does not achieve the absolutely smallest possible norm for a given misfit, but its effectiveness is enhanced by a factor that cannot be appreciated from a purely abstract analysis: the range of eigenvalues encountered in most practical problems is truly enormous. This is the cause of the very large condition number associated with the Gram matrix so often seen in practice (see section 2.08, for example); it is a simple exercise to show that the condition number of A in the spectral norm is given by λ_1/λ_N. Furthermore, in a large inverse problem with many data, not only the smallest eigenvalue λ_N, but frequently the great majority of them (say 95%) are smaller than λ_1 by factors of 10^6 or more. Terms in the expansion associated with such small eigenvalues may be capable of making only minor reductions in misfit but from (41) we see that they may increase $\|m\|$ dramatically.

Let us compare the results of the above heuristic recipe with those of the true minimum-norm model derived in the previous section. The optimal model can be found just as well starting from the transformed problem (33) as from the original equations.

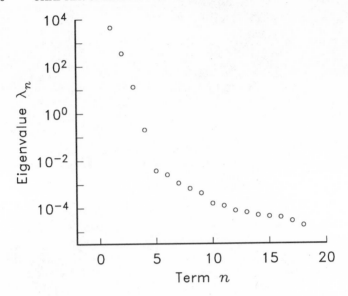

Figure 3.03a: Spectrum of the scaled Gram matrix $A = \Sigma^{-1}\Gamma\Sigma^{-1}$ in L_2 for the problem discussed in section 3.06. The great range of eigenvalues is not unusual in poorly conditioned problems.

Its coefficients in (38) are given by the solution to equation 3.02(13) for the proper value of the Lagrange multiplier ν_0; in the present context 3.02(13) reads

$$(\nu_0^{-1}I + \Lambda)\tilde{\mathbf{x}} = \tilde{\mathbf{e}} \tag{42}$$

where I is the $N \times N$ unit matrix, which is the appropriate covariance matrix for the transformed data $\tilde{\mathbf{e}}$. The matrix on the left is diagonal so that solving the equation is trivial:

$$\tilde{x}_n = \frac{\tilde{e}_n}{\nu_0^{-1} + \lambda_n} = \frac{w_n \tilde{e}_n}{\lambda_n}, \quad n = 1, 2, 3, \cdots N . \tag{43}$$

where

$$w_n = \frac{1}{1 + (\nu_0\lambda_n)^{-1}} . \tag{44}$$

We have written $\tilde{x}_n$ as a weighted version of $\tilde{e}_n/\lambda_n$; if we had wanted an exact fit, (39) and (40) show us that all the weights should be unity. Interpreted in terms of w_n the spectral recipe is equivalent to taking unit weight until $n = J$ and zero thereafter. Figure 3.03a is the spectrum for the mantle dissipation problem

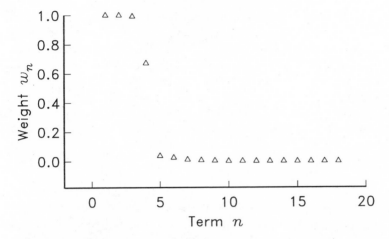

Figure 3.03b: The weight series corresponding to the spectrum of figure 3.03a for the norm-minimizing model with $v_0 = 10$. The resemblance of this weight series to a complete cutoff at $n = 5$ is obvious.

under the 2-norm; this will be discussed again in section 3.06. Notice the great range of eigenvalues. Figure 3.03b shows how the optimal weight series for a true minimum norm model behaves in the presence of many small eigenvalues. It is apparent that simply cutting off the expansion after some suitable value of n (in this case $n > 3$) has a very similar effect to that of solving the norm-minimization problem exactly, provided the distribution of the $\tilde{e}_n$ is moderately uniform. Another common feature of most inverse problems encountered in practice is that the basis functions ψ_n become increasingly oscillatory as n grows (see figure 3.04f, for example). This strengthens the analogy of (38) as a kind of Fourier expansion. Both norm minimization and spectral cutoff can then be interpreted as filtering procedures that attenuate "high frequency" features of the solution. This fits comfortably with our general guideline for building a model that it should contain as little extraneous detail as possible.

A comparison of computational costs between the spectral approach and exact norm minimization is revealing: to obtain the factors U and Λ numerically requires about $5N^3$ computer operations, although this is difficult to state with certainty because of an iterative phase in any algorithm (see Golub and Van Loan, 1983). As described in section 3.02, each Newton iteration demands $N^3/6$ operations if Cholesky factorization is used to solve 3.02(13). Thus, unless it takes more than thirty iterations to reach ν_0, the spectral approach is inferior from a numerical perspective. About ten iterations may be typical, but plain Newton iteration with X^2 is itself far from optimal as we shall see in the next section.

Is there an explanation for the almost universally observed concentration of eigenvalues near zero and the inverse correlation of basis function wiggliness with eigenvalue? We cannot be rigorous, of course, because it is perfectly possible to devise artificial examples that exhibit quite different patterns; nonetheless, I shall attempt to offer some plausible speculation. Many forward problems can be seen to be the application of Green's function for a differential equation. For example, the computation of the gravitational or magnetic field of a body is just the integration of a point source function over a region, and the response of the point source is in fact Green's function for the differential equation of the appropriate field. Viewed as an operator, Green's function is just the inverse of the differential operator, and it will therefore share the same eigenfunctions; but the *eigenvalues* of the operator associated with Green's function are the *reciprocals* of those for the differential equation (chapter 5, Lanczos, 1961). The spectrum of almost any differential equation of physics contains a smallest eigenvalue, but there is no largest member because the eigenvalues increase without bound (chapter 6, Morse and Feshbach, 1953). Thus the reciprocal spectrum of Green's operator is bounded above but has an accumulation point at zero, where the eigenvalues cluster ever more densely. To the extent that the finite-data forward problem approximates the underlying continuous operation of Green's function, we might expect to see the same kind of behavior in the eigenvalues of Γ or A. That there is an intimate relation between the spectrum of the Gram matrix and that of the forward operator can be seen most clearly from the numerical decomposition described in section 2.08 or 3.05.

Looking back over our progress in solving inverse problems by minimizing the model norm in a function space, we recognize a progression. Initially, one might suppose the search for a suitable model should be conducted in a space of infinite dimension because that is after all where the most general solution lies. We gained an important insight when we found the search may be confined to a finite-dimensional space, the one spanned by the representers. In the previous section we saw that the simplification is not lost when observational uncertainty is allowed for because the solution remains in the same subspace. Somewhat unexpectedly perhaps, one may go further: the spectral approach reveals that to a high degree of approximation, the minimum-norm model is contained in an even smaller subspace. Obviously N noisy data contain useful information about fewer than N model parameters but, as we have seen, it requires a delicate calculation to discern which parameters are the favored ones.

Exercises

3.03(i) Develop a spectral approximation for seminorm minimization. Perform the spectral factorization of the Gram matrix and thus obtain the orthogonal basis ψ_j for the representer subspace G. The model m_0 is replaced by a set of trial approximations with different numbers of terms in the spectral expansion. Follow the development at the end of section 2.05; notice how simple it is to calculate projections onto G.

3.03(ii) For the misfit functional defined in 3.03(32), prove that $dX^2/d\nu < 0$ and $d^2X^2/d\nu^2 > 0$ when $\nu > 0$ using spectral factorization. Prove, moreover, that all derivatives of odd order are negative and those of even order are positive.

3.04 The Magnetic Anomaly Problem Revisited

We come to a fairly straightforward application of the techniques described in the foregoing sections. The story is contained almost entirely in the pictures. We begin with the artificial magnetic anomaly profile of figure 2.01a and we add to the thirty-one magnetic field values an equally phony noise series consisting of members drawn from a computer-generated approximation to zero-mean uncorrelated Gaussian random numbers with standard deviation 10 nT; the data series is plotted in figure 3.04a. There are several reasons why the "noise" is oversimplified: we know its mean and variance exactly and that it is uncorrelated. In practice none of these things can be known with certainty and indeed, for the marine magnetic field, the noncrustal contributions to the signal are surely correlated—these are mainly fields from the core and ionosphere which have large length scales. The example is, however, a simple training exercise, so let us continue. The 2-norm of the noise is about 56 nT while that of the original data set is 915 nT so that in terms of the Euclidean norm we have

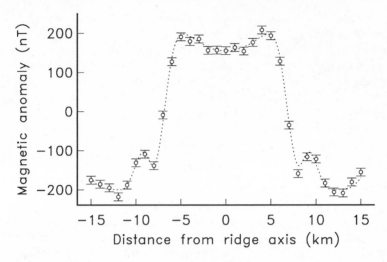

Figure 3.04a: An idealized marine magnetic anomaly at a mid-ocean ridge of figure 2.01a with uncorrelated Gaussian noise, $\sigma = 10$ nT, added. The dots are the noisy values with error bars at $\pm 1\sigma$; the dashed line indicates the original, noise-free anomaly.

added only about 6 percent uncertainty; the figure confirms this is a fairly small perturbation. Let us apply the theory of section 2.06 and fit the noisy data exactly, deriving the smallest model in C^1L_2 where size is the ordinary 2-norm of the first derivative. The resultant model, shown as the solid line in figure 3.04b, is very rough. The ill-posed nature of this inverse problem was discussed in 2.04—a small perturbation in the data does not necessarily lead to a small change in the solution. As later examples will illustrate, the present problem is relatively benign in its response to noise.

Obviously, when the observations are inexact, we should not attempt to match them precisely with model predictions. Instead we seek the smallest model that fits the data adequately. We turn to section 3.01 for guidelines on the question of an adequate fit: let us agree to accept a model if the standardized 2-norm of the misfit has the expected value in the statistical sense. With thirty-one independent Gaussian random variables 3.01(6) produces a tolerance $T = 5.523$ where σ is taken to be unity because in the relevant test equations, for example 3.02(5), each datum has been normalized by its standard deviation. We must proceed to solve 3.02(13) for the unknown Lagrange multiplier ν_0; with

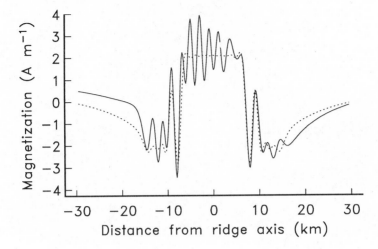

Figure 3.04b: The magnetization model with the smallest norm in the space W_2^1 exactly fitting the noisy anomaly values shown in figure 3.04a by the dots. Also plotted as a dashed line is the solution for noise-free values reproduced from figure 2.06c.

such a small number of observables it is the calculation of a few seconds in a modern computer to find the misfit functional $X^2[\nu]$ for a wide range of the parameter ν. The results appear in figure 3.04c. We have already described a foolproof iterative scheme for locating the desired Lagrange multiplier—it is just Newton's

Table 3.04A: Convergence of Newton's Method

n	ν_n	$X^2[\nu_n]$	$X[\nu_n]$
1	0.001	1822.7	42.693
2	0.002094	909.89	30.164
3	0.004229	483.71	21.993
4	0.008996	275.40	16.595
5	0.020952	162.80	12.759
6	0.049448	96.801	9.8387
7	0.10396	60.232	7.7609
8	0.18230	41.485	6.4409
9	0.25429	33.176	5.7599
10	0.28477	30.747	5.5450
11	0.28812	30.506	5.5232
12	0.28816	30.504	5.5230

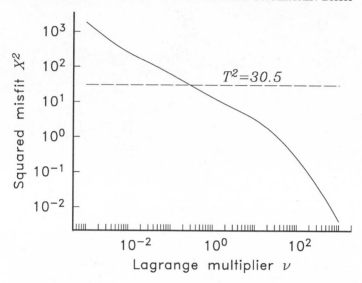

Figure 3.04c: The misfit functional $X^2[\nu]$ for the anomaly problem. The dashed line shows a tolerance level corresponding to the expectation of the norm of misfit: $T = 5.523$, or equivalently, $T^2 = 30.50$ for 31 standardized data. Solving $X^2[\nu_0] = T^2$ gives $\nu_0 = 0.2882$.

method, that is, 3.02(16) starting at any guess value below ν_0. Table 3.04B is a record of the sequence of iterates, starting with $\nu_1 = 0.001$. Except in the final stages the convergence rate of the process is unimpressive. Such a slow pace is not a critical problem with the small matrices of our simple illustration, but it would become a serious matter if the data numbered several hundred or more. The cause of the difficulty is the large magnitude of the second derivative of X^2 for small ν, invalidating the approximation of the curve by its tangent, which is at the heart of Newton's method. The reason this failure is not apparent in figure 3.04c can be attributed to the log scales, used to make a clear presentation of the large range of numbers. We may use the spectral approach to shed light on the question of convergence. If spectral factorization is performed in equation 3.02(12), in the notation of section 3.03 we find

$$X^2[\nu] = \sum_{n=1}^{N} \frac{\tilde{e}_n^2}{(1 + \lambda_n \nu)^2} . \tag{1}$$

When $\nu = -1/\lambda_n$ the functional X^2 is evidently singular; this

Table 3.04B: Convergence of Modified Scheme

n	v_n	$X^2[v_n]$	$X[v_n]$
1	0.001	1822.7	42.693
2	0.094618	64.053	8.0033
3	0.29564	29.984	5.4758
4	0.28816	30.504	5.5230

explains the forest of singularities seen in figure 3.02b when v is negative. The undesirably large second derivative of X^2 arises from the proximity of the singularity associated with λ_1, which is just below $v=0$. The gentle curvature of X^2 drawn on log axes suggests the idea of performing the solution in logarithmic variables, say $\mu = \ln v$, which generates the equation

$$\ln(X^2[\exp(\mu_0)]) - \ln T^2 = 0 .\tag{2}$$

The Newton step for this equation, rewrittten in terms of v, is

$$v_{k+1} = v_k \exp\left[\frac{-X^2[v_k]\,\ln(X^2[v_k]/T^2)}{v_k\,dX^2/d\,v_k}\right] .\tag{3}$$

Convergence of this scheme cannot be guaranteed in general but the results are encouraging as table 3.04B makes plain. Exercise 3.04(iii) suggests a way of obtaining further gains in computational speed.

Having thus determined that $v_0 = 0.28816$ we can compute the expansion coefficients in the representer basis and find a smooth model. This is presented in figure 3.04d. Most of the objectionable overshoot present in figure 3.04b has been successfully suppressed. The magnetization has lost the fine detail visible in the noise-free solution, but that is to be expected—uncertainty in the data must reduce the information content. The model produced when $T = E[\,\|\mathbf{X}\|\,]$ differs in a significant way from the smallest norm model found with exact data. In the latter case we knew every other model fitting the data would possess a larger norm; here we know only that every model fitting the data *equally well* is larger in the norm. It is fairly probable that the misfit of the true model exceeds the one we selected as adequate; in fact from 3.01(7) the chance of excluding the true data vector is about 50 percent. To be more confident that no unnecessary structure is present in the model we must weaken the fit criterion, so as to decrease the probability of rejecting the true data vector; this

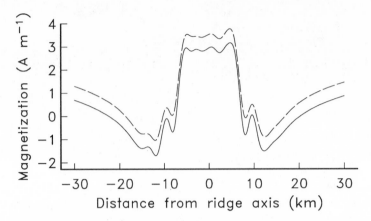

Figure 3.04d: Magnetization models with the smallest norm in the space W_2^1 approximately fitting the noisy anomaly values of figure 3.04a. The misfit level of the solid curve corresponds to the expected value of the norm of the noise, that is, $T = 5.523$. The dashed line shows a smoother model based upon a more generous fitting criterion, $T = 6.708$. Recall that the absolute level of the magnetization is undetermined in these solutions.

means T must be increased to obtain a solution that is the smoothest or smallest one compatible with observation rather than simply a typical model. Suppose we choose the 95 percent probability region; then from standard χ^2 tables we find that $T = 6.708$. The corresponding super smooth model is shown dashed in figure 3.04d; the gain in smoothness is in fact barely perceptible.

Next we generate the spectral solution to the problem. The complete eigensystem for the scaled Gram matrix is obtained numerically and the spectral factorization thereby performed. The eigenvalues arranged in decreasing order are shown in figure 3.04e. The range of values covered is moderately large, about five decades, with a cluster near zero as predicted in the previous section. The remarkable thing about the graph is the regularity of the distribution—taking the logarithmic scale into account we recognize an almost exact geometric progression. An explanation is given in exercise 3.04(i); a key ingredient here is that the forward solution for the equivalent continuous operator is a convolution (see 2.04(8)). The reader will appreciate the rarity of such simplicity. In the next picture, figure 3.04f, we have plotted a few of the orthogonal basis functions ψ_n that go into the spectral

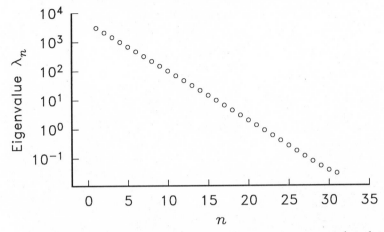

Figure 3.04e: Spectrum of the scaled Gram matrix $A = \Sigma^{-1}\Gamma\Sigma^{-1}$ for the magnetic anomaly problem. The unusually regular progression of eigenvalues can be ascribed to the particularly simple form of the forward problem.

expansion of the model 3.03(14). Again the expected pattern emerges: the largest eigenvalues are associated with the smoothest basis functions, the small ones with highly oscillatory functions.

The prescription of the previous section was to accept only the first J terms in 3.03(14), using 3.03(15) to calculate the nonzero coefficients. For each value of J we obtain a different model with its own misfit and norm. Thus, because $N = 31$, there are exactly thirty-one different models in the family of spectral solutions, not the continuum obtained when the Lagrange multiplier ν is varied. In figure 3.04g the circles indicate the positions of members of this family in the misfit/model-norm plane. (The exactly fitting member, with $J = 31$ and norm $\|m\| = 14.20$ is not plotted.) The spectral recipe calls for the particular solution whose misfit just falls within the preferred tolerance $T = 5.523$. In this problem that solution (with $J = 21$) has a misfit considerably below the critical level, and therefore the norm might be unnecessarily large; instead it makes more sense here to take a slightly worse fitting model. There are four models with misfits very close to, but above, the target value; we choose $J = 20$, giving 5.650 for δ. The corresponding magnetization function is plotted in figure 3.04h; the optimal solution at this misfit level is visibly smoother.

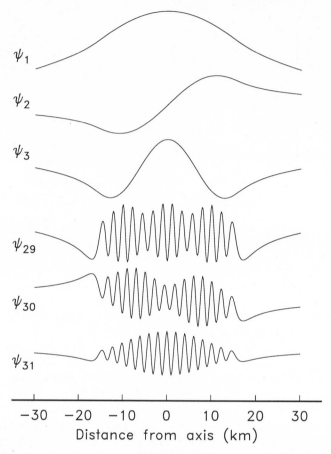

Figure 3.04f: The first and last three members of the complete set of orthogonal basis functions ψ_n associated with the spectral expansion of the model. The functions have been scaled for ease of plotting because properly normalized functions vary greatly in amplitude.

Figure 3.04g is an example of a trade-off diagram, showing how goodness of fit can be exchanged for small model size and conversely. The smooth curve is derived from the exact minimum norm solution; every conceivable model can be represented as a point in the plane lying on or above this line. The family of spectral expansion models lies reasonably close to the line, attesting to the validity of the expansion idea. Unfortunately for the purposes of qualitative comparison, the norm of the model space W_2^1, which

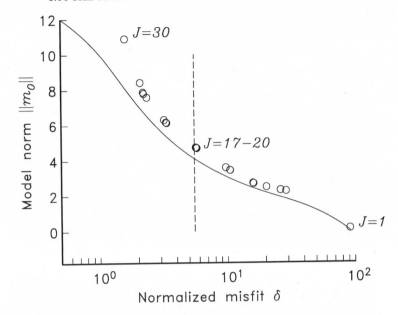

Figure 3.04g: Trade-off between data misfit δ and minimum model norm $\|m_0\|$. Results from the spectral expansion are shown by the circles, where J is the number of terms retained in 3.03(14). The smooth curve is the optimal relation obtained by the theory of section 3.02. The preferred value of misfit, corresponding to the tolerance $T = 5.523$, is indicated by the dashed line.

measures 2-norm of the slope, conveys little to the intuition: its units are A m$^{-3/2}$, which have been deliberately omitted from the graph to avoid confusion. Perhaps the differences in the models of figures 3.04d and 3.04h provide the best kind of illustration. Practical experience suggests that perhaps this magnetic anomaly example may be unusual for the relatively poor performance of the spectral technique and that the optimal solution and the one found by truncation of the spectral expansion are often indistinguishable.

Our next topic is the appearance of the theory when the Gram matrix cannot be obtained simply in elementary functions. While it is not in principle necessary to do anything new, considerable improvements can be made in exposition and stability if the numerical approach is integrated into the treatment from the beginning.

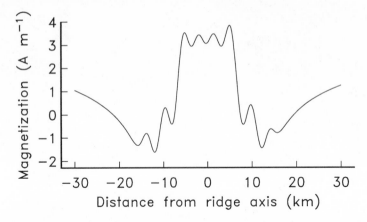

Figure 3.04h: The magnetization model with the smallest norm in the space W_2^1 approximately fitting the noisy anomaly values obtained by discarding the last ten terms of the spectral expansion 3.03(14). This corresponds to a misfit 2-norm of 5.65, slightly greater than that of the model shown solid in figure 3.04d; the model roughness is also noticeably larger.

Exercises

3.04(i) Because the continuous version of the forward problem 2.04(8) is a convolution, sines and cosines are eigenfunctions for the forward operator. Verify this. The Toeplitz form of the Gram matrix 2.06(30) means that $\sin \mu j$ and $\cos \mu j$ are approximate eigenvectors for Γ; use this to estimate the spectrum in figure 3.04e.

3.04(ii) If you have access to a computer with a user-friendly matrix arithmetic package, repeat the spectral factorization of this section but working in L_2. Use the Gram matrix 2.06(12) with $h = 2$ km and $x_{j+1} - x_j = 1$ km; $N = 31$. Note the completely different behavior of the eigenvalues in the spectral expansion. Calculate and plot the orthogonal functions associated with the largest eigenvalues; why do they fail to follow the usual pattern of increasing wiggliness with decreasing eigenvalue?
Hint: See the previous question.

3.04(iii) Numerical analysts will squirm to see that we repeatedly solve 3.02(4)

$$(\Sigma^2/\nu_0 + \Gamma)\alpha_0 = d$$

by Cholesky factorization for a set of different values of ν_0. Investigate the Householder tridiagonalization of Γ given, for example, in Golub and van Loan (1983) and show how this matrix factorization might reduce the numerical work needed to find v_0. Show that this approach is favorable if more than four Newton iterations are required for convergence.

3.05 The Theory in Numerical Dress

The theory of sections 3.02 and 3.03 will be reworked in a finite-dimensional form suitable for numerical computation. This is the inevitable shape assumed by most practical inverse problems because, as we have already mentioned, exact expressions for the Gram matrix are rarely obtainable. Naturally, one can simply evaluate the Gram matrix and other quantities numerically and proceed as before, but there are several reasons why it is worthwhile investigating alternatives. One is that we can give an elegant and uniform treatment of norm and seminorm minimizing solutions. We saw in 2.08 that slavish adherence to the exact theory for roughness minimization is unnecessarily awkward, and that a purely numerical realization is simpler and numerically more stable. These advantages also attend the theory when uncertainty is taken into account.

For concreteness our exposition is in terms of a model of one variable, dissipation in the mantle as discussed in 2.08, but the general theory closely parallels the specific case. Recapitulating that discussion, we recall that the model is represented by the vector $\mathbf{q} \in E^L$, whose components are samples of the dissipation as a function of radius in a spherically symmetric Earth. The observations are contained in $\mathbf{d} \in E^N$ and the solution to the linear forward problem is written

$$\mathbf{d} = G\,W\,\mathbf{q} \tag{1}$$

$$= B\,\mathbf{q} \tag{2}$$

where $W \in M(L \times L)$ is a diagonal matrix of weights providing a numerical quadrature approximation for functions sampled like $\mathbf{q}$, and $G \in M(N \times L)$ is a matrix whose rows are samples of the representers in the problem; obviously $B = G\,W$ and has the same row and column dimensions as G. In the earlier discussion when the data were alleged to be exact it was natural to suppose that $L > N$ to permit the mathematical indulgence of arbitrary improvement in the numerical approximation of the integrals; then G and B are arrays wider than they are tall. With noisy data, however, the situation is not so clear cut: the spectral theory shows us an excellent approximation to the norm minimizing solution can usually be found in a subspace of low dimension. With this in mind it seems unlikely that when N is large, L would normally need to be larger still. We may anticipate that good solutions can be obtained when $L < N$ in many practical cases.

The distinct problems of minimizing the model norm or a seminorm can be given a completely uniform treatment: in the seismic problem we minimize the functional

$$\|\hat{R}\,\mathbf{q}\|^2 = \mathbf{q}\cdot\hat{R}^T\hat{R}\,\mathbf{q} \tag{3}$$

where $\hat{R} \in M(L \times L)$ is a suitably chosen matrix, called the *roughening* or *regularizing* matrix; the ordinary Euclidean norm is used in E^L and E^N. When L_2-norm minimization of the model is contemplated

$$\hat{R} = \mathrm{diag}(w_1^{\frac{1}{2}},\,w_2^{\frac{1}{2}},\,w_3^{\frac{1}{2}},\,\cdots\,w_L^{\frac{1}{2}}) \tag{4}$$

$$= W^{\frac{1}{2}} \tag{5}$$

where the coefficients w_i are the weights of a quadrature rule. In this way the minimized functional is the numerical approximation to $\|q\|^2$. When we wish to make the model smooth, the 2-norm of the first or second derivative must be kept small and then $\hat{R}$ is a more complicated matrix composed of appropriate products of $W^{\frac{1}{2}}$ and ∂: for seminorm minimization in $W_2^1(0, a)$ for example

$$\hat{R} = W^{\frac{1}{2}}\partial \tag{6}$$

where

$$\partial = \frac{1}{h}\begin{bmatrix} 0 & & & & \\ -1 & 1 & & \text{\Large 0} & \\ & -1 & 1 & & \\ & & -1 & 1 & \\ \text{\Large 0} & & & \cdots & \\ & & & -1 & 1 \end{bmatrix} \tag{7}$$

which is the matrix for approximating differentiation by differencing. In a problem where the model is defined in a plane or on a sphere, like the second one in 2.07, $\hat{R}$ might be an approximation to ∇^2. In general, $\hat{R}$ provides a projection onto the orthogonal complement of the set of favored elements; recall, only the part of the model *not* lying in the subspace K is penalized in the solution. This means that in strict seminorm minimization $\hat{R}$

is a singular matrix and we should not rely upon the existence of an inverse in our manipulations.

Following the plan of 3.02 we seek the model $\mathbf{q}$ minimizing (3) subject to the constraint that the model predictions adequately match the observations, that is:

$$\|\Sigma^{-1}(\mathbf{d} - B\,\mathbf{q})\| \le T \tag{8}$$

where Σ is a diagonal matrix of standard deviations, and T is the tolerance level, both familiar from extensive use earlier in the chapter. By the arguments given in 3.02, inequality can arise in (8) only if the criterion can be met by $\mathbf{q}=0$ for norm minimization, or a vector in the subspace of ideal elements for the seminorm problem. Let us proceed with the most likely but more complicated case when equality holds in (8). To solve the minimization problem we use a Lagrange multiplier v just as before and seek the stationary values of

$$U[\mathbf{q}, v] = \mathbf{q} \cdot \hat{R}^{T}\hat{R}\,\mathbf{q} - v(T^{2} - \|\Sigma^{-1}(\mathbf{d} - B\,\mathbf{q})\|^{2}) \,. \tag{9}$$

Since this finite-dimensional problem is a special case of those treated in 3.02 we need not repeat the argument to show that there is only one stationary point with $v>0$ and it is the global minimizer of (3) under (8). Let $\mathbf{q}_0$ and v_0 be the arguments that make the functional U stationary. Differentiating U with respect to $\mathbf{q}$ and making the gradient vanish, we find vector $\mathbf{q}_0$ obeys

$$(B^{T}\Sigma^{-2}B + v^{-1}\hat{R}^{T}\hat{R})\,\mathbf{q}_0 = B^{T}\Sigma^{-2}\mathbf{d} \,. \tag{10}$$

As usual v_0 can only be discovered by solving (10) a number of times, searching for the value of v that makes the 2-norm misfit exactly equal to T. Defining the squared misfit function X^2 exactly analogous with the one in 3.02(12)

$$X^{2}[v] = \|\Sigma^{-1}(\mathbf{d} - B\,\mathbf{q}_0)\|^{2} \tag{11}$$

we can calculate the derivative

$$\frac{dX^{2}}{dv} = -2v^{-3}\mathbf{q}_0 \cdot \hat{R}^{T}\hat{R}(B^{T}\Sigma^{-2}B + v^{-1}\hat{R}^{T}\hat{R})^{-1}\hat{R}^{T}\hat{R}\,\mathbf{q}_0 \,. \tag{12}$$

Newton's method (perhaps in $\ln f$ and $\ln v$) can be applied to find the optimal value v_0 that solves

$$X^{2}[v_0] = T^{2} \,. \tag{13}$$

As promised these equations are valid for both norm and seminorm minimization and they apply whatever the relative magnitudes of L and N.

The use of (10) and Newton's method employing (12) can be quite satisfactory in many cases. There are however a number of alternative ways of writing or solving these equations that improve numerical stability or efficiency. First we describe how it is possible to increase the resistance of the numerical system to computational errors. Let the condition number of the matrix in parentheses on the left of (10) be κ^2; then, as in 2.08, we can exploit QR factorization to produce a linear system whose condition number is only κ. The reader will easily verify that the left matrix in (10) can be factorized thus:

$$B^T \Sigma^{-2} B + v^{-1} \hat{R}^T \hat{R} = C^T C \tag{14}$$

where the matrix $C \in M(L+N \times L)$ is given by

$$C = \begin{bmatrix} \Sigma^{-1} B \\ v^{-\frac{1}{2}} \hat{R} \end{bmatrix} . \tag{15}$$

Then (10) may be rewritten as follows

$$C^T C \, \mathbf{q}_0 = C^T \mathbf{d}' \tag{16}$$

where

$$\mathbf{d}' = \begin{bmatrix} \Sigma^{-1} \mathbf{d} \\ 0 \end{bmatrix} . \tag{17}$$

But (16) is exactly in the form of the normal equations appearing in the old-fashioned solution to a linear, overdetermined least squares problem; it is as if $\mathbf{q}_0$ had been defined as the solution of the problem

$$\min_{\mathbf{q} \in E^L} \| C \, \mathbf{q} - \mathbf{d}' \| . \tag{18}$$

Therefore (16) can be solved more stably by QR factorization of C as sketched in 1.13 and described in detail by Lawson and Hanson (1974) and Golub and Van Loan (1983). Writing $C = Q \, R$ in the usual way, the expression for the derivative is gratifyingly simple:

$$\frac{dX^2}{dv} = -2v^{-3} \| R_1^{-1} \hat{R}^T \hat{R} \, \mathbf{q}_0 \|^2 \tag{19}$$

where R_1 is the upper square matrix of R, below the diagonal of which there are only zeros.

A price has been paid for the improved stability: it is numerically more expensive to solve the least squares minimization (18) by QR than the algebraically equivalent symmetric linear system (10) by Cholesky factorization. With the figures mentioned in 1.13, the number of arithmetic operations needed to solve the square $L \times L$ system in (10) is about $L^3/6$, while QR solution of (18) requires $L^2(N + 2L/3)$. Thus QR is about four times more expensive when $L \gg N$ and about ten times when $L = N$. The inefficiency is exacerbated by the fact that we must solve (16) in the iterative cycle for v_0. The situation is not as bad as this naive analysis makes it seem. The action of the QR algorithm is to transform C into a matrix containing only zeros below the diagonal. If the matrix $\hat{R}$ is based upon a differencing operator, C is already filled with zeros in a large lower triangle. This fact can be recognized in a QR program modified to skip the unnecessary operations to transform blocks known to contain zeros and even to save the associated memory: the amount of arithmetic can be reduced to $\frac{1}{2}L^2N$ operations. The process is similar to the one for updating a large least-squares system described in chapter 27 of Lawson and Hanson (1974).

There are several more numerical tricks that we might use to speed up the calculations. For example, when $L > N$ and $\hat{R}$ has an inverse, it may be advantageous in (10) to apply a well-known (in numerical analysis, at least) matrix identity, the Sherman-Morrison-Woodbury formula (Golub and Van Loan, 1983). This states that

$$(A + U W^T)^{-1} = A^{-1} - A^{-1}U (I + W^T A^{-1}U)^{-1}W^T A^{-1} \quad (20)$$

where A is any square, nonsingular matrix, U and W are matrices of the proper sizes (but not necessarily square), and the left matrix must exist. When $\hat{R}$ is diagonal we can identify it with A and then the only computational intensive work is the inversion of the $N \times N$ matrix on the right. For small-scale systems with a few hundred data points or less, modern computers are fast enough that the required fancy programming is probably wasted effort. For very large numbers of observations, a radical revision of the computing techniques is necessary and we defer discussion of that until a later section.

The spectral approach to model construction also has a pleasing numerical realization. To understand it we need another kind of matrix factorization, closely related to the spectral factorization which we encountered in 3.03; this is the *singular value decomposition* or the SVD of a matrix. Every matrix $A \in M(M \times N)$ with $M \geq N$ can be expressed as the product of three factors:

$$A = U S V^T . \tag{21}$$

Here $U \in M(M \times M)$ and $V \in M(N \times N)$ and they are both orthogonal matrices; $S \in M(M \times N)$ and is assembled from a diagonal and a zero matrix:

$$S = \begin{bmatrix} S_1 \\ O \end{bmatrix} \tag{22}$$

where O is a matrix of the appropriate size filled with zeros and

$$S_1 = \text{diag}(s_1, s_2, s_3, \cdots s_N) . \tag{23}$$

The diagonal entries of S_1 are all nonnegative and are called the *singular values* of A. We shall assume they are ordered thus:

$$s_1 \geq s_2 \geq s_3 \geq \cdots s_N \geq 0 . \tag{24}$$

When $M < N$ an equivalent factorization is obtained by transposing (21). The possibility of performing the SVD factorization follows from the easily proved proposition that the nonzero members of the eigenvalue spectra of the (square, symmetric, positive semidefinite) matrices $A^T A$ and $A A^T$ are identical, except that when $M > N$ there are $M - N$ additional zero eigenvalues in the spectrum of the second matrix. Indeed we have that $s_n = \lambda_n^{\frac{1}{2}}$, where $\lambda_1, \lambda_2, \lambda_3, \cdots \lambda_N$ are the eigenvalues of $A^T A$. Furthermore, V is an orthogonal matrix of the spectral factorization of $A^T A$ and U is one for that of $A A^T$. (The reader may recall from 3.03 that the orthogonal matrix in the spectral factorization is not unique because of possible sign permutations of the columns.) To calculate the explicit SVD of a given matrix, it is not necessary to solve the pair of spectral factorizations—it can be obtained directly from A. There are two remarkable aspects of the numerical algorithms: first, they are extremely stable numerically; second, they are surprisingly efficient. See Golub and Van Loan (1983) for details. Incidentally, it is normal for computer programs to omit the last $M - N$ columns of U because they are always multiplied by zeros, not only in (21) but everywhere else in the theory. For more on the interpretation and application of SVD see Lanczos (1961) and Strang (1980).

Returning to the linear inverse problem, we combine the essential ideas of the spectral theory in section 3.03 and the numerical treatment in 2.08. As before we first scale the data by the standard errors, introducing $e_j = d_j/\sigma_j$, weighted "data" each with unit variance; in vector language

$$\mathbf{e} = \Sigma^{-1}\mathbf{d} .\tag{25}$$

Multiplying both sides of (1) by Σ^{-1}

$$\mathbf{e} = \Sigma^{-1}G\ W\ \mathbf{q} .\tag{26}$$

From now on, we shall assume that $\hat{R}$ is nonsingular. To treat the seminorm problem, which leads naturally to singular matrices for $\hat{R}$, we shall make an approximation: we select a true norm that places only insignificant emphasis on the members of K, the subspace of favored elements. This was the approach set out in 2.08. For example, we might choose the norm

$$\|q\| = [\gamma^2 q(a)^2/a^2 + \int_c^a (dq/dr)^2\ dr/a\ ]^{\frac{1}{2}}\tag{27}$$

where $\gamma > 0$, but very small. This true norm approximates the seminorm in which only the integral term contributes to the minimized functional. Then in its finite-dimensional equivalent, the matrix $\hat{R}$ will be invertible for the approximate seminorm minimization. Define $\mathbf{y} \in E^L$ by

$$\mathbf{y} = \hat{R}\mathbf{q}\tag{28}$$

and then from (26)

$$\mathbf{e} = \Sigma^{-1}G\ W\ \hat{R}^{-1}\mathbf{y}\tag{29}$$

$$= A_1\mathbf{y}\tag{30}$$

where

$$A_1 = \Sigma^{-1}G\ W\ \hat{R}^{-1} .\tag{31}$$

Recall that $\hat{R}$ is the matrix such that, to the accuracy of the numerical approximations, $\|\hat{R}\mathbf{q}\| = \|q\|$, where the norm on the left is the Euclidean norm, and the one on the right is appropriate for the model space. But $\|\hat{R}\mathbf{q}\| = \|\mathbf{y}\| = \|q\|$ and thus by minimizing $\mathbf{y}$ in the Euclidean norm we minimize q in the norm of its space.

In fact we do not require that (26) be met, but only the weaker condition (8). The misfit to the measurements of the model predictions in the new variables is

$$\delta = \| \mathbf{e} - A_1 \mathbf{y} \| . \tag{32}$$

(Although δ and $\mathbf{e}$ mean the same things here as they did in 3.03, the nonsquare A_1 is *not* the same as the square matrix A in that section.) We shall assume that the number of model sample points, L, is greater than the number of observations, N; therefore, to use (21) we must factorize A_1^T rather than A_1 itself: then

$$A_1 = V S^T U^T . \tag{33}$$

Substituting this SVD into (32) and exploiting the orthogonality of V, we see

$$\delta = \| \mathbf{e} - V S^T U^T \mathbf{y} \| \tag{34}$$

$$= \| V^T \mathbf{e} - S^T U^T \mathbf{y} \| \tag{35}$$

$$= \| (V^T \mathbf{e}) - [S_1 \; O^T] (U^T \mathbf{y}) \| \tag{36}$$

$$= \| \tilde{\mathbf{e}} - [S_1 \; O^T] \tilde{\mathbf{y}} \| \tag{37}$$

where the definition of the new rotated vectors $\tilde{\mathbf{e}}$ and $\tilde{\mathbf{y}}$ should be self-evident. Notice that $\| \tilde{\mathbf{y}} \| = \| \mathbf{y} \|$, so that this vector too has the magnitude of $\| q \|$. Because S_1 is diagonal, the last equation above can be written

$$\delta^2 = \sum_{n=1}^{N} (\tilde{e}_n - s_n \tilde{y}_n)^2 . \tag{38}$$

The last $L - N$ columns of S^T are all zeros, and so the last $L - N$ components of $\tilde{y}_n$ are irrelevant to the misfit. On the other hand, they do contribute to the norm of q, for

$$\| q \|^2 = \sum_{n=1}^{N} \tilde{y}_n^2 + \sum_{n=N+1}^{L} \tilde{y}_n^2 . \tag{39}$$

It follows of course, that if we wish to keep the norm of q as small as we can, we should take all the members of the second sum to be zeros; this is precisely the same as confining attention to models that lie in the subspace spanned by the representers. In applying the spectral recipe of 3.03, we annihilate even more of the components of $\tilde{\mathbf{y}}$: the first J terms in the sum (38) are arranged to vanish but the remaining components in the model vector are set to zero

$$\tilde{y}_n = \tilde{e}_n / s_n , \quad n = 1, 2, \cdots J$$
$$= 0, \qquad n = J+1, J+2, \cdots N . \tag{40}$$

The singular values s_n decrease (usually very rapidly) as we advance along the sequence and therefore we choose J to be just large enough to satisfy $\delta \leq T$, in this way adding only the smallest contributors to the first sum in (39).

To compute the model itself, we backtrack through the transformations and obtain

$$\mathbf{q} = \hat{R}^{-1}\mathbf{y} = \hat{R}^{-1}U\,\tilde{\mathbf{y}}\,. \tag{41}$$

We can interpret this equation as an expansion in a set of basis vectors:

$$\mathbf{q} = \sum_{n=1}^{J} \tilde{y}_n \boldsymbol{\psi}_n \tag{42}$$

where $\boldsymbol{\psi}_n \in E^L$ are the columns of $\hat{R}^{-1}U$

$$\hat{R}^{-1}U = [\,\boldsymbol{\psi}_1,\ \boldsymbol{\psi}_2,\ \boldsymbol{\psi}_3,\ \cdots\ \boldsymbol{\psi}_N\,]\,. \tag{43}$$

The relationship of this SVD approach to the spectral factorization of 3.03 is straightforward. It is readily verified that the product

$$A_1 A_1^T = \Sigma^{-1}\Gamma\Sigma^{-1} = A \tag{44}$$

where A is the scaled Gram matrix whose spectral expansion is sought. It follows that the eigenvalues of A are identical to the *squares* of the singular values of A_1. Notice how the squares of s_n will appear in the denominators of the first sum in (39) when (40) is substituted there, while λ_n appears in 3.03(17), the equivalent sum in the spectral theory. In a naive version of the SVD implementation (mysteriously referred to as "generalized" inverse theory in the literature) it is usually asserted that the columns of U form an orthogonal basis for solutions with expansion coefficients $\tilde{y}_n$; we see from (41) that is strictly true only when $\hat{R}$ is proportional to a unit matrix. Instead, the solution is built from the vectors $\boldsymbol{\psi}_n$, which are not usually orthogonal in E^L under the ordinary dot product. These vectors are, of course, the finite-dimensional analogs of the eigenfunctions ψ_n that we encountered earlier in the spectral theory.

Finally, we illustrate why the SVD approach is superior in its numerical behavior to that of a straightforward application of the spectral theory, particularly when seminorm minimization is undertaken. To keep everything as simple as possible we consider a primitive finite-dimensional approximation to the norm mentioned in (27): let the function q be sampled evenly on the interval, replace the differential by a first-difference, and integration by a simple sum. Instead of (27) we have, upon squaring

$$\|q\|^2 = \|\hat{R}_\gamma \mathbf{q}\|^2 = \gamma^2 q_1^2 / a^2 + \sum_{n=2}^{L} (q_{n-1} - q_n)^2 / ha . \tag{45}$$

We allow the small parameter γ to appear explicitly in the notation. For ease of comprehension we shall set out an illustrative matrix with $L = 6$, suppressing the sample spacing h and the Earth's radius a; it will be quite obvious how the situation generalizes. Let

$$\hat{R}_\gamma = \begin{bmatrix} \gamma & 0 & 0 & 0 & 0 & 0 \\ -1 & 1 & 0 & 0 & 0 & 0 \\ 0 & -1 & 1 & 0 & 0 & 0 \\ 0 & 0 & -1 & 1 & 0 & 0 \\ 0 & 0 & 0 & -1 & 1 & 0 \\ 0 & 0 & 0 & 0 & -1 & 1 \end{bmatrix} . \tag{46}$$

Then from (31)

$$A_1 = (\Sigma^{-1} G\ W) \hat{R}_\gamma^{-1} \tag{47}$$

$$= B_1 \hat{R}_\gamma^{-1} . \tag{48}$$

As in the discussion above we take $L > N$ so that B_1 is wider than it is tall. The SVD is performed upon A_1^T

$$A_1^T = (\hat{R}_\gamma^{-1})^T B_1^T \tag{49}$$

where we can compute the first factor on the right explicitly: it is an $L \times L$ matrix in which only the first row is influenced by γ:

$$(\hat{R}_\gamma^{-1})^T = \begin{bmatrix} \gamma^{-1} & \gamma^{-1} & \gamma^{-1} & \gamma^{-1} & \gamma^{-1} & \gamma^{-1} \\ 0 & 1 & 1 & 1 & 1 & 1 \\ 0 & 0 & 1 & 1 & 1 & 1 \\ 0 & 0 & 1 & 1 & 1 & 1 \\ 0 & 0 & 0 & 0 & 1 & 1 \\ 0 & 0 & 0 & 0 & 0 & 1 \end{bmatrix} . \tag{50}$$

As γ is made small in our effort to diminish the influence of the leading term on the right of (45), only the first row of this matrix becomes large, and similarly only the first row of A_1^T is heavily weighted. Because the numerical algorithms for SVD work by Householder triangularization into the upper part of the array, these large numbers in A_1^T do not become mixed up with those in the lower portion and overall accuracy of the SVD is preserved. See chapter 22, Lawson and Hanson (1974) for a complete analysis of a similar problem. In contrast, suppose we attempt to find the spectral factorization of A when γ is very small compared with unity. Then we must first compute A:

$$A = A_1 A_1^T \tag{51}$$

$$= B_1^T [\hat{R}_\gamma^{-1}(\hat{R}_\gamma^{-1})^T] B_1 \tag{52}$$

$$= B_1^T [\hat{R}_\gamma^T \hat{R}_\gamma]^{-1} B_1 . \tag{53}$$

The inner matrix is

$$[\hat{R}_\gamma^T \hat{R}_\gamma]^{-1} = \begin{bmatrix} \gamma^{-2} & \gamma^{-2} & \gamma^{-2} & \gamma^{-2} & \gamma^{-2} & \gamma^{-2} \\ \gamma^{-2} & \gamma^{-2}+1 & \gamma^{-2}+1 & \gamma^{-2}+1 & \gamma^{-2}+1 & \gamma^{-2}+1 \\ \gamma^{-2} & \gamma^{-2}+1 & \gamma^{-2}+2 & \gamma^{-2}+2 & \gamma^{-2}+2 & \gamma^{-2}+2 \\ \gamma^{-2} & \gamma^{-2}+1 & \gamma^{-2}+2 & \gamma^{-2}+3 & \gamma^{-2}+3 & \gamma^{-2}+3 \\ \gamma^{-2} & \gamma^{-2}+1 & \gamma^{-2}+2 & \gamma^{-2}+3 & \gamma^{-2}+4 & \gamma^{-2}+4 \\ \gamma^{-2} & \gamma^{-2}+1 & \gamma^{-2}+2 & \gamma^{-2}+3 & \gamma^{-2}+4 & \gamma^{-2}+5 \end{bmatrix} . \tag{54}$$

Observe how *every entry* of this matrix contains the term γ^{-2}. If γ^2 falls a little below the relative floating-point precision of the computer representation of numbers, no trace of the difference operator remains in A, because perturbations of order unity associated with the differences are too small to be recorded in the computer words that must hold the large number γ^{-2}. Then the numerical spectral factorization of A could not possibly approximate the desired problem. We can expect that even if this extreme loss of significance does not arise, the spectral factorization nonetheless incurs a severe penalty in accuracy compared with SVD when seminorm minimization is approximated by norm minimization.

Exercises

These exercises ask the reader to set out the numerical theory for seminorm minimization by simply following the rules of section 3.02; perhaps surprisingly, the matrix calculations are completely different, even though the final solutions are identical. The Hilbert space of models is E^L equipped with the inner product

$$(\mathbf{p}, \mathbf{q})_R = (\hat{R}\,\mathbf{p}) \cdot (\hat{R}\,\mathbf{q}) = (\hat{R}^T \hat{R}\,\mathbf{p}) \cdot \mathbf{q} = \mathbf{p} \cdot (\hat{R}^T \hat{R}\,\mathbf{q}) .$$

According to (1) the jth datum is given by

$$d_j = (W\,\hat{\mathbf{g}}_j) \cdot \mathbf{q}$$

where $\hat{\mathbf{g}}_j \in E^L$ are the rows of the matrix G.

3.05(i) Express the linear functional for the forward product as an inner product and write down the corresponding representers $\mathbf{g}_j \in E^L$, which are not just $\hat{\mathbf{g}}_j$! Show the Gram matrix is

$$\Gamma = G \ W \ \hat{R}^{-1} (G \ W \ \hat{R}^{-1})^T \ .$$

3.05(ii) We know from 3.02 that the seminorm minimizing solution fitting the data with tolerance T can be expanded thus:

$$\mathbf{q}_* = \sum_{j=1}^{N} \alpha_j \mathbf{g}_j + \sum_{k=1}^{K} \beta_k \mathbf{h}_k$$

where $\mathbf{h}_k \in E^L$ are special vectors to be exempted from the norm minimization. The equation for the expansion coefficients is 3.02(17); show that the submatrix A of that equation (not to be confused with its homonym $\Sigma^{-1} \Gamma \Sigma^{-1}$) is given by

$$A = G \ W \ H^T$$

where H is a matrix whose rows are the vectors $\mathbf{h}_k$.

3.05(iii) Give $\mathbf{h}_k$ and H for the seminorm obtained from (45) when $\gamma=0$. Estimate the number of numerical operations to solve the linear system 3.02(17). Can Cholesky factorization be used? Can the matrix on the left be factored as in (16)? What about the situation when a norm is minimized instead of a seminorm?

3.06 Return to the Seismic Dissipation Problem

The geophysical inverse problem to be discussed next is how to determine the seismic dissipation function from estimates of the decay constants of certain free oscillations. Two characteristics of this problem were prominent in our first encounter in 2.08. First, the representers are not available as simple formulas in elementary functions; instead they are given only as numerical values at a fixed set of radii—the result of calculations of the velocities in the Earth when it undergoes free vibration in its various eigenstates. Because of this the elements of the Gram matrix cannot be expressed exactly as was possible in all the earlier examples, and so we introduced purely numerical means of evaluation. Second, the representers (plotted in figure 2.08b), though strictly linearly independent, can be closely approximated by a linearly dependent set of functions, which leads to extreme difficulty in the numerical solution of the linear equations when an exact fit to observations is demanded. Now we shall replace the eighteen artificial data with the actual observations, complete with their

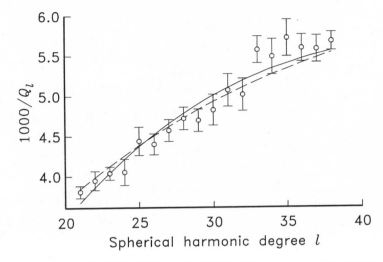

Figure 3.06a: Observed decay parameters and uncertainties for a series of 18 fundamental spheroidal modes of free oscillation $_0S_{21}$, $_0S_{22}$, $\cdots$ $_0S_{38}$. The error bars correspond to $\pm 1\sigma$ estimates. Also shown as lines are the predictions of two models, both fitting with tolerance $T = 4.184$.

one-standard-deviation uncertainties: these are shown in figure 3.06a, taken from the original paper of Masters and Gilbert (1983). The errors are governed by a Gaussian distribution and the correlations are small. We shall not show any model fitting the noisy data without misfit; we could predict from our preliminary study in 2.08 that even the smallest norm model must be extremely large and oscillatory: it is sufficient to note that q_0 varies between about -1700 and 1800. Since negative dissipation values contravene the second law of thermodynamics we must reject this model as unphysical. In fact, no positive attenuation model can exactly match the observations, although we must wait until chapter 4 for the techniques to demonstrate this fact.

Naturally a precise fit to the data is unnecessary; using the results in section 3.01 (equation 3.01(6) in particular) we find that $N = 18$ is associated with an expected value for the standardized 2-norm of misfit of 4.1841, and this will be the tolerance T to be regarded as an adequate misfit in 3.05(3) and the SVD models. Employing the same numerical framework as in 2.08 we sample the interval from the base of the mantle, $r = c = 0.547a$, to the surface, $r = a$, at 101 evenly distributed points; thus $L = 101$.

Table 3.06A: Convergence of Newton's Method

n	v_n	$X^2[v_n]$	$X[v_n]$
1	0.001	737.52	27.157
2	0.0016293	341.42	18.478
3	0.0026037	164.57	12.828
4	0.0041488	85.334	9.2376
5	0.0067025	49.543	7.0387
6	0.0111855	32.935	5.7389
7	0.0193010	24.546	4.9544
8	0.0312892	19.908	4.4618
9	0.0403148	17.913	4.2324
10	0.0425129	17.522	4.1859
11	0.0426006	17.507	4.1841

Simpson's rule provides the quadrature formula; evidently fanatical precision in the numerical integration is uncalled-for in view of the quite large uncertainties in the data.

The first problem to be tackled will be the construction of the smallest model in $L_2(c,a)$ with the stated misfit. Then $\hat{R}$ is just the square-root of the diagonal Simpson weight array:

$$\hat{R} = W^{\frac{1}{2}} = \text{diag}((h/3)^{\frac{1}{2}}, (4h/3)^{\frac{1}{2}}, (2h/3)^{\frac{1}{2}}, (4h/3)^{\frac{1}{2}}, \cdots (h/3)^{\frac{1}{2}}) \quad (1)$$

where $h/a = 0.00453$ is the spacing of the sampling. The easiest system to set up is 3.05(4), which leads to a set of 101 linear equations in as many unknowns. It hardly need be said that the model $q = 0$ does not satisfy the tolerance, and so we must set about finding v_0 iteratively. One possible way to guess the order of magnitude of v_0 is to compare the sizes of the two matrices on the left of 3.05(4): it is plausible to assume that the true value of v_0 will not be very far from the one that causes the two terms to

Table 3.06B: Convergence of Modified Scheme

n	v_n	$X^2[v_n]$	$X[v_n]$
1	0.001	737.52	27.157
2	0.0111503	33.002	5.7447
3	0.0296952	20.349	4.5110
4	0.0425000	17.524	4.1862
5	0.0426006	17.507	4.1841

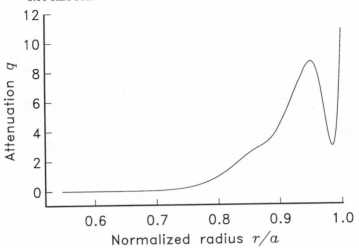

Figure 3.06b: Attenuation model with smallest L_2 norm satisfying the decay parameters shown in figure 2.08a with tolerance 4.184. The solid curve in the earlier figure represents the predictions of this model.

be comparable in size. For efficiency I used the crudest matrix norm, the Frobenius norm, which ignores the structure of the matrix layout and just computes the Euclidean norm of the string of numbers in the array. This rough calculation yields a value of 0.00215; since Newton's method is guaranteed to converge only if the initial guess lies below the actual answer, I took the initial value $v_1 = 0.001$; subsequent approximants are shown in table 3.06A. We observe the same behavior as in our earlier numerical example: although strictly convergent from below as it must be, Newton's method advances relatively slowly until the approximation is quite good. Since each iteration involves the solution of a fairly large linear system, this poor rate of improvement is a nuisance. But as the next table illustrates, transforming into logarithmic variables as described in 3.04 effects a considerable acceleration. There seems to be no good reason for even trying the unmodified Newton method in such calculations.

The smallest model in this norm is plotted as the solid curve in figure 3.06b. The dissipation function is positive and smooth, exhibiting a general increase in the upper mantle. The solution is of reasonable magnitude in the shallow regions, at least consistent with seismic losses estimated by other means. The extremely low values deep in the mantle should be viewed with caution—this is after all the *smallest* solution. Also somewhat disturbing is the local minimum near the Earth's surface: is this oscillation a real feature, or an accidental wiggle? One way to test the reality of the dip is to construct a model by a method that discriminates against large gradients or curvatures, both of which are needed to generate it. This leads naturally then to the seminorm minimizers that have occupied so much of our attention before.

The next natural choice is the functional for the 2-norm of slope, a seminorm in the space $W_2^1[c, a]$:

$$[\int_c^a (dq/dr)^2 \, dr/a]^{1/2}. \tag{2}$$

We may call the model that minimizes this functional the "flattest" solution in a 2-norm sense. A numerical approximation is needed for this integral: the simplest approach is to replace the derivative by the first difference and the integral by a sum,

$$\|\hat{R}\mathbf{q}\| = [(h/a) \sum_{n=2}^{L} (q_{n-1} - q_n)^2/h^2]^{1/2} \tag{3}$$

which is 3.05(17) with $\gamma = 0$. We are in effect applying the less accurate rectangle rule for numerical integration instead of the Simpson formula. It is awkward to apply Simpson's rule consistently to (2) because the difference approximation is an estimate of the slope halfway between the function samples, not at the samples themselves, and this means the integrand is missing at the ends of the intervals. The problem may be solved with a more elaborate approximation to differentiation (see Ralston, 1965, for example) but, as we have remarked before, there is no particular virtue in obtaining a very accurate evaluation of the minimized functional, provided the linear functionals that solve the forward problem are tended to carefully. The new $\hat{R}$ is the lower triangular array

$$\hat{R} = (ha)^{-1/2} \begin{bmatrix} 0 & & & & \\ -1 & 1 & & \mathbf{0} & \\ & -1 & 1 & & \\ & & -1 & 1 & \\ & \mathbf{0} & & \cdots & \\ & & & -1 & 1 \end{bmatrix}. \tag{4}$$

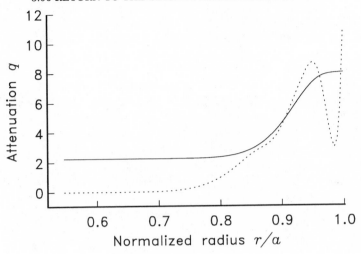

Figure 3.06c: Solid curve: the attenuation model with the smallest 2-norm of slope, the flattest model. This model predicts the values on the dashed curve in figure 3.06a. For comparison the model of figure 3.06b is also shown lightly dashed.

Attempting to start the iteration for v_0 as before, by finding the ratio of the two matrix norms, leads to difficulties because the guess $v_1 = 1171$ is *larger* than the correct value, evident from the fact that the misfit norm is too small. The next iterate of the modified (that is, logarithmic) Newton scheme is very small, far below the proper v_0, but the corresponding linear system is now so ill-conditioned that Cholesky factorization of 3.05(4) is impossible because accumulated inaccuracy makes the stored matrix non-positive definite! We shall return to the question of numerical stability in a moment. Clearly the initial guess in this case was not good enough, even for the modified iterative solution. To obtain a suitable starting value, the original crude guess was repeatedly divided by ten until the value fell below the true root; starting from that point the logarithmic scheme converged in three steps to $v_0 = 6.0668$.

The resultant flattest solution is plotted in figure 3.06c. The small values in the lower mantle have disappeared and so has the minimum near the surface. The model is almost featureless—it would be hard to imagine a solution with much less structure other than constant attenuation, which does not fit the data (How do we know this?). This solution suggests only that the material of the upper mantle is more dissipative than that in the lower. Evidently a simple model is consistent with the eighteen observations.

Although there is little hope of obtaining anything simpler, we can ask for a solution that minimizes the 2-norm of the second derivative. The corresponding $\hat{R}$ matrix is the one given as $W_1^{1/2}\partial^2$ in 2.08(25). The smoothest model is one that gently increases with positive second derivative in the whole interval. Unfortunately, the solution is unacceptable because, although it agrees with the other two solutions near the surface, in the lower mantle the solution is negative. The completely discordant behavior of three simple models in the lower mantle seems to say that the data contain little information about that region, a fact that could be guessed from the shape of the representers in figure 2.08a. The job of the present chapter is merely the construction of models, not the analysis of their content; in the next chapter we shall discuss methods to help us in that most important task.

Before giving the SVD version of this problem, let us briefly discuss the question of numerical stability. When an exact fit was demanded, we found it necessary to go to some lengths to obtain a reliable numerical solution. QR solution of 3.05(6) is the most stable way to solve the corresponding problem when exact fitting is not desired, but so far we have stayed with 3.05(4) which we repeat for convenience:

$$(B^T \Sigma^{-2} B + v^{-1}\hat{R}^T \hat{R})\,\mathbf{q}_0 = B^T \Sigma^{-2}\mathbf{d}\,. \tag{5}$$

Figure 3.06d is a summary of an empirical investigation: we plot the condition number κ in the spectral norm of the matrix $B^T \Sigma^{-2} B + v^{-1}\hat{R}^T \hat{R}$, the matrix on the left of (5) as the Lagrange multiplier varies over a wide range. The solid curve gives the behavior for the smallest model in L_2. When v is very small, so that large misfits are expected, the condition number tends to 4, and the matrix is extremely stable. At this end of the range the term $\hat{R}^T \hat{R}$ dominates in the matrix. For 2-norm minimization $\hat{R}^T \hat{R}$ is just the diagonal Simpson weight array, and it is obvious that the ratio of the largest to the smallest eigenvalue (which is the spectral condition number) is then 4. At the other end of the range the condition number increases indefinitely as greater accuracy in fitting is requested; the graph shows that in fact (5) cannot be used to obtain a precise fit, because the corresponding matrix is singular. At the actual value of v_0 needed in our problem, the matrix is clearly stable enough even for inversion in

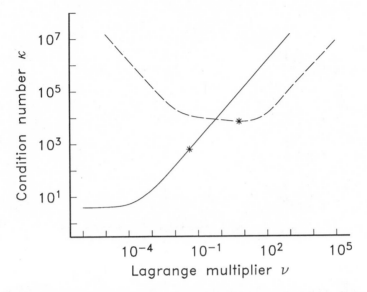

Figure 3.06d: Solid curve: spectral condition number for matrix in (5) when $\hat{R}$ gives the smallest model in L_2. Dashed curve: condition number when $\hat{R}$ is associated with the flattest model in W_2^1. The two emphasized points correspond to the Lagrange multipliers for the models in figures 3.06b and 3.06c.

"single precision" on my computer (representation error about 1 part in 10^7), and even had the data been much more accurate, stability does not seem to be a serious problem in this particular case.

The situation is somewhat different for the seminorm minimizer. Since the matrix $\hat{R}$ for this problem is the one in (4), which is obviously singular, the term $\hat{R}^T\hat{R}$ in (5) is also singular, and so when ν is small the condition number must rise; but as we just noted the other term in the matrix is singular too, and so we see the U-shaped dashed curve of κ rising for large or small Lagrange multiplier values. By good fortune, the desired level of misfit gives rise to a matrix that is nearly optimally conditioned in our case, although this can hardly be expected to be a general occurrence. The stability problem seems nonetheless to be very well in hand here, especially in comparison to the dreadful behavior exhibited with exact fitting. We have in reality a lot of stability in reserve, because by solving 3.05(6) with QR instead of (5) we can obtain a condition number of $\kappa^{1/2}$, and if this is done in

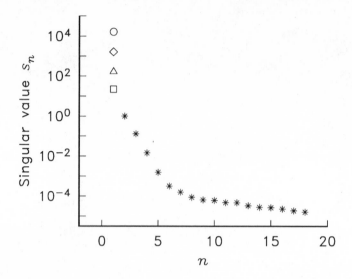

Figure 3.06e: Singular value spectra of dissipation problem in W_2^1 with the norm $\|\hat{R}_\gamma \mathbf{q}\|$ approximating a seminorm, and four different values of the parameter γ. The values are $\gamma = 0.1$ (square); 0.01 (triangle); 0.001 (diamond); 0.0001 (circle). For $n \geq 2$ the four spectra are coincident on the scale of the figure so that only the value of s_1 distinguishes them.

"double precision" we would expect to be immune from accuracy problems over an enormous range of v.

Let us now proceed to the application of SVD, the numerical implementation of the spectral theory. The eigenvalue spectrum has already been plotted in figure 3.03a for 2-norm minimization. We shall not treat this problem further, because it is quite uninteresting, but instead we move directly to construction of the flattest model, approximating the seminorm $\|dq/dr\|$ by $\|\hat{R}_\gamma \mathbf{q}\|$ according to 3.05(17). Obviously some value of γ needs to be selected; it must be small so that the derivative term in the definition of the norm receives most of the emphasis, but too small a value may lead to problems associated with the finite arithmetic precision of the computer. We find the SVD of the matrix $A_1^T = (\Sigma^{-1} G \, W \, \hat{R}_\gamma^{-1})^T$; the singular values are plotted in figure 3.06e for a number of values of the parameter γ. Notice how large s_1 is: roughly speaking the first singular value is inversely proportional to γ, while the rest of the spectrum is almost unaffected. The SVD has apparently decoupled the

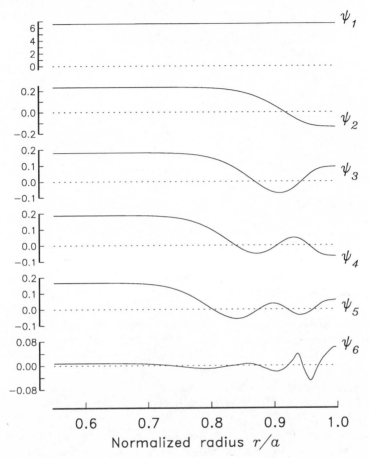

Figure 3.06f: The first six basis functions in the mantle dissipa-
tion problem, orthogonal in W_2^1 under the norm $\|\hat{R}_\gamma \mathbf{q}\|$. Note that
these are obviously not orthogonal in L_2.

seminorm approximation, and its dependence on γ, from the solution of the forward problem. One may predict from the figure
that the solution will be almost independent of γ provided that
$s_1 \gg s_2$; this is confirmed in the actual computations. For simplicity the SVD results given will be those corresponding to $\gamma = 0.001$.

Figure 3.06f shows a collection of the first eigenfunctions of
the spectral factorization, or in the finite-dimensional model, of
the vectors ψ_n; these are orthogonal under the inner product
associated with $\|\hat{R}_\gamma \mathbf{q}\|$. The first vector, companion of the largest
singular value s_1, is nearly constant. The constant function

Table 3.06C: Summary of the SVD Expansion

k	s_k	$\tilde{e}_k$	$\tilde{y}_k$	$\|\hat{R}_\gamma \mathbf{q}(k)\|$	$\delta(k)$
1	1649.2	142.87	0.08663	0.08663	18.794
2	1.0089	−18.489	−18.326	18.326	3.3744
3	0.12930	1.7138	13.254	22.617	2.9068
4	0.01446	0.7942	54.927	59.402	2.7962
5	0.00155	1.0858	700.43	702.94	2.5768
15	2.657×10^{-5}	0.48500	18253	53201	1.1536
16	2.281×10^{-5}	−0.80342	−35211	63798	0.8278
17	1.867×10^{-5}	0.15291	8187	64321	0.8135
18	1.615×10^{-5}	0.81354	50346	81682	0

should be transparent to the seminorm and so we recognize ψ_1 as $\mathbf{h}_1$, the sole member of the subspace K of ideal elements that are to be ignored by the seminorm, or in our approximate treatment, to be assessed very lightly by the norm. If there were several such elements instead of just one, they would appear (suitably orthogonalized) at the top of the list of basis functions, each associated with a large singular value. The remaining basis functions exhibit the familiar pattern of increasingly oscillatory behavior, the wiggles invading from the right.

Table 3.06C summarizes what happens as we undertake the expansion of $\mathbf{q}$ in this basis set, accepting further terms in the ordered sequence ψ_1, ψ_2, $\psi_3 \cdots$ with coefficients $\tilde{y}_n$ in 3.05(15). We denote the 2-norm of standardized misfit after accepting k terms by $\delta(k)$, and this is shown in the last column; similarly, $\mathbf{q}(k)$ stands for the solution vector at this stage. Not shown in the table is the misfit when $\mathbf{q}=0$, that is, the misfit norm of the data; this is 144.10. From this information we see that the addition of just one term in the series, the constant function, accounts for most of the observations. We also see that the first term contributes next to nothing in the norm, which is exactly what we want from a good approximation to a seminorm. The target misfit is $T = 4.184$, and the inclusion of only one further term in the sequence takes $\delta(k)$ below that tolerance. We must choose whether to over-fit slightly in accepting two terms, or to misfit rather more seriously in retaining only one term; the discrete nature of the spectral recipe means precise matching of the tolerance is impossible (see figure 3.04g). In view of the large misfit of the one-term model, the two-term model seems the better choice

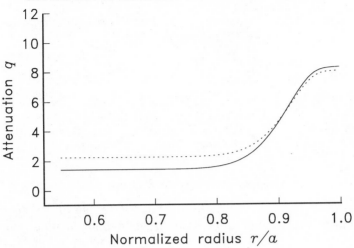

Figure 3.06g: Solid curve: SVD solution in W_2^1 with misfit closest to desired tolerance of 4.184; in fact $\delta = 3.3744$. Only two terms in the spectral expansion were needed, so this model is a combination of ψ_1 and ψ_2 in figure 3.06f. The dashed curve is the flattest model, with misfit 4.184, replotted from figure 3.06c.

here. That solution is plotted in figure 3.06g. The flattest model obtained by true seminorm minimization is shown on the graph lightly dashed; there is little qualitative difference between the two. If the reader is disturbed by the inability to meet the desired misfit in the spectral method, it is simple enough to perform the norm or seminorm minimization in a transformed problem, just as we did at the end of section 3.03; then, instead of truncating the spectral expansion at J terms, one includes all the members of the series, but with appropriate decreasing weights.

At the risk of belaboring the obvious, it must be emphasized that the proper choice of norm, one that discriminates against gradients, helps to eliminate unnecessary structure: in this example we avoid being taken in by an intriguing dip of the attenuation function near the surface found in the 2-norm minimizing solution. A fallacy has arisen in the geophysical literature that the 2-norm is generally satisfactory because, if a spectral or an SVD solution is performed, this automatically prevents the generation of spurious structure through the exclusion of the higher-order basis functions with their small-scale oscillations. We can see that is in fact not the case by noting that a solution almost exactly like the one shown in figure 3.06b can be produced by a two-term SVD expansion in L_2. A dramatic real-life illustration is given in the paper by Widmer et al. (1991).

Exercise

3.06(i) Find the matrix $\hat{R}$ corresponding to the seminorm minimization in $L_2(c, a)$ when h_1 is a constant function. Now find $\hat{R}$ when constant and linearly varying functions are exempt from norm penalization. Write down suitable norm approximations for use with SVD.

3.07 Large Numbers of Observations

The lessons of the magnetic anomaly example have yet to be exhausted. Anyone familiar with underway marine magnetic surveying will notice the sparseness of the field measurements in the earlier figures: there is an anomaly value every kilometer. At a vessel speed of ten knots and a data recording rate of one field value per minute, both typical values, a normal measurement density would be between three and four magnetic anomaly values per kilometer of ship track. Furthermore, it is sometimes necessary to analyze several hundred kilometers of track to cover a broader time span in the paleomagnetic reversal sequence. Both these factors point to a much larger data set than those we have treated so far, several hundred to perhaps a thousand numbers. Numerical solution of systems with a thousand unknowns, which would be necessary if the methods of the previous sections are employed, is a costly and time-consuming process because the number of arithmetic operations rises as the cube N. If the idealization of exact lineation of the anomalies is abandoned and an area is covered in the survey, the data set may contain tens of thousands of observations at which point the methods we have discussed may become prohibitively expensive. The present section will describe some ideas that can help to overcome the computational bottleneck. We cannot hope to provide a complete account of the subject because it is one that seems to occupy an ever-increasing proportion of the research geophysicists' time, particularly as seismologists and others begin to explore the construction of models whose properties may vary in an arbitrary way within the Earth instead of being confined to variation with depth or radius alone.

One way to slow the inexorable rise in computational cost with increasing problem size is to find a special trick to solve the problem in hand. This is hardly a subject that could be covered properly in the space available to us here: if such a trick exists at all, it may require a wide knowledge of analytical and numerical techniques to discover it. One class of measures that can dramatically improve efficiency is the exploitation of particular regularities in the Gram matrix that allow fast inversion. For example, we have already noted that if, as is usually the case, the magnetic data are taken evenly spaced in distance, the Gram matrix is Toeplitz, that is the entries are constant along lines parallel to the diagonal; such matrices can be inverted in $O(N^2)$ operations instead of the usual $O(N^3)$ (Golub and Van Loan, 1983). Spline interpolation is another place where a special matrix property is always exploited: set up in the appropriate way (de Boor, 1978), the matrices needed in exact cubic spline interpolation are pentadiagonal, that is, every row consists entirely of zeros, except for the diagonal and its two nearest neighbors on either side. If proper advantage is taken of the sparseness the interpolation can be accomplished in only $O(N)$ steps. When the interpolation is not expected to pass exactly through the data, *smoothing splines* result; we can obtain them by applying the theory of this chapter but the standard computational process given by Reinsch (1967) generates a linear system with a sparse array that also requires only $O(N)$ arithmetic operations. A remarkable example of a special matrix has been given by Henry et al. (1980); in their inverse problem, which involves the combination of many seismic time series to construct a so-called "wave field," they find that in a suitable Hilbert space the Gram matrix takes the form:

$$\Gamma_{ij} = \begin{cases} p_i q_j, & i \le j \\ q_i p_j, & j < i \end{cases}. \tag{1}$$

They show that such a matrix possesses a tridiagonal *inverse*, and it can be computed in only $8N$ operations.

Often to obtain important computational savings, it is necessary to depart completely from the norm-minimization strategy advocated in this book and make use of analytic results derived for idealized exact or continuous data. Probably the most valuable of the analytical methods is the Fourier transform. The Fourier theory briefly set out in section 2.04 for the magnetic anomaly problem applies equally well to any forward problem based upon convolution. Suppose the solution to the forward problem can be written

$$d(x) = \int_{-\infty}^{\infty} g(x - v) m(v) \, dv \tag{2}$$

where d, the data and m, the unknown model, are functions in $L_2(-\infty, \infty)$ and g is somewhat better behaved (bounded, for example). We saw in 2.04 that, when m is real, the Convolution Theorem gives

$$\hat{d}(\lambda) = \hat{g}(\lambda) \hat{m}(\lambda) \tag{3}$$

where ^ is the Fourier transform. Thus to obtain m we need only divide $\hat{d}$ by $\hat{g}$ and take the inverse transform. The method works in exactly the same way for models that vary in two or three dimensions. The Fourier approach is made even more attractive because there is a discrete approximation to the Fourier integral, the "Fast Fourier Transform" or FFT, as it is universally abbreviated, which can be calculated in only $N \log_2 N$ operations (Strang, 1986). Naive use of this seductively simple solution presents a number of serious difficulties. In almost all cases the transform $\hat{g}(\lambda)$ tends to zero rapidly with increasing λ; in the magnetic anomaly problem, for example, the fall-off is proportional to $\lambda \exp - |c \lambda|$. To find $\hat{m}$ one must effectively multiply the transformed data $\hat{d}$ by a function that increases extremely rapidly. With actual, noisy observations, $\hat{d}$ is dominated by noise terms at high λ because the true signal has fallen to low values there. Thus dividing by $\hat{g}$ has the effect of greatly amplifying errors at the small scale and this gives rise to spurious, short-wavelength oscillations. Of course such behavior is a manifestation of the inherent instability of the inverse problem. The usual way of coping with the instability is to filter the solution, that is, to multiply the transformed model by a function decreasing so rapidly that the undesirable growth is suppressed; the resemblance of this process to the truncation of the spectral recipe or the weighting of the norm-minimizing regularization is more than a coincidence. A troublesome aspect of approximations involving the FFT is that the real line in the convolution integral must be replaced by a finite interval. The approximate version of (2) is topologically *circular* convolution, which means that model values near one end of the interval influence the data predictions at the other end. To avoid this embarrassment, artificial buffer zones must be introduced to separate the end points close to one another in the metric of the convolution, but far apart in the original space. These devices and others have been applied in the practical inversion of marine magnetic data, originally by Schouten and McCamy (1972) and with increasing elaboration, to permit topographic roughness of the source layer and irregularity of the observation track, by Klitgord et al. (1975), and to treat practical data covering an area, by Macdonald et al. (1980).

We shall confine the rest of this section to methods of general applicability. The following is based upon the paper by Parker and Shure (1987) in which they suggested building a solution from a *depleted basis*. The key is the observation made at the end of section 3.03: the spectral theory shows us that, even though the norm-minimizing solution is known to lie in G with dimension N (the space spanned by the representers), an excellent approximation to it can usually be found in a low-dimensional subspace of G. When the ordered eigenvalues of the scaled Gram matrix (or singular values of the linear mapping) fall very rapidly, the dimension of the subspace containing the significant components may be very much smaller than the number of data; we saw this on a small scale in the previous section, where only two eigenvectors were selected for the expansion of the solution from the eighteen available. If we agree to be satisfied with a reasonable approximation to the regularized solution, we can exploit this property to cut down the numerical work. We return to the Hilbert space setting: the solution vector $m \in H$ is expanded in a basis set $b_1, b_2, b_3, \cdots b_L \in H$:

$$m = \sum_{l=1}^{L} q_l b_l \tag{4}$$

where some of these elements may belong to K, the subspace of ideal elements free from norm penalization; we make no special distinction in the notation now. The solution to the linear forward problem then becomes

$$d_j = (g_j, m), \quad j = 1, 2, \cdots N \tag{5}$$

$$= \sum_{l=1}^{L} B_{jl} q_l \tag{6}$$

where

$$B_{jl} = (g_j, b_l), \quad j = 1, 2, \cdots N; \quad l = 1, 2, \cdots L . \tag{7}$$

In vector language the prediction of the data values is simply

$$\mathbf{d} = B \, \mathbf{q} \tag{8}$$

where $B \in M(N \times L)$ and $\mathbf{d} \in E^N$ and $\mathbf{q} \in E^L$ have their obvious meanings. The norm of the model is calculated directly from the expansion (4)

$$\| m \|^2 = \sum_{l=1}^{L} \sum_{k=1}^{L} (b_l, b_k) q_l q_k \tag{9}$$

which we also write in vector form as

$$\| m \|^2 = \mathbf{q} \cdot \hat{R}' \mathbf{q} \tag{10}$$

with $\hat{R}' \in M(L \times L)$ defined by

$$\hat{R}_{lk}' = (b_l, b_k), \quad l = 1, 2, \cdots L; \quad k = 1, 2, \cdots L . \tag{11}$$

The familiar task is to minimize (10) subject to

$$\| \Sigma^{-1}(\mathbf{d} - B \mathbf{q}) \| \leq T . \tag{12}$$

When $\hat{R}'$ is substituted for $\hat{R}^T \hat{R}$ (10) becomes 3.05(2), equation (8) is identical to 3.05(1) without change and we have an optimization problem that looks exactly like the one treated by the theory of section 3.05! It should also be clear that if, instead of the norm in (10), $\hat{R}'$ is derived from a seminorm, most of the subsequent calculations remain the same. We may therefore apply all the equations in that section up to the beginning of the discussion of SVD. Specifically we must solve the linear system

$$(B^T \Sigma^{-2} B + v^{-1} \hat{R}') \mathbf{q} = B^T \Sigma^{-2} \mathbf{d} \tag{13}$$

with the appropriate Lagrange multiplier, so that when $\mathbf{q}$ is substituted into (12), equality is obtained. To apply the QR method of equation 3.05(7) one must first factorize $\hat{R}'$; the factors may be obvious if $\hat{R}'$ has been found by numerical quadrature. On the other hand, numerical factorization—by the Cholesky scheme for norm minimization or by SVD for the seminorm problem—may be necessary. If we write the factorization of $\hat{R}'$ thus:

$$\hat{R}' = \hat{R}_1^T \hat{R}_1 \tag{14}$$

then the QR equation tells us that the expansion we seek is the solution of the overdetermined least squares problem

$$\min_{\mathbf{q} \in E^L} \left\| \begin{bmatrix} \Sigma^{-1} B \\ v^{-\frac{1}{2}} \hat{R}_1 \end{bmatrix} \mathbf{q} - \begin{bmatrix} \Sigma^{-1} \mathbf{d} \\ 0 \end{bmatrix} \right\| . \tag{15}$$

The lower submatrix $\nu^{-\frac{1}{2}}\hat{R}_1$ applies the regularization that mitigates against unnecessary roughness. If that submatrix were omitted we would be seeking the coefficients of an expansion for the best-fitting model; this is *least-squares collocation*, a technique with little to recommend it but still widely used in geophysics.

The main difference from the discussion of 3.05 is that here the dimension of the space of basis elements L is much smaller than N, not greater than or equal to it. Because the matrix on the left of (13) is of order L the computational labor in solving the linear system grows only as L^3, which in many cases will be orders of magnitude smaller than N^3, provided the basis elements have been wisely selected; more about this in a moment. We see from (7) and (9) that there is a cost that grows directly with N and that is associated with the formation of B and of $\hat{R}'$. For an application of (15) in which all the details are exposed and considerable care is taken with computational costs, the reader should consult Constable and Parker (1988b).

A major question as yet unanswered is, How shall we choose the deleted basis? There is no hard and fast answer, although we can offer a number of suggestions. These are best described in terms of some concrete illustrations. Once again we turn to the old workhorse, the magnetic anomaly profile: now in figure 3.07a we have increased the data density by a factor of ten, leaving the noise level the same. To make our point we have generated too many data for this to be a realistic simulation of a magnetic profile—as we noted earlier, three or four data per kilometer of ship track is more normal. There are 301 data here, a number not outside limits of practical computation using ordinary methods. Actually, the complete solution with smallest norm in W_2^1 can be calculated very quickly (in about three minutes of computer time, including six Newton iterations) because, as we have just noted the Gram matrix is Toeplitz, and the computer program exploited the fact; therefore, in this case the exact solution is available for comparison. Based on 3.01(6), the expected value of the norm of standardized misfit gives the tolerance $T = 17.335$ for $N = 301$. The corresponding smallest possible model in W_2^1 for the given data has norm 6.415, a number with no intrinsic interest, but useful in assessing the smoothness of our approximations.

The first choice for a basis that comes to mind is surely the set of representers g_j themselves. If we take them all we simply

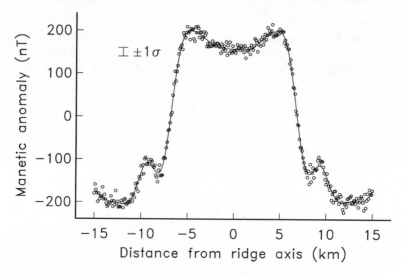

Figure 3.07a: The idealized magnetic anomaly sampled at a much higher rate, namely, ten field values per kilometer. The standard error of the noise remains 10 nT. The solid line is the predicted anomaly of the smoothest model with the expected misfit.

get back the theory of 3.02 thereby obtaining the exact solution to the optimization problem. But this set is much too large to be handled and so we must prune it. In the anomaly example every representer g_j is an identical function, translated so that its peak lies directly under the observation position; see figure 2.06a. Because in our particular illustration the anomaly sites are evenly distributed, an obvious way to thin the representer basis is to keep every n th one in a sequence ordered according to x_j. We cannot know without calculation what value to give n, so we compute a series of models with L increasing from a small number. When the dimension of the expansion subspace is much too small we will be unable to meet the misfit criterion. After it becomes possible to satisfy the tolerance level, increasing L decreases the norm, until at $L = N$, the minimum norm is reached. We expect, though, that the norm will reach a plateau not far above the minimum value for values of L that are some modest fraction of N. In the numerical illustration we make the trivial generalization to an arbitrary integer L, by creating basis functions according to 2.06(28) for any evenly spaced set of x_j on the interval $[-15, 15]$. Thus only when $L - 1$ evenly divides 300 does the model actually lie in the subspace G.

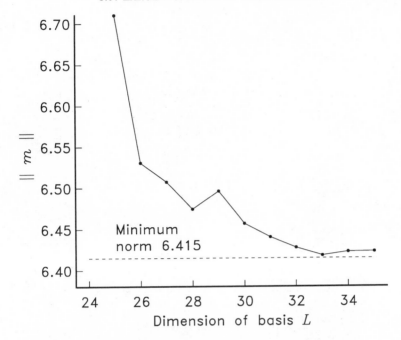

Figure 3.07b: Minimum norm of the model $\sum q_l b_l$ with misfit 17.335 as the number of elements in the basis increases. When $L \leq 24$ the models have misfits larger than the required one.

Until $L = 25$ the expansion is unable to match the required tolerance. In figure 3.07b we see how the model norm behaves as the number of basis elements is increased. There is no reason to expect a strictly monotonic decrease with L, and indeed, though there is a strong tendency for the norm to decrease, there are minor ripples in the function. The remarkable feature is how small the norm is for every one of the models. Figure 3.07c depicts three solutions: the one obtained when $L = 25$; model obtained with $L = 33$, a local minimum in figure 3.07b; and the true optimum solution found by solution of the complete system, effectively for $L = 301$. The model obtained with 25 basis functions is visibly rougher than the others, but the optimal model and the one found for $L = 33$ are virtually identical. The number of arithmetic operations to solve the linear systems for the optimal model would, in the absence of the Toeplitz symmetry, be more than 750 times greater than that needed for the model expanded in the thirty-three basis functions b_l. This is an

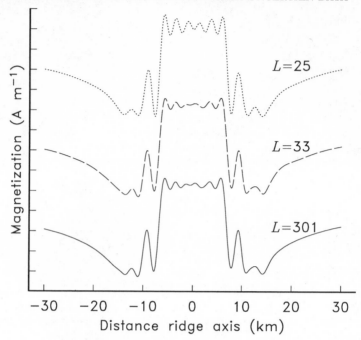

Figure 3.07c: Magnetization models with minimum norm and misfit 17.335; the dimension of the expansion subspace is indicated. When $L = 301$ the model is in fact the smallest one of any fitting to the requested tolerance. The models are defined only up to an arbitrary constant, so that only the variation in magnetization is significant, not the absolute level.

overestimate of the relative cost, because it does not take into account the cost of forming B and $B^T B$, nor can we ignore the fact that we must compute solutions at a number of different values of L, but the advantage of the method for large systems is obviously very great, and it would be considerably greater if many more data were involved.

What other basis functions might be sensible instead of a subset of representers? An apparently attractive alternative is to draw from the orthogonal set generated by SVD analysis. Clearly one could not perform the decomposition on the complete system —this is just too expensive. Perhaps a suitable starting point might be a subproblem making use of substantially fewer data. But obviously the subspace of basis functions thus made available is the same whether or not SVD has been performed and the only reason in support of the extra expense is a possible improvement in numerical stability.

A reasonable requirement is that the basis functions should in principle be capable of approaching the true optimal solution if enough are used; clearly the subset of representers satisfies this requirement. Another class meeting the demand is any complete basis for the whole space. An orthogonal set of functions has obvious attractions, since this leads to a diagonal matrix $\hat{R}'$. An additional useful property is simplicity of evaluation of the inner product with the representers, although this is unlikely in all but the most elementary inverse problems. The application of a complete orthogonal basis to the magnetic anomaly problem is straightforward in principle, but messy in practice. The awkwardness arises from the use of the whole real line $(-\infty, \infty)$ as the domain of the model; none of the familiar orthogonal families of functions in $L_2(-\infty, \infty)$, for example the Hermite functions, has a simple, closed-form inner product with the representers and that means the evaluation of B entails a lot of heavy numerical work, something the author was hoping to avoid. One way out would be to make the interval on which m is defined finite, in which case the ordinary Fourier basis becomes available, except that then a direct comparison with what we have already done becomes impossible.

Instead of flogging the magnetic anomaly example to death, for an illustration of the utility of orthogonal basis functions we return briefly to a problem considered in 2.07, spline interpolation on the sphere. Now, of course we include observational uncertainty, and we shall also assume that a very large number of data must be smoothly interpolated. Recall that by our use of the 2-norm of $\nabla_s^2 f$ to measure roughness of models we found that the optimally smooth solution is a linear combination of the constant function and dilogarithms, special functions defined by 2.07(60); the equation for the interpolation, 2.07(66), can be rewritten

$$f(\hat{r}) = \beta_1 + \sum_{j=1}^{N} \alpha_j \, \mathrm{dilog}(\tfrac{1}{4} \, |\hat{r} - \hat{r}_j|^2) \tag{16}$$

if we rescale α_j and absorb into β_1 the various constants under the sum in 2.07(66). Here $\hat{r}_j \in S^2$ are the points at which the function has been sampled, that is, the places where data are given. It seems natural to identify the basis elements b_l of (4) with a subset of the dilogarithm representers in (16); the inner products needed for B and $\hat{R}'$ can all be evaluated as dilogarithms using the expression for Γ_{jk} in 2.07(59). We can proceed exactly as in the magnetic anomaly case, choosing a subset of the observation sites as points on which to center the basis functions.

In this case, however, there is an alternative, perhaps even more natural basis for expansion—the spherical harmonics. Suppose we order the (complex, fully normalized) spherical harmonic functions as we did in B_2, the space used in 2.07; with the arguments θ and ϕ omitted, the sequence is

$$Y_0^0, Y_1^{-1}, Y_1^0, Y_1^1, Y_2^{-2}, Y_2^{-1}, Y_2^0, Y_2^1, Y_2^2, \cdots . \qquad (17)$$

We identify $b_1, b_2, b_3, \cdots$, with these in turn; they are now complex basis functions and q_l must be complex too if f is to be real as it usually is. We expand the interpolation according to (4):

$$f = \sum_{l=1}^{L} q_l \, b_l . \qquad (18)$$

The roughness of the model function f remains the 2-norm of the surface Laplacian and so (9) becomes

$$\int_{S^2} |\nabla_s^2 f|^2 \, d^2\hat{\mathbf{r}} = \int_{S^2} |\sum_{l=1}^{L} q_l \, \nabla_s^2 b_l|^2 \, d^2\hat{\mathbf{r}} \qquad (19)$$

$$= \sum_{l=1}^{L} \sum_{k=1}^{L} [\, q_l^* q_k \int_{S^2} (\nabla_s^2 b_l)^* \nabla_s^2 b_k \, d^2\hat{\mathbf{r}} \,] \qquad (20)$$

$$= \sum_{l=1}^{L} \lambda_l^2 |q_l|^2 \qquad (21)$$

where λ_l are the (degenerate) eigenvalues of ∇_s^2, namely, the sequence $0, -2, -2, -2, -6, -6, -6, -6, -6$ and so on with sets of $2n+1$ values of $-n(n+1)$. Thus $\hat{R}'$ is diagonal. The vast simplification follows from the eigenfunction property of the basis functions and their orthogonality in $L_2(S^2)$. The diagonal property is kept even if we double the number of unknowns by splitting into real and imaginary parts, something inconvenient for theory, but economical in computation. Notice how the zero eigenvalue reflects the fact that this is a seminorm, causing $\hat{R}'$ to be singular. Its factorization (14) is trivial and (15) has the air of least-squares fitting to a spherical harmonic series, but with a diagonal tail on the matrix.

The orthogonal basis functions have the advantage that, once enough of them have been taken to meet the misfit criterion, adding more can never increase the roughness; this is somewhat paradoxical at first sight, because one is getting smoother solutions by adding functions to the basis that are more and more oscillatory. Equation (21) shows that it is easy in this expansion to regularize with operators other than traditional surface derivatives. For example, other polynomials may be chosen for λ_l, those with higher degree effectively having higher differential power; notice though, that to keep indifference to the choice of coordinate system, the grouping into sets of $2n + 1$ constant terms in the sequence is essential.

When the expansion technique or other approximate methods are needed to make a large problem practicable we must sacrifice something: we do not know how far the approximate solution is from the optimal one. We cannot give even a useful upper bound on the distance between our approximation and the exact answer. This is an unsatisfactory state of affairs that pertains to many large-scale computer calculations. For example, when fluid flows are simulated by numerical solution of the partial differential equations, it is usually impossible to say rigorously how far the numerical solution differs from the exact one; still less can it be said how well the numerical result resembles physical reality. But the same fundamental uncertainty applies to the models we have constructed, even when we can avoid approximation in the optimization problem. So far in this book we have concentrated on the quest to build geophysical models that match the observations. The intrinsic nonuniquess has not been ignored; we have deliberately selected special kinds of models from the infinitely large family of solutions. It is now time to ask a much harder question: What can the data tell us about the Earth?

CHAPTER 4

RESOLUTION AND INFERENCE

4.01 The Central Question

Up until this point we have concentrated upon the discovery of the simplest kind of model consistent with a given, possibly noisy, data set. A simplification that we shall retain in the present chapter is the treatment of linear forward problems, that is, those where the predictions of the mathematical model take the form of linear functionals. But the issue of nonuniqueness can no longer be set aside: we must address the central question, What can be said about the Earth from our data? We shall describe a number of different approaches.

A natural extension of the process of model building suggests the notion of resolution: we attempt to assess the degree to which a particular model is reliable. Intuitively, it seems plausible that our limited data might be able to discern the grosser features of the real Earth, but that structure on a scale finer than some critical level would be invisible. The critical scale is called the resolution length, which would be expected to vary from place to place within the model. There are two ways of calculating the scale: one is intimately connected with construction of smooth models, and yields a qualitative idea of the intrinsic detail contained in the constructed models; the other, the famous method of Backus and Gilbert, is a more quantitative scheme that computes a scale length from the data and their errors. The general idea of resolution has become a firmly established part of geophysical folklore, and is frequently invoked in a qualitative manner without supporting calculations. Yet upon close examination we find that even quantitative resolution calculations do not provide a satisfactory solution to the problem of making rigorous deductions from the data about the Earth.

To approach this ideal we must at last give up the comfort of a single model as a basis for interpretation of the observations; instead we must come to grips with the existence of an entire universe of solutions in accord with observation. A powerful strategy is to seek common properties of the complete class of adequate solutions. Insofar as the mathematical model for the forward problem is satisfactory, the common properties shared by all solutions to the inverse problem must be properties of the real Earth. To make the term "property" concrete we replace it with the word "functional." For example, we might ask, How large is the *average* seismic velocity in this particular region? Or, because of the ambiguity imposed by data incompleteness and measurement noise, What are upper and lower limits on that average, given the data? Averages are examples of linear functionals and our attention will first turn to the mathematical problem of maximizing or minimizing a linear functional of the model, a problem that can be framed, perhaps not surprisingly, as an optimization. The Backus-Gilbert resolution theory asks essentially the same question, but, because the space of linear functionals made accessible to exact treatment turns out to be so limited, the theory chooses to approximate the true averaging functional. If we decide to do without that approximation, no further progress can be made unless further assumptions about the model are introduced.

The class of models that formally satisfies the observations, but has no other constraints placed upon it, is too large. It contains models that certainly lie outside geological experience and outside the realms of physical plausibility too. If we choose an appropriate norm for our space, a Hilbert space for the moment, we can remove the quote marks from the word large, and simply say that all the acceptable models must have a norm smaller than a specified amount—thus all models acceptable on the grounds of fitting the data, but larger than a stipulated amount in the norm, are rejected. If such a bound can be agreed upon, it can be translated into a bound upon any continuous linear functional of the model, like the average value in an interval or a volume. We can extend the bounding idea to finding the shape of the allowed region in a space made up of a collection of linear functionals taken together. The machinery for this translation will be given in section 4.03. Whether or not it is possible to state a sensible upper bound on the norm from available geophysical or geological evidence is a question that can be answered only on a case-by-case basis; whether or not the bound on the model translates into an interesting or useful constraint also cannot be decided in advance without detailed calculations. Nonetheless, the author's experience suggests that it is comparatively rare for this approach to yield any rewarding insight, if Hilbert spaces are the setting.

When we enlarge our perspective to norms other than those associated with inner products, the effectiveness of the bounding strategy improves considerably. The most convenient and physically natural norm is the uniform norm, the bound on the maximum value (for continuous models). Treatment of this kind of norm brings with it a bonus: it automatically allows the handling of an additional constraint, that of positivity. The unknown model often represents a property, like density or seismic attenuation, that cannot be negative on physical grounds. Being able to insist upon this requirement in all acceptable models sometimes improves our ability to make useful inferences. Many parameters are expected to rise with increasing depth in the Earth, for example temperature or density; if there is a departure from this trend something interesting is going on that deserves special attention. We can use the constraint on the sign of a property (in this case a derivative) to determine whether a given data set requires a departure from the expected tendency, by asking, Are there any models satisfying the data that increase with depth but never decrease? If the answer is no, we have rigorously proved the necessity of a property decrease. The inclusion of constraints on the sign enlarges the scope of our ability to make rigorous inferences about the Earth in a way that geophysicists are only now beginning to recognize.

A final popular approach is to introduce a statistical aspect to the space of models. So far in our treatment statistical descriptions have been relegated to a comparatively minor role in the measurement of goodness of fit of the theoretical prediction to the actual data. We can, however, if we choose, put the whole inverse problem into a statistical framework. To make the mathematics tractable and the answers practically useful requires the imposition of a large number of additional and often totally untestable assumptions about the nature of the model. In this book we shall deal briefly with the easiest kinds of problems, where it is assumed that the Earth can be reasonably taken as representative of a random stationary process. We find that many of the calculations that we have already been doing remain unchanged in form, but that the results can be given a new interpretation.

4.02 Resolution

When we are confronted with a solution to an inverse problem deliberately built to be as featureless as possible, for example the one shown in figure 3.06c, we are entitled to ask whether the flat places are merely manifestations of our ignorance of the true structure. If the inversion process is in principle incapable of creating undulations or other structure with a scale finer than that of the constant plateau, this must cast doubt on the necessity of a featureless zone. The alternative is of course that the Earth really is quite smooth in those regions. The distinction between the two can be defined only in terms of a characteristic scale: we cannot expect actual observations ever to be capable of generating accurate models on the scale of molecules or individual crystals; on the other hand, we may hope to perceive with confidence variations on the scale of the whole mantle, or of one hemisphere of the Earth's surface. The true characteristic scale lies between these extremes (though usually much nearer the upper end!). We are of course discussing the meaning of the *resolution* of our observations, their ability to "see" small features in the Earth. We shall show that the regularized model may be interpreted as the result of smoothing the real structure using a family of narrowly peaked functions, the resolving functions. In this section we shall adopt a fairly informal characterization of these ideas, hoping the examples provide insight and justify the general approach.

By the way in which the norm- and seminorm-minimizing models are constructed it is clear that even if a very sharp jump occurs in the Earth, our models will be smooth, if the feature is detected at all. Keep in mind the fact that conservative smoothing is a highly desirable property, because it protects us from creating totally spurious details that would otherwise arise. We can calibrate the inversion and discover just what would happen to a very small-scale disturbance in the real Earth. Let us start with an artificial model chosen to be as concentrated as possible, a δ function at some position. Then, by solving the linear forward problem, we generate the data values associated with it. Next we invert those "data" and thereby exhibit the smallest-scale single feature allowed to appear in any model based upon the regularization. The image of the δ function after inversion will be called the *resolving function*. Ideally a δ function is mapped into a function with a simple maximum near the single point of support in the original; the overall width of the spread-out image tells us the resolution scale. The reader may wonder how we can allow an object like a δ function into a model space where it usually would not belong. In most cases we can create a sequence of positive, increasingly narrow functions, just as in section 1.08 where we discussed Cauchy sequences (although the sequence will seldom be Cauchy in this case); if each of these functions lies in the model space, the members of the sequence may be submitted to the inner product solving the forward problem and a sequence of data vectors results whose limit will be the data set for analysis, on the assumption the sequence of vectors converges to a limit. For this to work, the representers must have a certain amount of smoothness, at least continuity; to see this, the reader should experiment with the interpolation problem using equation 2.07(1). We shall ignore mathematical niceties for the time being, since the resolution theory is at best a qualitative matter, not an instrument for making rigorous inferences.

Let us first carry through the program for the magnetic anomaly example of 3.04, the one with only thirty-one noisy data. (Notice that the way we have described the process does not preclude its application to exactly known data.) Recall from 2.06(1) that the solution of the forward problem may be written

$$d_j = \int_{-\infty}^{\infty} \frac{-\mu_0}{2\pi} \frac{(x-x_j)^2 - h^2}{[(x-x_j)^2 + h^2]^2} m(x)\,dx \tag{1}$$

$$= \int_{-\infty}^{\infty} g_j(x)\,m(x)\,dx, \quad j = 1, 2, 3, \cdots N \tag{2}$$

where d_j are the magnetic anomaly values at the observation sites x_j and m, the unknown model, is the density of vertical dipoles in the source layer. From this expression, if a δ function source is placed at x_0

$$d_j = g_j(x_0), \quad j = 1, 2, 3, \cdots N. \tag{3}$$

The artificial data are samples of the L_2 representer pictured in figure 2.06a. To interpret the significance of the regularized solution, say the one shown in figure 3.04d, we choose a series of target values for x_0, then solve the corresponding inverse problems *with the same Lagrange multiplier that was used to construct the model*. In this way we find the resolving functions corresponding to the proper model. Figure 4.02a shows the results for several values of the center x_0. For the interval between −13 and 13 km the resolving functions are virtually identical copies of each other, translated appropriately. Their width measures around 2 to 3 km, and this confirms that the central plateau in magnetization between ±8 km is in some sense real, and not just the imposition of innocuous behavior by the norm-minimization process; of course, on scales less than 3 km, the evident smoothness is just such an artifact. On the wings of the model we see a very rapid

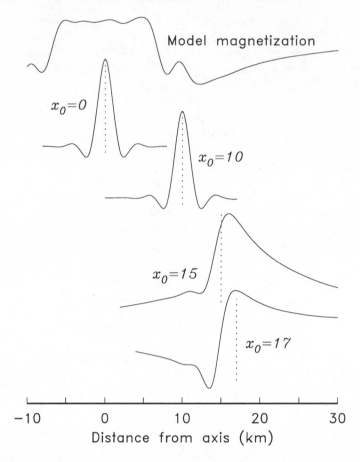

Figure 4.02a: Top, model shown in figure 3.04d with standard level of misfit. Below, four resolving functions for the 31-data magnetic anomaly problem for target centers as indicated. The behavior of the functions is symmetrical about $x_0 = 0$. Notice the rapid degradation in resolution as soon as $x_0 > 15$ km.

increase in the apparent width of the resolving functions, something presumably connected to the fact that all the anomaly observation sites lie within 15 km of the origin. Observe how the resolving function for $x_0 = 17$ resembles a step more than a δ function; the deterioration in resolution continues as the center moves even further away from the origin. Clearly the smoothness of the model more than 20 km from the ridge axis is the result of strong simplification by the regularization. The almost constant magnetization in the central anomaly region is quite different: the resolving length is considerably shorter than the width of the plateau. Conversely, we may tentatively accept as genuine the sharp transitions at ±7 km on the grounds that we have sought to suppress large gradients in the solution. These interpretations of the model do not, however, have the force of rigorous deduction; as we shall soon see, unless additional properties of the model are assumed, they could be completely misleading.

The numerical work for calculating the resolving functions is quite light because only one value of the Lagrange multiplier is needed and the matrix factorization (Cholesky, for example) need be done just once. The computational cost of a solution once a factorization has already been performed is $O(N^2)$ arithmetic operations, and so one can afford to examine many different centers. Another virtue of this approach to resolution is that it requires the writing of almost no new computer programs; given a program for solving the inverse problem by regularization, we can use it to compute resolving functions.

Let us quickly examine the application of this method to the seismic attenuation problem. Taking as our base solution the seminorm minimizing (flattest) model of figure 3.06c (reproduced at the top of figure 4.02b), we calculate the response of the linear system to δ functions at various depths in the mantle. The results are shown in figure 4.02b. As we might have already anticipated, the data of figure 3.06a are not capable of resolving fine detail anywhere in the mantle. An infinitely sharp perturbation, even at the optimum depth, say $r_0/a = 0.926$, is distorted into a function with width of about $0.1a$. Below $r_0/a = 0.9$ the resolving functions are all extremely broad and almost identical in shape. We must conclude the flat region in the attenuation model is there because the data contain almost no information about the lower mantle. This is in contrast to the magnetic anomaly example, where the central constant magnetization is situated where the data are capable of responding to relatively small-scale variations, but in that case none is detected.

The version of resolution theory we have developed is qualitative, some may feel excessively so: we merely inspect the resolving function and make a judgement about its width relative to wiggles in the model. There is another way to formulate the problem which offers quantitative estimates of the resolving scale, the celebrated theory of Backus and Gilbert. We shall describe this work in a quite simplified version because for almost all purposes

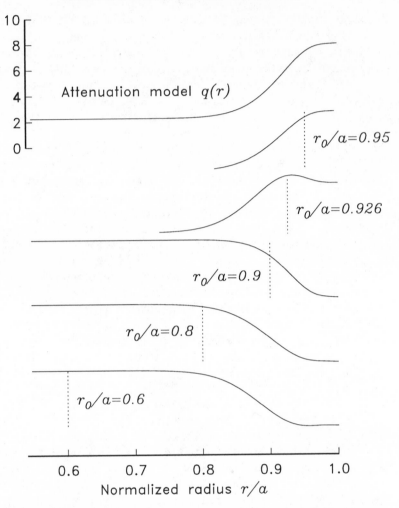

Figure 4.02b: *Top*, attenuation model shown in figure 3.06c. *Below*, five resolving functions centered as indicated. Observe the similarity in shape of all the resolving functions when $r_0/a < 0.9$, irrespective of the nominal center.

the qualitative treatment is perfectly satisfactory. Aside from its historical interest the Backus-Gilbert theory sheds light on the meaning of the regularized models and their relationship to the true structure; in addition the theory provides the first step in the transition we must make from the traditional business of model construction to the less familiar and largely unsolved problem of making logically sound deductions about the Earth from the observations. Initially, we return to the idealized world of chapter 2 where the measurement error is negligible. Then, because of the linearity of the solution to the forward problem, our knowledge of the unknown structure is contained entirely in the N numbers, the result of linear functionals applied to m:

$$d_j = L_j[m], \ j = 1, 2, 3, \cdots N . \tag{4}$$

If we restrict ourselves to values derived solely from numerical combinations of the d_j, we obtain unambiguous information about m. The nonuniqueness of the inverse problem means there is a vast family of equally good solutions but any model fitting the data shares the values of d_j. Let us choose a concrete version of this abstract form, somewhat like equation (2):

$$d_j = \int g_j(x)\, m(x)\, dx . \tag{5}$$

Further, let us agree to consider only *linear combinations* of the data; then a typical numerical property p of all models is

$$p = \sum_{j=1}^{N} \chi_j d_j \tag{6}$$

$$= \int [\sum_{j=1}^{N} \chi_j g_j(x)]\, m(x)\, dx \tag{7}$$

where the coefficients χ_j may be chosen as we wish. When we attempt to construct a model, we are again interested in linear functionals, specifically in the form:

$$m(x_0) = \int \delta(x - x_0)\, m(x)\, dx . \tag{8}$$

If it were possible to choose the coefficients χ_j in (7) so that the function in large parentheses becomes a δ function centered on x_0 we would have solved the problem of exactly recovering the true solution, but of course this usually cannot be done. We can, however, select χ_j so that the sum makes an approximation to $\delta(x - x_0)$, and this is what the Backus-Gilbert theory does. Let us write (7) as follows

$$p_{x_0}[m] = \int \tilde{\delta}(x, x_0)\, m(x)\, dx \tag{9}$$

where

$$\tilde{\delta}(x, x_0) = \sum_{j=1}^{N} \chi_j g_j(x) . \tag{10}$$

For each x_0 we find the best approximation we can, and the corresponding value of p_{x_0} is a smoothed, or averaged version of the true $m(x)$, smeared out by the *averaging kernel* $\tilde{\delta}(x, x_0)$. We know that when the true function m is appropriately averaged it must yield p_{x_0} because that number is obtained directly from measurement via (6) and every model that fits the data must obtain the same value.

One more element is needed to complete the picture: we must decide the sense in which we want to approximate $\delta(x - x_0)$. Backus and Gilbert (1968) examine several approximation techniques; for a complete discussion the interested reader should consult the original paper and the article by Backus (1970b), who adopts a Hilbert space framework similar to ours. A distinctive feature of the Backus-Gilbert approach is their use of norms or seminorms that give direct estimates of the scale of resolution, or *averaging length* as they call it. In our treatment, which is only one of those discussed by Backus and Gilbert, we draw upon our experience with norm minimization in Hilbert spaces: we introduce a Hilbert space containing δ functions as valid elements and write the solution to the forward problem in terms of its inner product. Thus we seek the approximation closest in the norm to the true δ function, namely, we look for the element $\tilde{\delta}$ that solves

$$\min_{\tilde{\delta} \,\in\, G} \| \tilde{\delta} - \delta_{x_0} \| \tag{11}$$

where G is the linear subspace spanned by the representers and δ_{x_0} is a δ function centered at x_0. This minimization problem is the one solved in section 1.12 by applying the Projection Theorem. In section 1.07 we mentioned a norm based upon indefinite integration, the 2-norm of the antiderivative:

$$\| f \|_I = [\int_a^b \{ \int_a^x f(t)\, dt \}^2 \, dx]^{\frac{1}{2}} \tag{12}$$

$$= [\int_a^b f^{(-1)}(x)^2 \, dx]^{\frac{1}{2}} \tag{13}$$

recalling the notation of 2.06 for indefinite integration. We showed in 1.08 that one of the classical sequences for a δ function was a Cauchy sequence under this norm. It follows that completion of an appropriate pre-Hilbert space produces a Hilbert space of the kind we need, one containing the δ function; we shall call the space $W_2^{-1}(a, b)$ for obvious reasons. The inner product is clearly

$$(f, g)_I = \int_a^b f^{(-1)}(x)\, g^{(-1)}(x)\, dx \; . \tag{14}$$

To understand more intuitively what the norm of W_2^{-1} measures, consider its application to the discontinuous boxcar function:

$$B_w(x) = \begin{cases} 1/w, & |x - x_0| \leq w/2 \\ 0, & \text{otherwise} \end{cases} \tag{15}$$

where $w > 0$. We may regard B_w as the archetype for functions of width w centered at x_0 with unit area under them. A short calculation shows that

$$\| B_w - \delta_{x_0} \|_I^2 = \frac{1}{12} w \; . \tag{16}$$

Thus the squared norm of the difference, a quantity minimized in the Backus-Gilbert program for resolution, is proportional to the width of a function in an idealized case. It is an acceptable generalization to define the *averaging length* about x_0 of an arbitrary element of W_2^{-1} as

$$A_{x_0}[f] = 12 \| f - \delta_{x_0} \|_I^2 \tag{17}$$

since A_{x_0} is measured in units of length and can be made small only by functions closely resembling a true δ function at the proper place. Therefore (11) can be interpreted as saying that we seek the narrowest function in the sense of (17) from those in the set of all linear combinations of the representers. Exercise 4.02(i) provides a list of scales of familiar functions under this width estimator.

Compared with the theory for exact data, the Backus-Gilbert treatment for noisy data is a somewhat clumsy affair. An individual model average p_{x_0}, which is just a single real number, obviously cannot satisfy the measurements. The function produced by the union of evaluation of all such functionals for every x_0 can be regarded as a model but, because that function is just a collection of averages, it does not in general provide a solution to the inverse problem by fitting the original observations. Therefore the usual method of demanding a fit within the expected uncertainty cannot be applied when the theory is extended to handle noisy data. Backus and Gilbert (1970) return to (6) for the key to this dilemma; if independent Gaussian behavior is assumed for the uncertainties in the data, the number p becomes a random variable whose standard deviation is

$$\Delta p = [\sum_{j=1}^{N} \chi_j^2 \sigma_j^2]^{1/2} \tag{18}$$

where σ_j^2 is the variance of the measurement d_j. (Correlated errors could be treated if necessary.) The smaller Δp can be made by suitable choice of the coefficients χ_j the more accurately p, or more specifically p_{x_0}, is known. But at the same time we want A_{x_0} small in order to focus on the smallest possible region within the Earth. The two functionals cannot be minimized simultaneously; instead, we must explore the resolution-uncertainty plane to discover the best compromise between loss of detail and statistical imprecision. Mathematically, this is another minimization problem: we ask, How small can A_{x_0} be made if we demand that Δp is less than a fixed amount? Putting (10) into (17), we see that we must choose χ_j to minimize

$$A_{x_0} = 12\| \sum_{j=1}^{N} \chi_j g_j - \delta_{x_0} \|_I^2 \tag{19}$$

subject to

$$\sum_{j=1}^{N} \chi_j^2 \sigma_j^2 \le T^2 . \tag{20}$$

As in 3.02 we can show that *equality* must hold for the smallest averaging length to be achieved, so a Lagrange multiplier ρ may be introduced and stationary values sought for the functional

$$V[\chi, \rho] = A_{x_0}[\chi] - \rho(T^2 - \Delta p[\chi]^2) \tag{21}$$

where $\chi \in E^N$, vector of the expansion coefficients. Expanding all the inner products we obtain

$$V[\chi, \rho] = 12(\chi \cdot \Gamma \chi - 2\mathbf{v} \cdot \chi + \| \delta_{x_0} \|_I^2) - \rho(T^2 - \chi \cdot \Sigma \chi) . \tag{22}$$

Here Γ is the familiar Gram matrix of representers in W_2^{-1} and Σ is the diagonal matrix of variances; the components of $\mathbf{v}$ are

$$v_j = (g_j, \delta_{x_0})_I . \tag{23}$$

Differentiating to find the stationary argument χ_0, we obtain the linear system:

$$(\Gamma + \frac{1}{12}\rho\Sigma)\chi_0 = \mathbf{v} \tag{24}$$

and as usual ρ must be chosen to achieve the desired variance T^2.

The behavior of the system is so similar to the one examined at great length in 3.02 that we shall simply skip ahead to the results, which are probably easily anticipated. The Lagrange multiplier ρ is nonnegative. No matter how large the tolerance T is allowed to be (corresponding to the limit as ρ tends to zero), there is a value below which the averaging length cannot be driven, the minimum value obtained by the theory for exact data; as T is reduced A_{x_0} increases monotonically, attaining its greatest value when $T = 0$ (which occurs for infinite ρ). As ρ is varied one sweeps out the most famous trade-off curve in all of inverse theory: good resolution in the estimated linear functional can be obtained only at the expense of reduced statistical reliability.

Let us briefly compare the classical Backus-Gilbert theory with the more qualitative approach given earlier. When one traces through the equations for finding the regularized model, it becomes apparent that, if the Lagrange multiplier is fixed, the value of $m_0(x)$ for any particular x depends linearly upon the measurements d_j, and therefore the value is a universal property, an example of a linear functional p which may be expressed as in (6) or (7). Thus there is an averaging kernel, call it $\tilde{\delta}_0$, for the norm- (or seminorm-) minimizing model:

$$m_0(y) = \int \tilde{\delta}_0(x, y) \, m(x) \, dx \ . \tag{25}$$

The coefficients χ_j and hence the function $\tilde{\delta}_0$ are constant once the Lagrange multiplier has been chosen. Now we construct the resolution function by the recipe described earlier—we solve the inverse problem for data generated by the artificial model

$$m(x) = \delta(x - x_0) \ . \tag{26}$$

Putting this into equation (25) shows that the regularized solution of the δ function model is the averaging kernel, in other words, the resolution function and the averaging kernel are the same thing. Thus we have shown that the regularized solution can be viewed as the true model smoothed through a family of averaging kernels. Those kernels are not optimally narrow in the sense of minimizing (17) as the Backus-Gilbert kernels are, but as we shall argue in a moment, it is wise not to take the width estimate too seriously.

With this interpretation in mind, look again at figures 4.02a and 4.02b: our models, shown at the top, can be thought of as the true solutions averaged with the functions below. A very strong impression is gained of the effect of nonuniqueness: we sense that all detail below a certain length scale must be invisible. In figure 4.02a we feel that if the true magnetization undulates with a wavelength of less than 5 km near the ridge axis, we would not be aware of this fact, but we could detect an oscillation with a 10-km scale; 20 km from the axis the scale of invisible detail is very large. Conversely, we may wish to express confidence in the existence of the transition from higher to lower magnetization at $x = 6$ km; but before becoming too confident, the reader should look at exercise 4.02(iii). For the seismic problem, figure 4.02b tells us that throughout the deeper three-quarters of the mantle the data can give us a rough estimate of the average value of q but little else. The upper mantle is known only through an average value in the top one-fourth.

We shall not provide a numerical illustration of the Backus-Gilbert approach; there are several drawbacks to the classical theory that have led to its having received more citation than actual application. On a practical level the computations are extensive: a trade-off curve must be computed for every location of interest in the model. This a large number of operations and it produces a great volume of material for inspection. A considerable investment must be made in additional programs to perform the computations, as well. It is also necessary to invent new norms and seminorms to handle δ functions defined on a plane or within the solid Earth, when modeling in higher dimensions is undertaken. Backus (1970b) gives the skeleton of a general approach. Next, a reasonable balance must be struck between resolution and variance for every location and it is by no means clear what the proper criteria ought to be. Having made these choices, one is expected to plot a series of points with horizontal and vertical error bars to express the uncertainty in our knowledge of the structure, one bar indicating the resolving scale over which the functional p_{x_0} averages, the other giving the statistical uncertainty of the estimate. At a fundamental level this picture is misleading. The averaging length is only a crude measure of the scale and does not give the true interval over which the actual structure has been averaged, even though one is very tempted to treat it as such. The approximations $\tilde{\delta}(x, x_0)$ normally depart a long way from the boxcar shape B_w and the value of p_{x_0} will always depend on the value of the profile at places outside the interval indicated, particularly strongly if the true solution varies widely. To be protected from making the mistake of taking the averaging length too literally, we should always plot $\tilde{\delta}$ —we simply cannot rely on the numerical width estimator. This raises the question of whether there is any real advantage at all in calculating *a number* for the averaging length. The author's own conclusion is that there is none and that the resolving function treatment we gave earlier does everything that can be expected in a resolution analysis and, since it is so much more easily calculated, it is generally to be preferred to the Backus-Gilbert treatment.

Exercises

4.02(i) Each of the following functions is an element of $W_2^{-1}(-\infty, \infty)$ and there is unit area under each one, since it is to be assumed that $a > 0$. Using the definition (17) verify the following width estimators

$$f_1(x) = \frac{e^{-|x/a|}}{2a}, \qquad A_0[f_1] = 3a$$

$$f_2(x) = \frac{e^{-x^2/a^2}}{a\,\pi^{1/2}}, \qquad A_0[f_2] = \frac{6a\,(2-2^{1/2})}{\pi^{1/2}} = 2.8245a$$

$$f_3(x) = \frac{a}{\pi(x^2+a^2)}, \qquad A_0[f_3] = \frac{24a\,\ln 2}{\pi} = 2.2997a$$

$$f_4(x) = \frac{\sin(x/a)}{\pi x}, \qquad A_0[f_4] = \frac{12a}{\pi} = 3.8197a$$

$$f_5(x) = \frac{a\,\sin^2(x/a)}{\pi x^2}, \qquad A_0[f_5] = \frac{6a}{\pi} = 1.9099a \ .$$

By choosing suitable values for a rescale these functions to make them of unit width and then plot them on the same scale.

4.02(ii) Starting with equation 2.08(12) apply the Backus-Gilbert theory to the problem of interpolation on the real line with exact data. Find the averaging length according to (17) for an interpolated point lying between two sample points x_j and x_{j+1}. As x varies on this interval, how does $p_{x_0}[x]$ behave? This is not the cubic spline interpolator; why not?

4.02(iii) Describe a method for constructing a magnetization model fitting the anomaly data with the property that m is precisely constant on the entire interval shown in figure 4.02a. We have asserted that the average of this solution taken through the family of resolving functions yields the highly variable model at the top of the figure; explain how this is possible.
Hint: Recall the model is defined on the whole real line.

4.03 Bounding Linear Functionals in Hilbert Space

The theory of resolution strongly suggests that only certain averages of the solution can be learned from the data, not the values at points in space. In any case, the traditional continuum mechanics or electrodynamics of geophysics are merely approximations obtained by *averaging* more fundamental laws of physics —on the atomic level the very idea of the value of the seismic velocity or of the electrical conductivity *at a point* is devoid of meaning. We shall now describe a theory for making estimates of model averages, or more generally of any bounded linear functional of the unknown parameter distribution. We return to a strict Hilbert space setting and first of all we treat the more elementary situation in which N error-free observations are known. In other words we are given the N numbers

$$d_j = (g_j, m), \quad j = 1, 2, 3, \cdots N \tag{1}$$

where $g_j \in H$ are linearly independent representers. From this information we wish to make a series of M predictions about the model m; they are the numbers associated with certain bounded linear functionals:

$$p_k = (\tilde{g}_k, m), \quad k = 1, 2, 3, \cdots M. \tag{2}$$

We shall refer to the elements $\tilde{g}_k \in H$ as *predictors*. In an appropriate space (but not in L_2) even the prediction of the model value at a point can be expressed like this, as we showed in section 2.07. If any of the predictors lie in the subspace G spanned by the representers, the corresponding prediction p_k is exactly computable from the data; this is just the Backus-Gilbert argument and leads to equations 4.02(6) and (6). What about all the other properties, not expressible as linear combinations of observations? If for some k, the predictor $\tilde{g}_k$ is not in G and nothing besides the values in (1) is known about m, then p_k *is completely unconstrained*. This result is implicit in the discussion of section 2.03—we can invent any value we like for p_k, which is after all simply another linear functional, and there is always a model m compatible with it and the demands of (1).

In the magnetic anomaly example, suppose we wish to ascertain the average value of magnetization in some interval (a, b); then working in L_2

$$p[m] = \int_a^b \frac{m(x)}{b - a} \, dx \tag{3}$$

$$= \int_{-\infty}^{\infty} B_w(x)\, m(x)\, dx \tag{4}$$

$$= (B_w,\, m)\,. \tag{5}$$

Here B_w is the boxcar function with support interval (a, b). Since it is easily shown that B_w, which is the predictor in this simple case, does not lie in G, it follows that the average value of magnetization in any finite interval is completely indeterminate.

To make any further progress it is necessary to introduce some kind of new information about the model. If we were to construct models with outrageous mean values but consistent with the anomaly values, we would find they have enormous oscillations and magnetization values. Our immediate reaction to these models would be that they are implausible, but to make mathematical use of this feeling it is necessary to have a quantitative statement about how large an acceptable model can be. For now let us assume it is possible to assign an upper bound on the *norm* of any reasonable model in the Hilbert space; we shall return in 4.04 to the question of how this might be done in practice. Thus in addition to (1) we shall assert that for acceptable models $m \in \mathrm{H}$

$$\|m\| \le \Xi_0 \tag{6}$$

where Ξ_0 is a positive number. This constraint immediately puts limits on the possible values bounded linear functionals can attain: by Schwarz's inequality we know from (2) and (6) that

$$|p_k| \le \Xi_0 \|\tilde{g}_k\| \tag{7}$$

without even having to make any use of the measurements.

We can equally well apply the inequality to the observations themselves; if (6) is to be consistent with (1) we must certainly have

$$|d_j| \le \Xi_0 \|g_j\|\,. \tag{8}$$

But (1) makes more severe demands of the observations than those of (8). In section 1.12 we discovered the element m of smallest norm consistent with a given finite collection of linear functionals like (1):

$$m_0 = \sum_{j=1}^{N} \alpha_j g_j \tag{9}$$

where the vector of coefficients $\alpha \in \mathbb{R}^N$ is given by

$$\alpha = \Gamma^{-1}\mathbf{d} \tag{10}$$

with $\mathbf{d} \in \mathbb{R}^N$ the vector of data values and of course Γ is the Gram matrix of representers. This norm-minimizing element was a mainstay of chapter 2. Since m_0 has the smallest possible norm of any element capable of satisfying (1), it follows that a necessary and sufficient condition for consistency of the bound (6) and the observations is

$$\|m_0\| \le \Xi_0 . \tag{11}$$

Squaring and substituting for m_0 we find

$$\|m_0\|^2 = (\sum_i \alpha_i g_i , \sum_j \alpha_j g_j) \tag{12}$$

$$= \sum_i \sum_j (g_i , g_j) \alpha_i \alpha_j \tag{13}$$

$$= \alpha \cdot \Gamma \alpha \tag{14}$$

$$= (\Gamma^{-1}\mathbf{d}) \cdot \Gamma (\Gamma^{-1}\mathbf{d}) \tag{15}$$

$$= \mathbf{d} \cdot \Gamma^{-1}\mathbf{d} . \tag{16}$$

Thus the strongest generalization of (8) is the following fundamental inequality

$$\mathbf{d} \cdot \Gamma^{-1}\mathbf{d} \le \Xi_0^2 . \tag{17}$$

This equation has a simple geometric interpretation in the Euclidean space $\mathbb{R}^N$: the region within which all acceptable data vectors must lie is the interior and surface of an N-dimensional ellipsoid. To show this notice that the expression on the left is a quadratic form, centered at the origin. Let us perform the spectral factorization of Γ (see 3.03). If

$$\Gamma = U \Lambda U^T \tag{18}$$

equation (17) becomes

$$\frac{y_1^2}{\lambda_1} + \frac{y_2^2}{\lambda_2} + \cdots + \frac{y_N^2}{\lambda_N} \le \Xi_0^2 \tag{19}$$

where $\mathbf{y}$ is a rotated version of $\mathbf{d}$:

$$\mathbf{y} = U^T \mathbf{d} . \tag{20}$$

Since Γ is positive definite all the eigenvalues λ_n are positive and (19) is clearly the equation of an ellipsoid in a rotated frame of reference.

The complete symmetry between the equations for the observations (1) and those for the predictions (2) suggests we can treat them in an entirely similar way. Suppose for the moment that we simply guess a set of values for p_k so now we have a collection of measurements d_j together with an additional M guesses for the properties. We can test consistency of this combined set of numbers as follows: form a vector $\mathbf{e}$ in $\mathbb{R}^{N+M}$:

$$\mathbf{e} = (d_1, d_2, \cdots d_N, p_1, p_2, \cdots p_M) \tag{21}$$

or equivalently

$$\mathbf{e} = \begin{bmatrix} \mathbf{d} \\ \mathbf{p} \end{bmatrix}. \tag{22}$$

Then, adopting the same ordering, define a family of $M+N$ "generalized" representers $\hat{g}_j \in H$:

$$\begin{aligned} &\hat{g}_1 = g_1, \quad \hat{g}_2 = g_2, \quad \cdots \quad \hat{g}_N = g_N \\ &\hat{g}_{N+1} = \tilde{g}_1, \ \hat{g}_{N+2} = \tilde{g}_2, \quad \cdots \quad \hat{g}_{N+M} = \tilde{g}_M. \end{aligned} \tag{23}$$

Then (1) and (2) can be combined into a uniform set of equations

$$e_i = (\hat{g}_i, m), \quad i = 1, 2, 3, \cdots M+N. \tag{24}$$

Evidently the theory used to derive (17) from (1) can be applied equally well to (24):

$$\mathbf{e} \cdot \hat{\Gamma}^{-1} \mathbf{e} \leq \Xi_0^2 \tag{25}$$

where $\hat{\Gamma}$ is the Gram matrix of inner products among the $\hat{g}_i$. Thus we see in the Euclidean product space of data and predictions, the bound (6) constrains acceptable vectors to an elliptical volume. For this analysis to hold we must have linear independence of the $\hat{g}_k$; it makes sense to insist that the predictions form a linearly independent subset of elements, and any of these that can be formed from linear combinations of the representers should be extracted and computed exactly as in 4.02(6).

We can now drop the pretense that values for p_k need to be guessed. Equation (25) defines the region of acceptable vectors consistent with the assigned bound on the norm of m even before observations are made. When those measurements are performed, numerical values are assigned via (1) and (21) to a subset

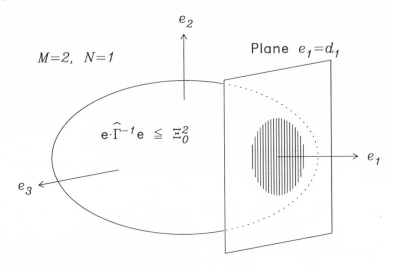

Figure 4.03a: Intersection of the region defined by the bound constraint (6), an ellipsoidal ball, and the single data constraint (1), a plane, gives the shaded elliptical zone. This zone gives the allowed set of predictions (2) consistent with both demands.

of the components of $\mathbf{e}$. Geometrically, this is the same as passing an M-dimensional hyperplane through $\mathbb{R}^{N+M}$; the set of points consistent with (1) and (25) is the intersection of these two regions. This is illustrated in a very low dimensional example by figure 4.03a. Here we imagine two predictions are to be made, that is, $M = 2$, after only one observation has been taken, so that $N = 1$. The single observation confines $\mathbf{e}$ to the plane shown, which intersects the ellipsoid in a shaded elliptical zone. We see that if Ξ_0 is chosen to be too small the plane will miss the ellipsoid entirely and then the assumed bound on m is simply inconsistent with observation. On the other hand, when Ξ_0 is very large, the size of the region within which the predictions are confined becomes correspondingly large; we see that as Ξ_0 tends to infinity, the values of predicted linear functionals become totally indeterminate, just as we described at the beginning of this section.

The algebraic statement of this pleasing geometrical picture is fairly concise. We partition the inverse of $\hat{\Gamma}^{-1}$ into four submatrices $G_{11} \in M(N \times N)$, $G_{12} \in M(N \times M)$, $G_{21} \in M(M \times N)$ and $G_{22} \in M(M \times M)$ in the obvious way

$$\hat{\Gamma}^{-1} = \begin{bmatrix} G_{11} & G_{12} \\ G_{21} & G_{22} \end{bmatrix}. \tag{26}$$

It is possible to express these submatrices as combinations of the original Gram matrix and the other inner products; see exercise 4.03(i). With this partition (25) can be rewritten

$$
\mathbf{e} \cdot \hat{\Gamma}^{-1} \mathbf{e} = \begin{bmatrix} \mathbf{d} \\ \mathbf{p} \end{bmatrix} \cdot \begin{bmatrix} G_{11} & G_{12} \\ G_{21} & G_{22} \end{bmatrix} \begin{bmatrix} \mathbf{d} \\ \mathbf{p} \end{bmatrix} \le \Xi_0^2 \tag{27}
$$

which may be expanded into

$$
\mathbf{p} \cdot G_{22}\mathbf{p} + 2\mathbf{p} \cdot G_{21}\mathbf{d} \le \Xi_0^2 - \mathbf{d} \cdot G_{11}\mathbf{d} . \tag{28}
$$

Here we have recognized the symmetry of $\hat{\Gamma}^{-1}$ which gives $G_{21}^T = G_{12}$. Since $\mathbf{d}$ is known, this inequality describes the region in which the desired prediction vector must lie. A little further massaging gives

$$
(\mathbf{p} - \mathbf{p}_0) \cdot G_{22}(\mathbf{p} - \mathbf{p}_0) \le \Xi_0^2 + \mathbf{p}_0 \cdot G_{22}\mathbf{p}_0 - \mathbf{d} \cdot G_{11}\mathbf{d} \tag{29}
$$

where

$$
\mathbf{p}_0 = -G_{22}^{-1} G_{21} \mathbf{d} . \tag{30}
$$

Equation (29) describes an ellipsoid centered at $\mathbf{p}_0$ in $\mathbb{R}^M$, provided that the real number on the right, which controls the size of the body, is positive; this is because G_{22} is always positive definite (why?). The condition is equivalent to the one that says the observations are consistent with the assumed bound; it follows that the right side of (29) must be

$$
\Xi_0^2 - \|m_0\|^2 \tag{31}
$$

and so

$$
\|m_0\|^2 = \mathbf{d} \cdot G_{11}\mathbf{d} - \mathbf{p}_0 \cdot G_{22}\mathbf{p}_0 . \tag{32}
$$

This can be shown directly (see Parker, 1977) but only after moderate algebraic effort.

The analysis can be extended in several ways. The first of these, to which we shall pay only cursory attention, is the counterpart of the seminorm-minimizing solution. It may be easier or more effective to place a bound on the size of a part of the model instead of on the whole thing. Let us express the model as a sum of two parts as we did in section 2.05:

$$
m = \sum_{k=1}^{K} \beta_k h_k + r \tag{33}
$$

where $h_k \in \mathrm{H}$ are special elements (for example, basis functions in an expansion) and we claim to know that

$$\| r \| \le \Xi_1 . \tag{34}$$

In 2.05 we described how to find the model with the smallest possible $\| r \|$ subject to exact agreement with measurement, that is, (1). Recall how we showed that the minimal length element, which we shall call r_*, must lie in G, the space spanned by the representers; then

$$r_* = \sum_{j=1}^{N} \alpha_j g_j \tag{35}$$

and expanding in the familiar way

$$\| r_* \|^2 = \boldsymbol{\alpha} \cdot \Gamma \boldsymbol{\alpha} . \tag{36}$$

We can now appeal to the equations of 2.05 for an explicit expression for $\boldsymbol{\alpha}$ in terms of the data values: we use 2.05(17) and 2.05(16) to arrive at

$$\| r_* \|^2 = \mathbf{d} \cdot H^T \Gamma^{-1} H \, \mathbf{d} \tag{37}$$

where

$$H = I - (A^T \Gamma^{-1} A)^{-1} A^T \Gamma^{-1} \tag{38}$$

and $I \in M(N \times N)$ is the unit matrix, while $A \in M(N \times K)$ is the matrix of inner products of g_j with h_k. Thus, since r_* has the smallest possible value consistent with (1), equation (34) leads to the following constraint system on possible measurements

$$\mathbf{d} \cdot H^T \Gamma^{-1} H \, \mathbf{d} \le \Xi_1^2 . \tag{39}$$

This can be extended to include predictions exactly as (17) was expanded into (25). We shall omit the details; Parker (1977) gives a fuller account. One significant difference arises: the region containing all the data and the predictions is not of finite volume in $\mathrm{I\!R}^{N+M}$—it is a cylinder, a degenerate ellipsoid with K infinitely long axes. We might expect something like this because by bounding $\| r \|$ we put no constraints on the coefficients β_k, which are themselves examples of linear functionals of m.

The most important extension is the treatment of uncertain data. Complete generalization becomes complicated so we confine ourselves to the simplest case, the prediction of a single linear functional. To obtain several averages we must find them one at a time, but this leads to an unnecessarily large volume in parameter space—we would obtain an enclosing rectangular box in the example of figure 4.03a not the elliptical region, which is clearly smaller. Before we engage in the serious computations we shall give a simple approximate treatment that sheds light on the general ideas. Instead of exactly matching the observations we permit misfit; we continue to use the form of 3.02(1), reproduced here as

$$\sum_{j=1}^{N} \left[\frac{d_j - (g_j, m)}{\sigma_j} \right]^2 \leq T^2 \,. \tag{40}$$

Our objective is to find the largest and smallest values of

$$p_1 = (\tilde{g}_1, m) \,. \tag{41}$$

Suppose that we use the Decomposition Theorem to split the predictor $\tilde{g}_1$ into a part in G, the representer subspace and $G^\perp$, its orthogonal complement; then with $\tilde{g}_\perp \in G^\perp$

$$\tilde{g}_1 = \tilde{g}_\perp + \sum_{j=1}^{N} \gamma_j g_j \tag{42}$$

where we can easily calculate the coefficients γ_j from the orthogonal projection operators given in 1.12. When we put this decomposition into (41) we have

$$p_1 = (\tilde{g}_\perp, m) + \sum_{j=1}^{N} \gamma_j (g_j, m) \,. \tag{43}$$

$$= (\tilde{g}_\perp, m) + s \,. \tag{44}$$

For a simple estimate of the range of the prediction we can bound the first term and the sum separately. The inner products in the sum s are just the same as those in (40); it is a simple exercise to show that subject to (40)

$$|s - \sum_j \gamma_j d_j| \leq T [\sum_j \gamma_j^2 \sigma_j^2]^{1/2} \,. \tag{45}$$

We can use Schwarz's inequality on the first term:

$$|(\tilde{g}_\perp, m)| \leq \Xi_0 \, \|\tilde{g}_\perp \| \,. \tag{46}$$

In this way we see that the ability of the representers to make a good approximation to the predictor, hence leaving only a small remainder $\tilde{g}_\perp$, is crucial in keeping the size of the interval down. This is the quantitative interpretation of Backus-Gilbert theory in the one case where $\tilde{g}_1$ is a δ function. We can usually expect the

Schwarz bound to be a very crude bound because a large part of m must lie in G in order to fit the data and hence $\tilde{g}_\perp$ will make a large angle with m and not be parallel with it. Much use is made of these geometrical considerations in Backus's development (1970a) for the exact data problem.

Our final task of this section is to improve these crude bounds on p_1. We return to the theory of section 3.02 for guidance; suppose we seek the smallest solution in the norm consistent with (40). We found in 3.02 that the norm minimizer always lies in the subspace G spanned by the representers. Next we include as an artificial datum p_1, the value to be predicted by (2), and we associate with it a tiny uncertainty. Clearly the smallest model compatible with these $N+1$ "data" is to be found in G_1, the union of G and the one-dimensional subspace containing $\tilde{g}_1$, and this is true whatever value is assigned to p_1. The largest allowable value of p_1, arguing as we did in the case of error-free observations, arises when the minimum norm exactly matches our bound, Ξ_0. Thus we can confine the search for a model achieving the upper (or lower) bound for p_1 to the subspace G_1, a space of dimension only $N+1$ (we assume of course that the predictor does not lie in G). So we can write

$$m = \sum_{j=1}^{N} \alpha_j g_j + \beta_1 \tilde{g}_1 \tag{47}$$

and ask for the model in this form that maximizes (or minimizes) $(\tilde{g}_1, m)$ subject to (6) and (40). For the sake of definiteness we shall consider maximizing p_1 and we assume that the upper limit on the prediction is positive. Substituting (47) into these equations we obtain the following optimization problem, in vector notation:

$$\max_{\beta \in E^{N+1}} \mathbf{u} \cdot \hat{\Gamma} \beta \tag{48}$$

subject to

$$\| S(\mathbf{e} - \hat{\Gamma}\beta) \| \leq T \tag{49}$$

and

$$\beta \cdot \hat{\Gamma}\beta \leq \Xi_0^2 . \tag{50}$$

where the norm is the Euclidean length. The notation is consistent with the earlier one in the analysis of error-free data: $\mathbf{e} \in E^{N+1}$ is defined as in (21) except that the last entry is zero—

we do not know the value of p_1 and it does not enter the definition of W; $\hat{\Gamma}$ is the augmented Gram matrix of inner products among g_j and $\tilde{g}_1$; the coefficient vector $\beta \in E^{N+1}$ is

$$\beta = (\alpha_1, \alpha_2, \alpha_3, \cdots \alpha_N, \beta_1) \tag{51}$$

the constant vector $\mathbf{u} \in E^{N+1}$ is

$$\mathbf{u} = (0, \cdots \ 0, \ 0, \ 0, \ 1) \tag{52}$$

and finally $S \in M(N{+}1{\times}N{+}1)$ is a projection matrix to insure that the predicted value of p_1 does not influence the misfit:

$$S = \begin{bmatrix} & & & & 0 \\ & \Sigma^{-1} & & & 0 \\ & & & & \vdots \\ 0 & 0 & 0 & \cdots & 0 \end{bmatrix}. \tag{53}$$

Let us assume for now that the inequalities (49) and (50) can be replaced by equalities and introduce the two Lagrange multipliers in the usual way: define the unconstrained functional to be made stationary over $\beta \in E^{N+1}$

$$W[\beta, \nu, \mu] = \mathbf{u} \cdot \hat{\Gamma}\beta - \nu(\|S(\mathbf{e} - \hat{\Gamma}\beta)\|^2 - T^2)$$

$$- \mu(\beta \cdot \hat{\Gamma}\beta - \Xi_0^2). \tag{54}$$

Both Lagrange multipliers μ and ν must be positive. To show this recall that we mentioned in 1.14 that the Lagrange multiplier is the derivative of the objective functional as the corresponding constraint varies. We can argue that the largest consistent value of p_1 can only increase with increasing norm or misfit. For a fixed pair μ and ν the functional W is stationary at β_0 found by taking the gradient of (54); after minor rearrangement we find this linear system for the vector

$$((\mu/\nu)I + S^2\hat{\Gamma})\beta_0 = \mathbf{u}/2\nu + S^2\mathbf{e}. \tag{55}$$

We can easily solve the system, and so our job is now to discover the Lagrange multiplier pair that gives the correct misfit T and norm Ξ_0. To reduce the numerical effort we systematically vary the ratio μ/ν: for some value of this ratio, ρ say, we solve the system and compute the corresponding value of $\|m\|^2$:

$$\|m\|^2 = \beta_0 \cdot \hat{\Gamma}\beta_0 \tag{56}$$

$$= (\nu^{-1}\mathbf{a} + \mathbf{b}) \cdot \hat{\Gamma}(\nu^{-1}\mathbf{a} + \mathbf{b}) \tag{57}$$

where **a** and **b** are the known vectors

$$\mathbf{a} = (\rho I + S^2 \hat{\Gamma})^{-1} \mathbf{u}/2$$
$$\mathbf{b} = (\rho I + S^2 \hat{\Gamma})^{-1} S^2 \mathbf{e} . \tag{58}$$

We select the Lagrange multiplier ν_0 by the demand that the norm match the specified bound, Ξ_0:

$$(\nu_0^{-1} \mathbf{a} + \mathbf{b}) \cdot \hat{\Gamma} (\nu_0^{-1} \mathbf{a} + \mathbf{b}) = \Xi_0^2 \tag{59}$$

which reduces to the simple quadratic equation

$$\nu_0^2 (\mathbf{b} \cdot \hat{\Gamma} \mathbf{b} - \Xi_0^2) + 2\nu_0 (\mathbf{a} \cdot \hat{\Gamma} \mathbf{b}) + \mathbf{a} \cdot \hat{\Gamma} \mathbf{a} = 0 . \tag{60}$$

On the assumption that one of the roots is real and positive, we find the corresponding β_0 from

$$\beta_0 = \nu_0^{-1} \mathbf{a} + \mathbf{b} \tag{61}$$

and hence the corresponding misfit

$$\| S (\mathbf{e} - \hat{\Gamma} \beta_0) \| . \tag{62}$$

We vary ρ over positive values, seeking the value that generates the desired misfit T with the largest p_1.

The simplest way to show there are no local maxima or minima which would lead to spurious solutions is by appealing to the notion of convexity, an idea we have tried to do without so far. We wish to appeal to the theorem concerning the minimization of convex functionals given in 1.15. Working in the finite-dimensional space E^{N+1}, when we *maximize p_1* we are *minimizing $-p_1$* or $-\mathbf{u} \cdot \hat{\Gamma} \beta$. A linear functional is convex. The minimization must be performed over elements β in the intersection of the two sets $\| S (\mathbf{e} - \hat{\Gamma} \beta) \|^2 \leq T^2$ and $\beta \cdot \hat{\Gamma} \beta \leq \Xi_0^2$. Each of these sets is convex because it is associated with a norm or seminorm. The intersection of two convex sets is always convex and so the constraint set inside which β lies is also. Thus the conditions of the theorem in section 1.15 are obeyed. This means a local minimum of $-p_1$, discovered by calculus and Lagrange multipliers, is a true, global minimum of the functional. The functional $-\mathbf{u} \cdot \hat{\Gamma} \beta$ is linear and therefore has no extrema in the interior of the constraint set, justifying the use of equalities in (49) and (50) when we set up the system of Lagrange multipliers for (54). There is an exception to the condition of equality in (49). The usual situation is that the upper and lower bounds arise at points on the boundary of the intersection of the misfit and norm constraints, but it may happen that the data uncertainty is so large that only the norm constraint limits the value of the functional. Then equality of the misfit constraint is not required and (49) can be dropped when solving (48). The solution to this much simpler optimization problem is, as the reader may easily verify,

$$\beta_0 = (0, \; \cdots \; 0, \; 0, \; 0, \; \Xi_0/\|\tilde{g}_1\|) \,. \tag{63}$$

If this vector can satisfy observation through (49), it generates the optimal solution and the upper bound on p_1 is $\Xi_0 \|\tilde{g}_1\|$. When this happens there is then no need to solve the more complex system with both constraints.

Another loose end concerns the other possibilities for p_1; for example, minimizing or maximizing it when the largest value is negative and so on. We just state the result, which is easily derived: μ and ν are always both positive or both negative. This means that that search is always over positive ρ and the each of the two roots of (60) corresponds to a bound on the functional (but for different misfits; see figure 4.04b). To obtain the lower bound in the degenerate case simply reverse the sign of β_0 in (63).

The search process over ρ is computationally expensive if the number of observations is moderately large (more than 100, say) and it will then be worthwhile to make the solution of (58) as efficient as possible. We could introduce Newton's method to solve for the value of ρ at which misfit and norm exactly meet the desired goals but we should also recognize that several different functionals may be under investigation. The best way to improve computational efficiency in the face of this variety seems to be by spectral factorization of Γ. The details are left to the reader in exercise 4.03(iii). We are ready now to try out this promising new tool for making proper inferences about Earth.

Exercises

4.03(i) Partition the augmented Gram matrix $\hat{\Gamma}$ as follows:

$$\hat{\Gamma} = \begin{bmatrix} \Gamma & J^T \\ J & \tilde{\Gamma} \end{bmatrix}$$

where Γ is of course the Gram matrix of representers and the entries of the other submatrices are

$$J_{kl} = (\tilde{g}_k, g_l), \;\; k = 1, 2, \; \cdots \; M; l = 1, 2, \; \cdots \; N$$

$$\tilde{\Gamma}_{jk} = (\tilde{g}_j, \tilde{g}_k), \;\; j = 1, 2, \; \cdots \; M; k = 1, 2, \; \cdots \; M \,.$$

Show that the submatrices in the partitioned inverse $\hat{\Gamma}^{-1}$ are given by

$$G_{11} = (\Gamma - J^T \tilde{\Gamma}^{-1} J)^{-1}$$

$$G_{22} = (\tilde{\Gamma} - J \, \Gamma^{-1} J^T)^{-1}$$

$$G_{12} = -G_{11} J^T \tilde{\Gamma}^{-1} = G_{21}^T$$

$$G_{21} = -G_{22} J \, \Gamma^{-1} = G_{12}^T \, .$$

Hint: Solve the 2×2 linear system of submatrices by elimination.

4.03(ii) Compare the theory for exact and uncertain observations. First, plot a diagram like figure 4.03a for the case, $N = 1$ and $M = 2$. How does the picture change when the data misfit criterion is (40) instead of (1)? Next, show that the results of the theory for bounding a single linear functional with exact data are identical to those of the theory allowing data uncertainty in the limit as the data errors tend to zero. Hint: Use the expressions obtained in 4.03(i).

4.03(iii) Because of the need to compute the vectors **a** and **b** for a range of positive ρ and the possibility that more than one $\tilde{g}_1$ might be examined for a given problem, it is important to reduce the numerical work in evaluating (58); as it stands $O(N^3)$ arithmetic operations are required for every new ρ. Perform spectral factorization of $\Sigma^{-1} \Gamma \Sigma^{-1}$ and convert (1) into the form of 3.03(33). Show $\hat{\Gamma}$ is transformed into a matrix that is diagonal except for its last column and bottom row; solve the linear systems (58) in $O(N)$ operations for such a matrix.

4.03(iv) Suppose that instead of a bound on the model norm, the actual value of the norm were known exactly. Would this mean that the inequality in (25) could be replaced by an equality? If not, is there any advantage in knowing the norm?

4.04 Bounding Functionals in Practice

The reader is surely waiting for an illustration of the theory described in the last section. If past experience is our guide, it would be natural to expect that the magnetic anomaly example or the mantle dissipation problem should receive first attention. Those eagerly anticipating the reworking of these familiar examples will, unfortunately, be disappointed. It is instructive to understand why we are unable to exploit our favorite expository vehicles; the difficulties encountered are not uncommon. Beginning with the magnetization, we are required to estimate an upper bound upon some Hilbert-space norm of the model. When we are penalizing oscillations of the solution we find it convenient to work in $W_2^1(-\infty, \infty)$ where the norm is

$$\| m \| = [\int_{-\infty}^{\infty} \left[\frac{dm}{dx} \right]^2 dx \,]^{\frac{1}{2}} . \tag{1}$$

(As described in section 2.06, the whole set of constant magnetization functions is equivalent to the zero element of the space, so that this really is a norm, despite appearances to the contrary.) Dredging and drilling give marine geologists a fairly complete knowledge of the petrology and magnetic properties of the rocks in the top ten meters or so of the young sea floor and there is a reasonable basis for speculation down to a kilometer or more, probably right through the magnetized zone of the oceanic crust. This kind of information leaves considerable doubt that the integral in (1) exists! On a fine scale discontinuities of rock magnetization arise where lava flows of different ages make contact; on a gross scale, reversals of the polarity of the magnetic dipole cause rapid transitions in the magnetization function. If hard pressed, perhaps one would be willing to come up with a minimum scale for a magnetic transition (a few millimeters for individual lava flows) and a maximum jump in intensity to be used to limit the magnitude of the derivative dm/dx. Then a second problem is encountered—the interval of definition of the function is whole real line. Even if $|dm/dx|$ could be bounded by appeal to laboratory measurements of rock samples, that does not bound $\|m\|$ in (1) because a finite $|dm/dx|$ at every point on $(-\infty, \infty)$ must still lead to a divergent integral unless the value decays to zero for x large. We used a Sobolev norm deliberately to build smooth solutions, ones that avoid the propensity for spurious oscillations and jumps. Since derivatives, even low-order ones, may not exist or be difficult to estimate from empirical data, norms involving derivatives are usually unsuitable for the theory of the previous section; we may expect spaces with less stringent smoothness requirements might be better.

If we decide to return to L_2 to avoid the difficulties surrounding estimation of gradients of magnetization, the problem of how to bound the norm of the function of the whole real line remains. There is no natural scale on which the function m might be supposed to tend to zero other than the width of the whole ocean; it scarcely needs to be said that bounds calculated from this scale are nugatory—the constraints they provide are utterly worthless. It seems impossible in the marine magnetic example to come up with a Hilbert-space norm which can be expected to provide useful restrictions based upon empirical knowledge of ocean-bottom rocks. The picture might be different if we could be certain the magnetic source material was confined to a relatively small volume, for example a seamount. In fact, even in this case results of the theory have not been encouraging either; see Parker et al. (1987). Alternatively, a norm that limited the magnetic intensity *at every point* might be very effective, but such norms are not Hilbert-space norms; we come to them in section 4.05.

Rather different considerations hinder the application of the Hilbert-space theory for the dissipation problem: unlike the sea floor, the mantle of the Earth is completely inaccessible to sampling and so there is very little direct evidence concerning the way in which mantle material dissipates seismic energy. Our present knowledge is based almost entirely upon models constructed to satisfy attenuation data. It would be logically circular to use those models to arrive at an upper bound for any norm of the function q. For example, there is no fundamental physical reason why the materials of the upper mantle might not be extremely lossy, leading to very large values of q there and hence large $\|q\|$. Something that does rest on fundamental physics, namely the Second Law of Thermodynamics, is the fact that q cannot be negative. This information, rather than a bound on a norm, offers the greatest promise for learning about the dissipation from attenuation observations. Positivity is not a property that fits comfortably in a simple Hilbert-space setting, and so for an application of the theory of the previous section we must look elsewhere; see section 4.07.

Let us revive the global interpolation problem of section 2.07. In our earlier treatment we made no use of the fact that the interpolated function was a magnetic field—it could just as well have been barometric pressure or annual mean temperature. Now we shall put some more physics into our model, which will enable us to come up with a plausible quantity to identify with a norm. As covered in every good introductory text (for example, Stacey, 1977), the main geomagnetic field of the Earth varies so slowly that the displacement current in Maxwell's equations can be dropped, and in a region $V \subset \mathbb{R}^3$ where the flow of electric currents is negligible, the magnetic field can be derived from a scalar potential:

$$\mathbf{B}(\mathbf{r}) = -\nabla \Omega, \quad \mathbf{r} \in V . \tag{2}$$

If, in addition V contains no magnetized materials, the potential is harmonic, that is, it obeys Laplace's equation

$$\nabla^2 \Omega = 0, \quad \text{in } V . \tag{3}$$

These are excellent approximations in the atmosphere below the ionosphere, but it is commonly assumed that we may continue to use this representation right through the mantle to the surface of the core, accounting for the fields associated with magnetization and induced currents by a small error term in the observed field. In other words, we split the observed field into two parts: the main one is the geomagnetic field produced by strong electric currents in the core of the Earth, and making their way to the surface under the control of (2) and (3); the other, much smaller part is due to crustal magnetization (like the fields of the marine magnetic example), currents flowing in the mantle, the iono-sphere, and magnetosphere. We shall regard the main field as our signal, and the remainder as noise, whose magnitude we shall estimate later. Thus V in (2) and (3) is the region of space out-side the core; then, because the sources of $\mathbf{B}$ are within the core, Laplace's equation can be separated in spherical polar coordinates so that Ω can be represented as a spherical harmonic expansion:

$$\Omega(r, \theta, \phi) = a \sum_{l=1}^{\infty} \sum_{m=-l}^{l} b_l^m \left[\frac{a}{r} \right]^{l+1} Y_l^m(\theta, \phi) \tag{4}$$

where r, θ, and ϕ are the spherical polar coordinates of a point in V centered to the center of the Earth, a is traditionally the Earth's mean radius, Y_l^m are the fully normalized, complex spheri-cal harmonic functions familiar from section 2.07, and b_l^m are complex coefficients that completely characterize the potential Ω and through (2), the magnetic field $\mathbf{B}$. Notice the sum over l begins at $l = 1$; the $l = 0$ term represents the part of the field due to magnetic monopoles, which may not exist at all, or if they do, are known to be exceedingly rare.

In 2.07 our observations were values of the so-called Z com-ponent of the magnetic field, the one measured vertically down-wards; to an adequate approximation this is $-\hat{\mathbf{r}} \cdot \mathbf{B} = -B_r$, observed on the sphere $r = a$. Assuming (4) is uniformly convergent for all $\mathbf{r} \in V$ so that we can rearrange the sum, we substitute (4) into (2) and exhibit the radial part:

$$B_r(\mathbf{r}) = -\frac{\partial \Omega}{\partial r} \tag{5}$$

$$= \sum_{l=1}^{\infty} \sum_{m=-l}^{l} (l+1) b_l^m \left[\frac{a}{r}\right]^{l+2} Y_l^m(\theta, \phi) . \tag{6}$$

For all the observations $|\mathbf{r}| = a$ so using this fact and reversing the sign of B_r we find the observed jth datum:

$$d_j = -\sum_{l=1}^{\infty} \sum_{m=-l}^{l} (l+1) b_l^m Y_l^m(\theta_j, \phi_j) \tag{7}$$

which closely resembles the expansion used in 2.07.

Just as before we choose an ordered sequence of coefficients b_l^m to be the model element in a linear vector space. The space must be equipped with a norm of a Hilbert type, one whose magnitude we might hope to bound. The density of electromagnetic energy in a magnetic field in a nonpolarizable medium like the atmosphere is given by $|\mathbf{B}|^2/2\mu_0$. By integration the total amount of energy stored in the magnetic field outside the core is

$$E = \frac{1}{2\mu_0} \int_V \mathbf{B} \cdot \mathbf{B} \, d^3\mathbf{r} \tag{8}$$

$$= \frac{1}{2\mu_0} \int_V \nabla\Omega \cdot \nabla\Omega \, d^3\mathbf{r} \tag{9}$$

$$= \frac{1}{2\mu_0} \int_V [\nabla \cdot (\Omega \nabla\Omega) - \nabla^2\Omega] \, d^3\mathbf{r} . \tag{10}$$

Here we have used the well-known vector identity for $\nabla \cdot (f \nabla g)$. Now using (3) and applying Gauss's Theorem we have

$$E = -\frac{1}{2\mu_0} \int_{\partial V} \Omega \frac{\partial \Omega}{\partial r} \, d^2\mathbf{r} \tag{11}$$

where ∂V is the surface of the core, which is taken to be a sphere of radius c. We have also needed the fact that there are no field sources outside the core, which guarantees that $\partial\Omega/\partial r$ tends to zero at great distances. Since Ω is real we may replace $\partial\Omega/\partial r$ with its complex conjugate; then introduction of (4) and (6) gives:

$$E = \frac{1}{2\mu_0} \int_{S^2} \sum_{l,m} b_l^m \left[\frac{a}{c}\right]^{l+1} Y_l^m(\theta, \phi) \times$$

$$\sum_{n,k} (n+1) (b_n^k)^* \left[\frac{a}{c}\right]^{n+2} Y_n^k(\theta, \phi)^* c^2 d^2\hat{\mathbf{r}} \tag{12}$$

$$= \frac{a^4}{2\mu_0 c} \sum_{l=1}^{\infty} \sum_{m=-l}^{l} (l + 1) \left[\frac{a}{c}\right]^{2l} |b_l^m|^2 . \tag{13}$$

The massive simplification results from the orthogonality of the spherical harmonic functions.

Equation (13) shows the electromagnetic energy of the field outside the core has a simple expression in terms of the expansion coefficients b_l^m, one suitable as a basis for a norm on the model space. Why should we prefer (13) to 2.07(35) which we used in the interpolation study? Because (13) represents the energy of a system, we know from profound physical principles that it must be finite. On the other hand, there is nothing to say the L_2-norm of $\nabla_s^2 B_r$ must exist for a magnetic field. The flow of energy from one place to another or from one form to another are the concerns of the First and Second Laws of thermodynamics and these can place strong constraints on the magnitudes of energies or energy densities in natural systems. In fact, from this perspective there is a better norm for our problem than one based on (13): it is related to the heat generated by electric current flow in the core, but its complicated nature makes it an inappropriate illustration for us; for further details see Shure et al. (1982).

With these considerations in mind we define the normed vector space E_2 with elements comprised of infinite sequences of complex numbers, doubly indexed like b_l^m which, as an element of E_2, is written

$$b = (b_1^{-1}, b_1^0, b_1^1, b_2^{-2}, b_2^{-1}, b_2^0, b_2^1, b_2^2, \cdots) . \tag{14}$$

The norm for the space is

$$\|b\| = \left[\sum_{l=1}^{\infty} \sum_{m=-l}^{l} (l + 1) \left[\frac{a}{c}\right]^{2l} |b_l^m|^2 \right]^{\frac{1}{2}} \tag{15}$$

$$(f, g) = \sum_{l=1}^{\infty} \sum_{m=-l}^{l} (l + 1) \left[\frac{a}{c}\right]^{2l} f_l^m (g_l^m)^* . \tag{16}$$

The inner product has been formed in the usual way; with it E_2 is a Hilbert space. The energy of a magnetic field model associated with $b \in E_2$ is

$$E = \frac{a^4 \|b\|^2}{2\mu_0 c} . \tag{17}$$

From (7) we may write the linear functional for an observation as an inner product (16)

$$d_j = (b, y_j) \tag{18}$$

where

$$y_j = ((c/a)^2 (Y_1^{-1})^*, \ (c/a)^2 (Y_1^0)^*, \ (c/a)^2 (Y_1^1)^*,$$
$$\cdots \ (c/a)^{2l} (Y_l^m)^*, \ \cdots \ ) \tag{19}$$

and we have suppressed the argument θ_j, ϕ_j. Because $c/a = 0.547$ it is easily verified that $\|y_j\| < \infty$, so that $y \in E_2$. We shall of course need the Gram matrix of these representers. By its definition and that of the inner product (16)

$$\Gamma_{jk} = (y_j, y_k) \tag{20}$$

$$= \sum_{l=1}^{\infty} \sum_{m=-l}^{l} (l+1) \left[\frac{c}{a} \right]^{2l} Y_l^m(\theta_j, \phi_j)^* Y_l^m(\theta_k, \phi_k) \tag{21}$$

$$= \frac{1}{4\pi} \sum_{l=1}^{\infty} (2l+1)(l+1)(c/a)^{2l} P_l(\hat{\mathbf{r}}_j \cdot \hat{\mathbf{r}}_k) \tag{22}$$

where we have applied the Addition Theorem for spherical harmonics to collapse the sum over m. To reduce this sum further we require the generating function for Legendre polynomials, a relationship we omitted to mention in 2.07; this is

$$\sum_{l=0}^{\infty} \xi^l P_l(\mu) = \frac{1}{R(\xi, \mu)} = \frac{1}{(1 + \xi^2 - 2\xi\mu)^{\frac{1}{2}}} . \tag{23}$$

Differentiating the sum above with respect to ξ twice and performing a little rearrangement gives

$$\sum_{l=0}^{\infty} (2l+1)(l+1)\xi^l P_l(\mu) = \left[2\xi^2 \frac{\partial^2}{\partial \xi^2} + 5\xi \frac{\partial}{\partial \xi} + 1 \right] \frac{1}{R} \tag{24}$$

$$= \frac{3(1 - \xi^2)^2}{2R^5} - \frac{1 + 3\xi^2}{2R^3} . \tag{25}$$

This provides an elementary expression for evaluating the Gram matrix; but we must remember to subtract one to remove the $l = 0$ term when substituting into (22).

We are nearly in a position to apply our theory for bounding linear functionals. Two more things are needed: first we require a definite number for Ξ_0, the upper bound on the norm; second, we must decide what linear functionals to estimate. The magnetic field of the year 1900 is poorly known compared with today's. Since the late 1960s artificial satellites have made possible global coverage of the main field and we shall assume that the spherical harmonic models based on ground stations and these measurements are reasonably reliable, adequate to give us a figure for E the electromagnetic energy of the core-generated field in V. Without a doubt the best-determined field models are those derived from the NASA Magsat mission; Magsat was a satellite with a vector magnetometer operating during 1980 for the sole purpose of measuring the Earth's magnetic field. Three main-field models were consulted: IGRF1980, the International Geomagnetic Reference Field; GSFC980, a Goddard Space Flight Center model; and PHS80, the Preliminary Harmonic Spline model, for 1980. These gave values for E of 7.1×10^{18} J, 6.8×10^{18} J, and 6.6×10^{18} J, respectively. Furthermore, the value of E is found to be decreasing by about 0.10 percent in models that include rates of change. Because the electromagnetic energy is unlikely to change drastically in as short a time as eighty years (main-field morphology has an apparent time scale of several thousand years) we may postulate 10^{19} J as a sensible upper bound on E in 1900; the corresponding bound on the norm is $231\,\mu T$. The careful (some would say pedantic) reader could correctly claim that the accuracy of even the modern models is quite unknown; they are just models fitting the observations (two of them found by collocation!) without defensible uncertainty bounds. Most geophysicists would be willing to accept the modern estimates for E, and our inability to give sound support for this belief is not atypical of the situation regarding models of the deep interior of the Earth. A case can be made for a bound on the norm based upon interior heat production by electric current flow; the reader is referred to the paper by Backus (1989).

Rather than treating an interpolation problem, let us attempt to estimate linear functionals of more general significance, the $l = 1$ or dipole components of the main field, which is an important inverse problem of geomagnetism. Traditionally, spherical harmonic expansions are not written fully normalized; instead the so-called *Gauss coefficients* are normally listed. We predict these more familiar numbers; their relationship to the fully normalized set for the dipole is

$$p_1 = g_1^0 = \quad (3/4\pi)^{\frac{1}{2}}\, b_1^0 \tag{26}$$

$$p_2 = g_1^1 = -(3/8\pi)^{\frac{1}{2}}[\, b_1^1 - b_1^{-1}\,] \tag{27}$$

$$p_3 = h_1^1 = -i(3/8\pi)^{\frac{1}{2}}[\, b_1^1 + b_1^{-1}\,]. \tag{28}$$

The predictors $\tilde{g}_k \in E_2$ for these values are

$$\tilde{g}_1 = \frac{c^2(3/4\pi)^{\frac{1}{2}}}{2a^2}(0,\ 1,\ 0,\ 0,\ 0,\ 0,\ 0,\ \cdots\ ) \tag{29}$$

$$\tilde{g}_2 = \frac{c^2(3/8\pi)^{\frac{1}{2}}}{2a^2}(1,\ 0,\ -1,\ 0,\ 0,\ 0,\ 0,\ \cdots\ ) \tag{30}$$

$$\tilde{g}_3 = \frac{c^2(3/8\pi)^{\frac{1}{2}}}{2a^2}(-i,\ 0,\ -i,\ 0,\ 0,\ 0,\ 0,\ \cdots\ )\ . \tag{31}$$

And finally, inserting these elements and (11) into (16)

$$(\tilde{g}_j, \tilde{g}_k) = \begin{cases} \dfrac{3c^2}{8\pi a^2}, & j = k \\[2ex] 0, & j \neq k \end{cases} \tag{32}$$

$$(\tilde{g}_1, y_j) = \frac{3c^2}{4\pi a^2}\cos\theta_j \tag{33}$$

$$(\tilde{g}_2, y_j) = \frac{3c^2}{4\pi a^2}\sin\theta_j\ \cos\phi_j \tag{34}$$

$$(\tilde{g}_3, y_j) = \frac{3c^2}{4\pi a^2}\sin\theta_j\ \sin\phi_j\ . \tag{35}$$

Notice how the predictors are mutually orthogonal. Now all the entries of the generalized Gram matrix $\hat{\Gamma}$ have been calculated.

We shall see that the theory for noise-free data cannot be applied to the 1900 data set, so let us first concoct an artificial example to explore how well the method might be expected to perform under almost ideal circumstances. Suppose that the values of B_r are known at a network of one hundred points roughly uniformly distributed on the sphere. (It is impossible to cover a sphere with this many points in a perfectly uniform manner, so a random spatial assignment was used.) The artificial field, roughly resembling today's geomagnetic field, is that from an axial dipole and has radial component $-60\cos\theta\ \mu T$; our task is to see how well we can recover the dipole structure. The actual values of the linear functionals to be predicted are $p_1 = -30\,\mu T$ and $p_2 = p_3 = 0$. For this field $\|b\| = 158.7425\,\mu T$ and, by (17), $E = 4.7 \times 10^{18}\,J$. The augmented Gram matrix $\hat{\Gamma}$ is computed as we have explained, then inverted and partitioned. Equation 4.03(29) describes an ellipsoid in $\mathbb{R}^3$; we reproduce the equation as follows:

$$(\mathbf{p} - \mathbf{p}_0) \cdot G_{22}(\mathbf{p} - \mathbf{p}_0) \leq \Xi_0^2 - \|b_0\|^2\ . \tag{36}$$

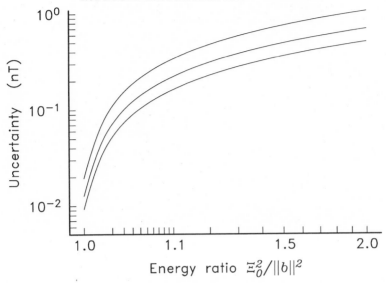

Figure 4.04a: Lengths of the principal axes of the uncertainty ellipsoid for the three predictions p_1, p_2, and p_3. The quantity $\|\Xi_0\|^2/\|b\|^2$ is the ratio of the bound on the electromagnetic energy of the field to the actual energy. The horizontal axis has been stretched to clarify the very rapid variation near unity.

For the artificial exact data we calculate

$$\mathbf{p}_0 = (0.0031, 0.0028, -29.989)\ \mu\mathrm{T} \tag{37}$$

and

$$G_{22} = \begin{bmatrix} 55491 & -5421 & 15563 \\ -5421 & 23418 & 4402 \\ 15563 & 4402 & 88836 \end{bmatrix}. \tag{38}$$

Notice how remarkably close the center $\mathbf{p}_0$ is to the exact point. The center and orientation of the ellipsoid is always the same but the size of the uncertainty region depends of course on our choice of Ξ_0, the bound on the energy norm. We find that $\|b_0\|$, the smallest possible norm for any exactly fitting model, is $158.7166\ \mu\mathrm{T}$, only slightly smaller than $\|b\|$, the true norm given earlier. To generate figure 4.04a we vary the value of the bound and plot the lengths of the principal axes of the body, found by spectral factorization of G_{22} just as Γ was factored in 4.03(19). The orientation of the axes is fortuitously close to the original axes of the prediction subspace: the shortest principal axis,

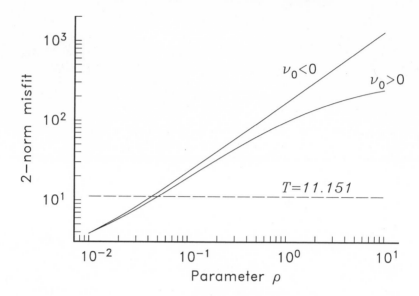

Figure 4.04b: Variation of misfit to observation as the parameter ρ is varied over positive real values and the bound on the norm is fixed at $\Xi_0 = 1.1 \|b\|$. The two curves correspond to the two roots of the quadratic equation 4.03(60); the upper curve gives the lower bound, and conversely. The dashed line is the desired misfit $T = 11.151$.

corresponding to the least uncertainty, makes an angle of 21° with the $p_1 = g_1^0$ axis and thus may be thought of as roughly aligned with it; the longest is only 14° away from the $p_2 = g_1^1$ direction. If by some miracle we knew a precise value of the energy, so that we could set $\Xi_0 = \|b\|$, we see from the figure that all the uncertainties in the estimated dipole coefficients are less than 0.02 μT. (When the model norm is assumed to be exactly known we are in fact solving a nonlinear inverse problem in which the value of the norm is an additional datum, along with the hundred other field values.) It is more realistic to assume that we could bound the energy to within 50 percent of its actual value. In figure 4.04a this corresponds to $\Xi_0^2/\|b\|^2 = 1.5$; we see for example, that now the axis roughly aligned with p_1 measures 0.36 μT; in other words the width of the interval containing p_1 is about 0.72 μT, while p_1 is about −30 μT. Although this represents a considerable increase in uncertainty the precision is still quite respectable.

Next we ascertain how sensitive these bounds are to the presence of noise in the data. As we have already discussed, the fields measured at an observatory may be divided into two parts: those generated by fluid motions in the core and everything else, such as fields caused by crustal magnetization or ionospheric electric currents. Fields not directly generated in the core will be regarded as noise for our purposes. Crustal magnetization always causes the largest of these and we shall assign such noise an RMS magnitude of 100 nT (or 0.1 μT), which represents a reasonable signal from crustal magnetization. Thus the field data in the artificial example are supposed to be known to a few parts in a thousand. Because of their height about the Earth's surface, the magnetometers in artificial satellites are far less susceptible to this source of noise; we can expect those measurements to have a standard error ten times smaller than that of ground stations. When we attempt to discover the extremes between which a prescribed functional must fall while satisfying the data, we should choose a more generous value for the tolerated misfit than the one we used when we constructed smooth models. This is because we want to be reasonably sure we have captured the true data vector in the allowed volume of data space. In the present example, where $N = 100$, I selected a value of T giving a confidence level of 95 percent; from the standard χ^2 tables $T^2 = 124.34$, or $T = 11.151$. We shall single out the axial dipole coefficient p_1 for study. In figure 4.04b we see how the misfit to the data varies as ρ is swept through positive values for a fixed value of the norm bound, $\Xi_0 = 1.1 \, \|b\|$. The two curves correspond to the two roots of the quadratic equation 4.03(60); the upper branch represents to the lower bound on p_1, the lower branch to the upper bound. Our principal interest is in the values of ρ where the misfit meets the tolerance level T. Figure 4.04c shows how the bounds are altered by the introduction of data uncertainty. Again we allow the assumed energy in the field to vary up to twice its true value, which in this artificial example is known. The degradation in the bounds is remarkably slight, compared with the results of previous analysis with exact data. For example, when $\Xi_0^2/\|b\|^2$ is assumed to be no more than 1.5, we find that p_1 is confined to an interval of 1.1 μT, in comparison with the value 0.72 μT obtained earlier. Because random noise has been added to the data, the smallest model in the energy norm that fits these numbers exactly has an energy much larger than the true one. If we disregard the errors in the data and use the theory for

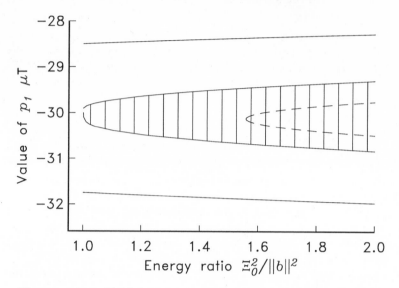

Figure 4.04c: Shaded region contains the axial dipole coefficient based upon 100 noisy, artificial data; the horizontal axis is the ratio of the assumed energy in the field to the true value. The dashed curve delineates the bounds obtained if the data were regarded as error free. The two outer lines represent crude bounds based on a primitive estimate.

exact measurements we obtain the dashed bounds in the figure. The outer lines in the figure result from using the simplified estimate based on projecting $\tilde{g}_1$ onto G. Recall from the previous section that we sought the best approximation for this element from linear combinations of the representers; the part unaccounted for by the approximation is the component called $\tilde{g}_\perp$ in the subspace $G^\perp$. In the present example $\|\tilde{g}_1\| = 0.189\,\mu T$ and after the standard fitting calculation we find $\|\tilde{g}_\perp\| = 0.0035\,\mu T$. Despite the small size of $\tilde{g}_\perp$ the associated bounds are rather wide compared with the best possible ones found from the complete theory.

    Finally we return to the historical record of the magnetic observatories in the year 1900. We take all thirty-three stations, but we continue to use just the Z fields, even though X and Y components were measured; this is to simplify the calculation of the Gram matrix (which is complicated enough as it is). Recall we selected $\Xi_0 = 231\,\mu T$. The fact that the smallest model of all those fitting the observations exactly has a norm of $22{,}770\,\mu T$ explains why we were unable to apply the theory for exact data to these

observations. The noise level of $0.1\,\mu T$ will be retained from the
artificial data. We shall predict only the axial dipole coefficient as
before. With thirty-three measurements at the 95 percent level
we obtain a tolerance for data misfit of $T = 6.888$. Running
through the calculations, fixing Ξ_0 at $231\,\mu T$, we find that -31.5
$\leq p_1 \leq -12.8\,\mu T$. This is a very broad interval. Notice, however,
that the lower bound is not very different from the 1980 value of
-30.0, and the dipole moment is known to be decreasing at about
0.05 percent per year at present. To understand the relatively
poor performance of the theory with the real observatory data let
us calculate the size of that part of the predictor lying in $G^{\perp}$; now
we find $\|\tilde{g}_{\perp}\| = 0.088\,\mu T$ while $\|\tilde{g}_1\|$ is the same as in the previous
illustration. Thus nearly half of the predictor lies outside the sub-
space spanned by the representers, in contrast to the mere 2 percent in the artificial example. The failure of the representers to
approximate the predictor with reasonable accuracy must lead to
disappointingly wide bounds. This inability is certainly caused by
the uneven distribution of observatories and not the smaller
number of representers: there are no measurement sites in the
Pacific Ocean and only two in the Southern Hemisphere. The
crowding together of the observer sites in Europe causes approximate linear dependence among the corresponding representers.
One way of seeing this is by calculating the condition number of
the Gram matrix: the spectral condition number κ for the observatory data is 6.1×10^{13} while that for the one hundred artificial
stations is only 1.7×10^7. If this example were to be pursued seriously, one would have to include the other magnetic components
but it is unlikely that this alone would result in a significant
improvement because nothing has been done to improve the poor
geographic coverage. That can be achieved by calling upon the
much more complete record of the magnetic field directions; but
then we would run into what is for the moment an insurmount-
able obstacle—the forward problem for those data is nonlinear.

Let us summarize the results of this section. First, the
kinds of norms that were useful in finding regularized solutions
are usually not suitable for obtaining bounds on linear functionals. This is because the estimation of upper limits on the norms
depending upon derivatives may be very difficult, or indeed, those
norms may be unbounded in reality. Norms based upon energy
are quadratic and hence Hilbert in character, and will virtually
always be finite for real physical systems, although in practice it

may be impossible to discover such a norm for many problems; indeed, the difficulty of finding a quadratic norm whose magnitude can be estimated is the biggest drawback of the theory. We showed with an artificial example that a suitable data set can lead to gratifyingly tight bounds for some interesting functionals. Good performance need not be terribly sensitive to the choice on the bound of the norm, which is probably always going to be the least reliable ingredient of our recipe.

Our first attempt at providing defensible inferences has not been an overwhelming success: while the theory is powerful, there is often difficulty in feeding it with the necessary parameters. This theory has been a natural first step, relying as it does on the familiar paraphernalia of Hilbert space; it is time to move on to something new.

Exercises

4.04(i) From laboratory measurements of the attractive force of gravity and observations of the motions of satellites it is possible to obtain the mean density of the Earth, $\bar{\rho} = 5517 \, \mathrm{kg\,m^{-3}}$, and the ratio $C/M_e a^2 = 0.33078$, where C is the polar moment of inertia, M_e is the mass of the Earth, and a is its mean radius. Show that under the approximation of a spherically symmetric density distribution

$$d_1 = \tfrac{1}{3}\bar{\rho} = \int_0^a \rho(r) \, r^2 \, dr/a^3$$

$$d_2 = \bar{\rho} C/2M_e a^2 = \int_0^a \rho(r) \, r^4 \, dr/a^5 \, .$$

Suppose the core radius c is known to be $0.547a$; consider the problem of predicting the mean densities of the core and the mantle using the two data d_1, d_2 above. Write the predictions as integrals and identify the representers and predictors in $L_2(0,a)$. Calculate the norm and $\|\tilde{g}_\perp\|$ for both of the predictors. Give the best possible constraints on these predictions if it is known that the RMS density of the Earth does not exceed $10{,}000 \, \mathrm{kg\,m^{-3}}$. In reality can the RMS density be estimated without at the same time knowing the mean densities of the core and mantle?

4.04(ii) The solutions to most linear forward problems in geophysics are written as integrals. This is possible for the global magnetic field problem used as an illustration in this section. Let the model be the radial field, B_r, evaluated on the surface of the core. Use the orthogonality of spherical harmonics (equations (6) and (14)) to write the energy norm in terms of an integral of B_r^2. Express the representers for B_r measured at the Earth's surface and the predictor for g_1^0 in terms of the associated inner product. Give the Gram matrix of the representers.

Hint: Expressions given by Shure et al. (1982) may be found useful.

4.05 Ideal Bodies

In geophysical prospecting work, particularly in gravity and magnetics, it has long been recognized that the observations are subject to a wide variety of interpretations. A broad gravity anomaly, for example, might be caused by a compact ore body at great depth, or a lower density mass spread over a large area. Even if the density of a buried mass is given, a number of strong additional assumptions are needed to guarantee the uniqueness of the shape based upon the exactly known gravity anomaly (Smith, 1961). This led in the early days to various formulas to help the field geophysicist estimate gross properties of the body, most importantly its depth below the surface with *maximum-depth rules*. The rules made no pretense of calculating a shape but gave instead a bound on an important functional; they are perfect examples of the strategy we are currently pursuing in this chapter. This section will present the background theory for the depth rules of potential theory; it represents a condensation of two papers: Parker (1974) and Parker (1975). Fundamentally the theory depends on the minimization of a norm, this time the uniform norm. Just as in our work with Hilbert space, we are able to harness norm minimization for the purposes of estimation of other, more geophysically significant quantities. For the simple problem we address first we are able to give fairly complete results in closed form, but they contain unknown parameters whose values can only be obtained by the solution of nonlinear equations. This means that the theory is of practical value only when the number of data is quite small, say no more than four. To handle larger systems we should turn to a numerical approximation via linear programming, with which we can solve a much wider variety of problems. This is the topic of the next section. We shall also postpone until then the treatment of uncertainty in the data.

We focus on one particular application, although alternatives may be readily invented. Gravity values are observed at the surface of a halfspace $D \subset \mathbb{R}^3$, defined by the points with $z \leq 0$, where z is the vertical coordinate, positive upwards. Within D density variations $\Delta\rho$ are distributed. It is easily shown that the (free-air) gravity anomaly measured by an observer at position vector $\mathbf{r}$ is

$$\Delta g(\mathbf{r}) = \int_D \frac{G\,\hat{\mathbf{z}} \cdot (\mathbf{r} - \mathbf{s})}{|\mathbf{r} - \mathbf{s}|^3}\, \Delta\rho(\mathbf{s})\, d^3\mathbf{s} \tag{1}$$

where G is Newton's gravitational constant, approximately $6.6726 \times 10^{-11}\ \mathrm{N\,m^2 kg^{-2}}$. We suppose N such measurements are made at the positions $\mathbf{r}_1, \mathbf{r}_2, \mathbf{r}_3, \cdots \mathbf{r}_N$, and we write them as

$$d_j = \int_D g_j(\mathbf{s})\, \Delta\rho(\mathbf{s})\, d^3\mathbf{s}, \quad j = 1, 2, 3, \cdots N \tag{2}$$

with

$$g_j(\mathbf{s}) = \frac{G\,\hat{\mathbf{z}} \cdot (\mathbf{r}_j - \mathbf{s})}{|\mathbf{r}_j - \mathbf{s}|^3}\,. \tag{3}$$

This is a linear forward problem of a very familiar kind. Among all the models $\Delta\rho$ satisfying (2) we shall seek the smallest one, but not under the L_2-norm. Let us restrict our attention to the space of bounded models and minimize the uniform norm $\|\Delta\rho\|_\infty$, that is, the least upper bound on $|\Delta\rho|$; for continuous functions this is of course just the same as the maximum value, but as we shall shortly see our answer is not in fact continuous. To make the problem more interesting we add another constraint: we suppose that the anomalous material is known to have a positive contrast with the local material in which it is embedded: we say that

$$\Delta\rho(\mathbf{s}) \geq 0, \quad \mathbf{s} \in D\,. \tag{4}$$

We shall relax this additional restriction later.

The solution we are about to give can be motivated in a number of ways. The way I first found it was by minimizing the L_p-norm and taking the limit as p grows (see section 1.06, and Parker, 1972). If we want the result without (4) another approach is by the more rigorous but less accessible theory of functional alignment (see chapter 5, Luenberger, 1969). In the next section we shall see by linear programming that the norm minimizing function takes on just two values. Rather than going into any of these approaches in detail, we merely state the result:

If real constants $\alpha_1, \alpha_2, \alpha_3, \cdots \alpha_N, \Delta\rho_0$ can be found so that the function defined by

$$\Delta\rho^0(\mathbf{s}) = \begin{cases} \Delta\rho_0, & \text{where } \sum_j \alpha_j g_j(\mathbf{s}) > 0 \\ \\ 0, & \text{where } \sum_j \alpha_j g_j(\mathbf{s}) \le 0 \end{cases} \tag{5}$$

satisfies (2) and (4), then $\Delta\rho_0$ *is the smallest norm* $\|\Delta\rho\|_\infty$ *of any function* $\Delta\rho$ *satisfying those two conditions.*

Before proving this statement let us discuss the solution briefly. First it is obvious that $\|\Delta\rho^0\|_\infty = \Delta\rho_0$, so that the function $\Delta\rho^0$ is a density-contrast model that achieves the smallest norm; it is called the *ideal body* for the problem. The ideal body consists of one or more regions of constant density on a zero density background, the surfaces of contact between the two being given by the equation

$$\sum_{j=1}^{N} \alpha_j g_j(\mathbf{s}) = 0 . \tag{6}$$

Because $g_j(\mathbf{s})$ is the potential of a vertical dipole placed at the observer location $\mathbf{r}_j$, these surfaces are the equipotential surfaces of a collection of point sources at the boundary of D, and are usually (though not always; see Parker, 1975) perfectly smooth. What makes this prescription so mysterious at first sight is the fact that the problem of actually finding the constants α_j is left open. In reality this is no different from the initial phase of the construction of the equivalent 2-norm minimizer, when we characterize the solution as a linear combination of representers, equation 1.12(9), for example; but in that case the next step, substituting the general form into the constraints, results in a system of linear equations, easily solved. Substitution of (5) into (2) has a much more forbidding aspect, because the equations for the unknown parameters are nonlinear. We shall return to this question later.

To prove the above proposition in italics, which we call P, we show that assuming the contrary leads to a contradiction. We consider the situation in which the constants, $\alpha_1, \alpha_2, \alpha_3, \cdots \alpha_N$ have been found so that the ideal body given by (5) satisfies (2) and $\Delta\rho_0 > 0$, in order that (4) is obeyed; but we assert that, contrary to the statement P, there exists a function $\Delta\sigma^0$ such that $0 \le \Delta\sigma^0(\mathbf{s}) < \Delta\rho_0$ for every $\mathbf{s} \in D$ and, furthermore, $\Delta\sigma^0$ satisfies (2). For convenience let us write

$$F(\mathbf{s}) = \sum_{j=1}^{N} \alpha_j g_j(\mathbf{s}) . \tag{7}$$

We subdivide D into two regions: V^+ where $F(\mathbf{s}) > 0$ and V^- where $F(\mathbf{s}) \leq 0$. From (5) V^+ delineates the ideal body. Now let us use the constants α_j to form the number X defined by

$$X = \sum_{j=1}^{N} \alpha_j d_j \, . \tag{8}$$

Because $\Delta\rho^0$ satisfies (2) we may write

$$X = \sum_{j=1}^{N} \alpha_j \int_D g_j(\mathbf{s}) \, \Delta\rho^0(\mathbf{s}) \, d^3\mathbf{s} \tag{9}$$

$$= \int_D [\sum_{j=1}^{N} \alpha_j \, g_j(\mathbf{s})] \, \Delta\rho^0(\mathbf{s}) \, d^3\mathbf{s} \tag{10}$$

$$= \int_{V^+} F(\mathbf{s}) \Delta\rho^0(\mathbf{s}) \, d^3\mathbf{s} + \int_{V^-} F(\mathbf{s}) \Delta\rho^0(\mathbf{s}) \, d^3\mathbf{s} \tag{11}$$

$$= \int_{V^+} F(\mathbf{s}) \Delta\rho_0 \, d^3\mathbf{s} \, . \tag{12}$$

The last line follows because $\Delta\rho^0(\mathbf{s})$ vanishes inside V^-. Next we calculate X using the fact that $\Delta\sigma^0$ also fits the data. In exactly the same way we obtain

$$X = \int_{V^+} F(\mathbf{s}) \Delta\sigma^0(\mathbf{s}) \, d^3\mathbf{s} + \int_{V^-} F(\mathbf{s}) \Delta\sigma^0(\mathbf{s}) \, d^3\mathbf{s} \, . \tag{13}$$

Subtract equation (13) from (12):

$$0 = \int_{V^+} F(\mathbf{s}) [\Delta\rho_0 - \Delta\sigma^0(\mathbf{s})] \, d^3\mathbf{s} - \int_{V^-} F(\mathbf{s}) \Delta\sigma^0(\mathbf{s}) \, d^3\mathbf{s} \, . \tag{14}$$

By the assertion that $\Delta\sigma^0 < \Delta\rho_0$ everywhere in D, the first integral in (14) must be positive because F is positive in V^+ by its definition. But the second integral in (14) is nonnegative because F is negative or zero in V^- while $\Delta\sigma^0$ is positive or zero. Thus (14) states that zero is obtained by adding a positive term to one that is either positive or zero. This is clearly impossible. We conclude that under the conditions assumed no such function as $\Delta\sigma^0$ exists, in other words, $\Delta\rho^0$ has the smallest least upper bound of all functions satisfying (2) and (4) and P is proved.

Let us look at a few examples. The one class of problems in which a simple solution is guaranteed is the set of one-datum problems, that is, when $N = 1$. We see from (5) that, depending only on the sign of α_1, we fill the region where $g_1 > 0$ with matter, or the region where $g_1 < 0$. In the case of simple gravity

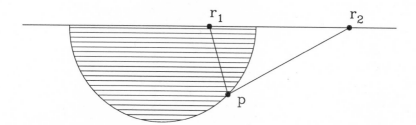

Figure 4.05a: Vertical cross section through the ideal body for two measurements of gravity made on a horizontal plane. The point **p** lies on the surface of the ideal body and the position vectors of the observation positions are $\mathbf{r}_1$ and $\mathbf{r}_2$. Then $|\mathbf{r}_1 - \mathbf{p}| = \beta \, |\mathbf{r}_2 - \mathbf{p}|$.

anomalies g_1 is positive throughout D and so we must either fill the whole region with matter or have none at all. Filling the halfspace with material of any positive density leads to a divergent integral for (2); we infer that an arbitrarily small density contrast can satisfy any single positive datum, and that is the solution to the bounding problem for one observation. This result is not very interesting but we shall see that not all one-datum problems are so dull. Next consider the system when $N = 2$. Referring again to (5) and the definition of g_j for gravity anomalies we find a point **p** on the surface of the ideal body obeys

$$\alpha_1 \frac{G \hat{\mathbf{z}} \cdot (\mathbf{r}_1 - \mathbf{p})}{|\mathbf{r}_1 - \mathbf{p}|^3} + \alpha_2 \frac{G \hat{\mathbf{z}} \cdot (\mathbf{r}_2 - \mathbf{p})}{|\mathbf{r}_2 - \mathbf{p}|^3} = 0 \,. \tag{15}$$

In the present configuration (see figure 4.05a) where the two observer positions are on the same level, $\hat{\mathbf{z}} \cdot (\mathbf{r}_1 - \mathbf{p}) = \hat{\mathbf{z}} \cdot (\mathbf{r}_2 - \mathbf{p})$ and so after rearrangement:

$$\frac{|\mathbf{r}_1 - \mathbf{p}|}{|\mathbf{r}_2 - \mathbf{p}|} = \left(\frac{-\alpha_1}{\alpha_2} \right)^{1/3} = \beta \tag{16}$$

where β is a constant, which must be positive. (Another, trivial solution to (15) is the plane $z = 0$, which coincides with boundary of D.) Equation (16) describes a point moving on a surface such that the ratio of the distances from two fixed points is constant. It has been known since the time of the Ancient Greeks that such a surface is a sphere. The reader may like to confirm in this case the radius is $\beta \, |\mathbf{r}_2 - \mathbf{r}_1| \, / \, |1 - \beta^2|$ and the center has the position vector $(\mathbf{r}_1 - \beta^2 \mathbf{r}_2)/(1 - \beta^2)$ which lies in the plane $z = 0$. Therefore,

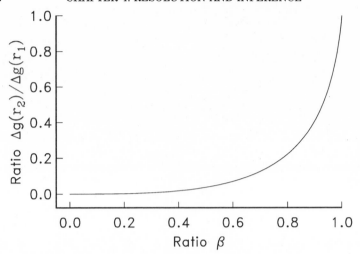

Figure 4.05b: Ratio of gravity anomalies associated with the hemispherical ideal body of figure 4.05a as β varies. The parameter $\beta = (-\alpha_1/\alpha_2)^{1/3}$.

the ideal body is one of a family of hemispheres, with their diametral planes coincident with the observation surface.

Another question left open by the prescription for the ideal body is whether there are any satisfactory constants; it merely asserts that when there are, the prescription works. Indeed, there may not be any density distribution obeying (4) and satisfying (2). This is the question of consistency of the conditions, or to put it another way, the problem of *existence* of solutions. We shall be able to treat this matter completely when we come to the linear programming approach. A case in point for the present example is that of negative gravity anomalies: because the representers g_1 and g_2 are both positive within D, a negative value for d_1 or d_2 is impossible under the assumption (4). Can every positive pair of values be generated by an ideal body? To answer this question we sweep through all positive values of β and compute the corresponding anomaly values. The absolute sizes of the anomalies can always be adjusted by suitable choice of $\Delta\rho_0$; the critical quantity is the ratio d_2/d_1, or equivalently $\Delta g(\mathbf{r}_2)/\Delta g(\mathbf{r}_1)$. This calculation is summarized in figure 4.05b. When $\beta < 1$ the matter fills a hemisphere under the observer at $\mathbf{r}_1$ and so the gravity anomaly there is larger than that at $\mathbf{r}_2$. As β tends to zero the mass concentrates in a tiny hemisphere under position $\mathbf{r}_1$ and the ratio of data tends to zero also. As β approaches unity, the radius of the associated hemisphere increases without bound, and its center moves far from the observation sites; the two observers experience virtually identical attractions from the body and hence d_2/d_1 tends to unity also. When β exceeds unity so does the ratio of attractions, but we need not discuss this case because there is no loss of generality in assigning d_1 to the larger of the two data when the values differ. Thus all ratios of attractions can be generated by an appropriate ideal body and so every legitimate pair of data can also be connected to one. The actual bound on density for this problem can be obtained from the curve labeled $h^* = 0$ in figure 4.05d.

If we move on to larger numbers of gravity measurements we find the nonlinear equations for the parameters α_j become increasingly difficult to solve. Since the conditions of (5) are homogeneous in α_j there are truly only N independent unknowns. As we have already mentioned, the approach is exactly the same as the one we took when we found the norm minimizer in Hilbert space: we substitute the ideal body into the constraint equations, and seek parameters that cause (2) to be satisfied. For every choice of α_j we will generate a body whose density is uniform and whose shape is determined by (5). We find the gravitational attraction of that body at each of the observation sites by evaluating the integrals, a challenging task in itself as N grows. Normally, a guess for the parameters will be made and this must be refined. Experience with a two-dimensional version of the problem (Parker, 1975) suggests a multivariable version of Newton's method is satisfactory provided $N < 6$. The necessary partial derivatives of $\Delta g(\mathbf{r}_j)$ as α_k changes are given in Parker (1975) for those who are interested, as well as detailed calculations involving many data. We shall not spend any more time on these complex calculations because linear programming provides a more reliable and general solution.

The smallest positive density contrast is not often a particularly valuable property of the unknown model. It is more likely in the real world that the density of a prospective ore mass is roughly known and the important issue is its location in space. For buried bodies the depth to the top seems to be a vital parameter. To use the norm bounding machinery we revise the definition of D, the region containing the anomalous material: let us say it is confined to the region $D(h)$ defined by $z \leq -h$, that is, the top of the anomalous material is at least h below the surface.

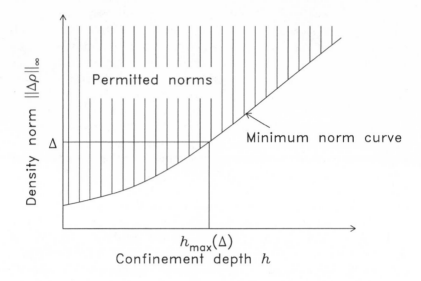

Figure 4.05c: Sketch of the plane of confinement-depth and density-norm pairs. The minimum norm curve is obtained from ideal body calculations. The shaded zone contains pairs consistent with (2) and (4). The depth $h_{max}(\Delta)$ is the greatest confinement depth compatible with the norm Δ.

Nothing else is changed in the development of the theory for bounding the norm. Imagine now performing a series of ideal body calculations for values of h increasing from zero. When $h_1 < h_2$, $D(h_2) \subset D(h_1)$ and the density bound obtained for h_2 must be a valid bound for h_1 also, since the ideal body for h_2 lies inside $D(h_1)$ and meets all the necessary conditions. Thus the best (that is, least possible) bound found by the ideal body calculation for each depth cannot decrease with h. (It can be shown that the ideal body for a given data set is unique and must touch the boundary of D [Parker, 1975]; from these facts it follows that the bound can never remain constant, but must grow as h increases.) In Figure 4.05c we sketch the plane of h and $\|\Delta\rho\|_\infty$ for all possible positive density contrasts consistent with the given gravity anomaly and a presumed confinement depth h. From our ideal body calculations every model is represented by a point lying in the shaded zone, on or above the curved line; no model can generate a point outside the zone for then it would possess a norm impossibly small for that depth. Therefore, if the norm of the density of the anomalous mass is known to be Δ, and we draw a horizontal line across Figure 4.05c at that level, we are certain that the point representing the body must lie to the left of the intersection of the line and the curve, that is, within the shaded region. Hence, a part of every body satisfying (2) and possessing the specified density norm must either touch the plane $z = -h_{\max}(\Delta)$ or be located above that level. We have obtained its greatest possible depth of burial, the so-called maximum depth.

It should be clear that the depth is not the only spatial parameter we can constrain by this kind of argument. One can imagine bringing up a horizontal plane from below, or inserting a vertical plane and asking for the body with the least extent to the North, and so on. See section 4.07 and exercise 4.05(iii) for examples.

Returning to the gravity anomaly example with $N = 2$, let us prepare a graph completely describing the solution, figure 4.05d. To do this we must reduce the number of parameters by introducing non-dimensional variables. As we mentioned earlier, one critical (and dimensionless) variable is the ratio of the two anomaly values which, as before, we take to be less than unity. Another scaling is the normalization of the confinement depth by the distance between the two observation sites: we write $h^* = h / |\mathbf{r}_2 - \mathbf{r}_1|$. We may render the density dimensionless by dividing $\Delta g(\mathbf{r}_2)$ by $G |\mathbf{r}_2 - \mathbf{r}_1|$, which has the units of density. To generate figure 4.05d we select arbitrary positive values for $d_2 = \Delta g(\mathbf{r}_2)$ and $|\mathbf{r}_2 - \mathbf{r}_1|$. We hold h constant, thus fixing $D(h)$ and the geometry. If β is given we have enough parameters to calculate the radius of the hemisphere and the ratio of attractions and thus, since we have assigned d_2, the other datum. We can also calculate the density bound. Now β is run through positive values and values of the gravitational attraction and the corresponding density bound are computed from the ideal bodies. The dimensionless ratios in the figure are calculated and one of the curves of constant h^* is plotted. Another h is chosen and the process repeated. The figure is constructed by finding the minimum density bounds for various depths of confinement but we have argued that there is an alternative interpretation as a series of curves of maximum depth, when density is known, and this interpretation is the more useful one.

Because of the evident complexity of finding maximum depths for multi-data problems, the traditional approach has been as follows. A single number is computed from a linear combi-

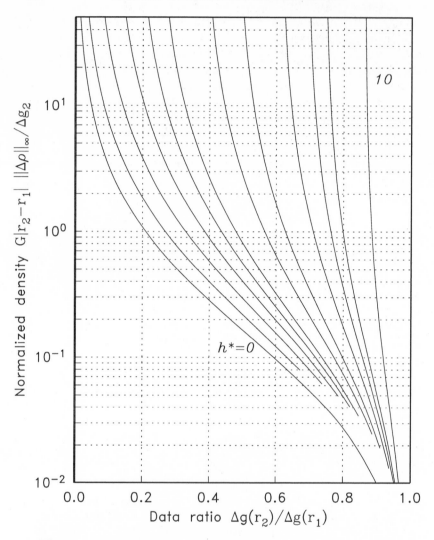

Figure 4.05d: Summary of the complete depth rule or density bound for two gravity measurements. The normalized depth bounds are for the values $h^* = h/|\mathbf{r}_2 - \mathbf{r}_1| = 0$, 0.2, 0.4, 0.6, 0.8, 1, 1.5, 2, 3, 4, 5, 10.

nation of actual measured values. A typical example might be an approximation for the second derivative:

$$\frac{\partial^2 \Delta g}{\partial x^2} \approx \frac{\Delta g\,(\hat{\mathbf{x}}\xi) - 2\Delta g\,(0) + \Delta g\,(-\hat{\mathbf{x}}\xi)}{\xi^2} \tag{17}$$

where we have set the coordinate origin at the central observation site. It is straightforward to compute the representer for the single datum by taking x-derivatives of the representer for a single gravity anomaly:

$$\frac{\partial^2 \Delta g}{\partial x^2} = \int_{D(h)} G\hat{\mathbf{z}} \cdot \mathbf{s} \left[\frac{15(\hat{\mathbf{x}} \cdot \mathbf{s})^2}{|\mathbf{s}|^7} - \frac{3}{|\mathbf{s}|^5} \right] \Delta\rho(\mathbf{s}) \, d^3\mathbf{s} \tag{18}$$

and to map out the surfaces on which the new representer vanishes. We have seen that these will be the boundaries of the ideal body. The integral is performed over this shape. The confinement depth h is necessary here from the beginning, otherwise the integral diverges from the singularity at the origin. Only linear equations arise and the result can be neatly summarized in closed form:

$$h_{\max} = \frac{48\pi}{5^{5/2}} \frac{G \, \|\Delta\rho\|_\infty}{|\partial^2 \Delta g / \partial x^2|} . \tag{19}$$

For details of the calculation see Parker (1974) or try exercise 4.05(ii). Of course one could equally well treat the single datum for the quantity on the right of (17) rather than the derivative that it approximates, but the integrals and subsequent equation for $h_{\max}$ cannot be written in an elementary way.

Depth rules have been derived for other kinds of measurements, particularly for magnetic field data. In this case it may be useful to remove the positivity condition (4) since reversely magnetized material frequently occurs even in relatively homogeneous geological units. The reader may wish to verify the revised version of (5), specifying the new ideal body, which corresponds precisely to pure norm minimization:

$$\Delta\rho^0(\mathbf{s}) = \begin{cases} \Delta\rho_0, & \text{where } \sum_j \alpha_j g_j(\mathbf{s}) > 0 \\ -\Delta\rho_0, & \text{where } \sum_j \alpha_j g_j(\mathbf{s}) \leq 0 . \end{cases} \tag{20}$$

Estimating the norm $\|m\|_\infty$ is much more straightforward than supplying the value of a Hilbert-space norm; one simply provides the largest anticipated value of the magnitude of the model. This gives the theory a tremendous advantage over the approach taken in sections 4.03 and 4.04. But the functionals for which we have obtained rigorous bounds based on the uniform norm have been somewhat peculiar; furthermore, the highly nonlinear nature of the equations for the unknown parameters puts severe limits on the size of the data set that can be treated practically. While the theory can be extended to cover other kinds of functionals (such as linear ones) the nonlinearity remains an insuperable obstacle for even medium-scale applications. In the next section we develop some powerful new methods, optimization with linear and quadratic programming, that enable us to overcome this and other limitations of the present approach.

Exercises

4.05(i) In the two-datum gravity anomaly problem treated in this section, consider bounding the depth when the density contrast tends to infinity. Show that then

$$h_{max} = |\mathbf{r}_2 - \mathbf{r}_1| \cdot |(d_2/d_1)^{2/3} - (d_1/d_2)^{2/3}| .$$

For the values in the table below, using an assumed density contrast of 150 $\mathrm{kg\,m}^{-3}$, calculate the best bound you can on the depth of burial of an ore body. Use figure 4.05d, taking all possible pairs of values. Also apply the rule (19) with the approximation of equation (17). The data were measured on a straight traverse and have been corrected to a single horizontal level. $(1 \text{ Gal} = 10^{-2}\,\mathrm{m\,s}^{-2})$.

Distance (m)	Anomaly (mGal)
0	2.85
500	1.92
1000	0.68
1500	0.26

4.05(ii) When a gravity anomaly is elongated in one direction it is convenient to approximate the system by a model in which the density is invariant in one direction, say the y-direction. Show that then

$$\Delta g(\mathbf{r}) = \int_b \frac{2G\,\hat{\mathbf{z}} \cdot (\mathbf{r} - \mathbf{s})}{|\mathbf{r} - \mathbf{s}|^2} \Delta\rho(\mathbf{s})\, d^2\mathbf{s} .$$

Here the vectors are contained in the x-z plane, normal to the axis of the system. Differentiate the representer twice and find the maximum depth rule for the single datum $\partial^2\Delta g/\partial x^2$. Would you expect the depth to be greater or smaller than the one found by (19)?

4.05(iii) Gravity values are measured in a vertical shaft and processed to give the second vertical derivative $\partial^2\Delta g/\partial z^2$. Some values are tabulated below. The uniform target material is known to have density contrast with the surrounding uniform sediment of 200 kg m^{-3}.

(a) Derive the maximum depth rule based on the second z derivative given in Parker (1974). Assuming that the deposit lies somewhere beneath the bottom of the shaft, find an upper bound on the depth of the top of the material below the deepest point in the borehole.

(b) There is reason to believe part of the anomalous material lies above the 1300 m level. Derive a new rule giving an upper bound on the distance *in any direction* of the high density material from a point at which $\partial^2 \Delta g / \partial z^2$ is known. Apply the rule to the data below and then find the greatest possible distance of some part of the deposit from the end of the shaft.

Depth (m)	$\partial^2 \Delta g / \partial z^2$ (μGal m^{-2})
1050	0.036
1200	0.035
1300	0.027

Hint: You may work out the solution in a full space, ignoring the effects from the interface at the Earth's surface.

4.06 Linear and Quadratic Programming

We introduce in this section an indispensable tool of optimization theory, linear programming. We shall describe only the finite-dimensional version of the theory. For applications to function spaces we simply invoke the naive approximation of a model space with a very large but finite dimension. Thus we shall have the equivalents for sections 1.13 and 3.05 of our Hilbert space theory. Some mathematically valid results can be obtained by considering a sequence of problems where the dimension of the model space increases systematically, but results about the uniqueness of solutions for example, obtained almost painlessly in Hilbert space, cannot be proved so easily here, if at all. To do justice to the general theory of optimization in Banach spaces requires a much larger investment in heavy mathematical machinery than we are prepared to make. For Hilbert space all we have required is the Decomposition Theorem and the Projection Theorem; with these we have been able to reduce every problem to a calculation in a finite-dimensional space, in principle without any approximation. To discover the requirements for more general spaces the reader should consult *Optimization by Vector Space Methods* by D. G. Luenberger (1969), a work cited several times already and a

model of clear but rigorous exposition: Luenberger devotes twice
as much space and proves twice as many theorems in his treatment of the optimization of functionals on Banach spaces as
opposed to Hilbert spaces. The richness of the general theory
arises chiefly from the fact that the space of bounded linear functionals, that is, the dual space, and the original space are no
longer identical. The symmetry between the two imposed in Hilbert space is responsible for the simplicity of its theory, but it is
somewhat artificial: for example, there seems no compelling reason why, if a model is continuously differentiable, the functions
combining with it to form inner products should be.

　　We shall state the famous Fundamental Theorem of Linear
Programming and examine some of its consequences. First we
must define the standard Linear Programming (LP) problem:

　　An unknown vector $\mathbf{x} \in \mathbb{R}^M$ satisfies the following linear constraints:

$$A\mathbf{x} = \mathbf{b} \tag{1}$$

where $\mathbf{b} \in \mathbb{R}^N$ with $N < M$ and of course A is a nonsquare
matrix in $M(N \times M)$. In addition the vector $\mathbf{x}$ obeys

$$\mathbf{x} \geq 0 \tag{2}$$

by which we mean every component of the vector is nonnegative. The first part of the standard LP problem is to ask,
(i) Are these conditions consistent? In other words, Can *any*
$\mathbf{x}$ satisfy (1) and (2)? If the answer is yes, we can go on to
the second part:
(ii) For vectors within the constraint set, what is the smallest
value of the penalty functional

$$P[\mathbf{x}] = \mathbf{p} \cdot \mathbf{x} \tag{3}$$

where $\mathbf{p} \in \mathbb{R}^M$ is a known vector? The answer to the second
part may be some finite number, or there may be no lower
limit on P.

To proceed we need some more terminology. The first new term
is *feasible solution*. A vector $\mathbf{x}$ is a feasible solution if it obeys
conditions (1) and (2). For the optimization problem only the
feasible vectors are important and no others. The second piece of
terminology, a *basic solution*, is more complicated. Let us write
the constraint matrix as an ordered set of vectors $\mathbf{a} \in \mathbb{R}^N$ as follows:

$$A = [\mathbf{a}_1, \mathbf{a}_2, \mathbf{a}_3, \cdots \mathbf{a}_M]. \tag{4}$$

The size of the largest set of linearly independent column vectors in A is called the *rank* of A; obviously the rank can never be more than N, but it might be less. We shall assume A is of full rank. (If the rank of A is less than N it is easily shown that either (1) is inconsistent or the constraints can be reduced in number to $N' < N$ and the new matrix A' will be of maximal rank, N'.) Equation (1) can be rewritten

$$\sum_{j=1}^{M} x_j \mathbf{a}_j = \mathbf{b}. \tag{5}$$

Suppose in equation (5) that we delete all but a subset of N linearly independent column vectors, and that (5) can still be satisfied by suitable choice of x_j. When this is possible, it is equivalent to setting $M - N$ components of $\mathbf{x}$ to zero; the corresponding vector $\mathbf{x}$ is called a *basic solution*. Notice positivity plays no part in the definition. The two terms can be combined to define the entity, a *basic feasible solution*; such a vector is of course simultaneously feasible and basic. Finally, we define an *optimal* solution as a feasible solution that minimizes the penalty functional with a finite value.

The Fundamental Theorem of Linear Programming can now be stated:

(I) If a feasible solution exists, there is a basic feasible solution.

(II) If an optimal solution exists, there is a basic solution that is optimal.

The proofs of these propositions can be found in every book on linear programming; see, for example, Gass (1975) or another book by Luenberger (1984). It is not immediately obvious that these two results really answer the original questions, but we shall see that they do, at least in principle. The crucial point of (I) is that there are only finitely many subsets of N column vectors in the matrix A. Imagine examining each such subset in turn. We throw out those that are linearly dependent. For the remainder we solve the linear system (5) for the subset of N components in $\mathbf{x}$, while the rest of the components are set to zero. Each such linear system has a unique solution because the column vectors of the square submatrix under consideration are linearly independent. In this way we can generate *all* the basic solutions. If any of the basic solutions is feasible, we immediately have a vector satisfying (1) and (2). If none of the basic solutions is feasible (which means every one has one or more negative components) part (I) of the Fundamental Theorem asserts there simply are no feasible solutions at all; in other words the constraints (1) and (2) are inconsistent. Thus we have answered question (i).

To answer the second question, that of finding the optimal solution, it is not enough simply to search the list of basic feasible solutions for the one with the smallest penalty functional P. This is because we do not know whether the functional is bounded below or not. We can appeal to (I) again for the solution to this problem. We construct the constraint system

$$\begin{bmatrix} A \\ \mathbf{p}^T \end{bmatrix} \mathbf{y} = \begin{bmatrix} 0 \\ -1 \end{bmatrix}. \tag{6}$$

Here $\mathbf{y} \in \mathbb{R}^M$ and the vector on the right consists of N zeros over a single entry of -1. If this system is compatible with $\mathbf{y} \geq 0$ the penalty functional $P[\mathbf{y}] = \mathbf{p} \cdot \mathbf{y}$ is not bounded below, that is, it may be made arbitrarily large in magnitude but negative while $\mathbf{y}$ remains within the constraints; conversely, if these constraints are incompatible the optimal solution is finite. This assertion is proved as follows. We know there is at least one feasible solution to the original LP problem found as described above—call it $\mathbf{x}_0$. Suppose now there is also a feasible solution to (6) called $\mathbf{y}_0$. Form the vector $\mathbf{z} = \mathbf{x}_0 + \alpha \mathbf{y}_0$ with $\alpha > 0$; then $\mathbf{z} > 0$ and

$$A\mathbf{z} = A\mathbf{x}_0 + \alpha A\mathbf{y}_0 \tag{7}$$

$$= \mathbf{b} + \alpha\, 0 \tag{8}$$

$$= \mathbf{b}. \tag{9}$$

So $\mathbf{z}$ is feasible for the original LP problem for every $\alpha > 0$. But

$$P[\mathbf{z}] = \mathbf{p} \cdot \mathbf{z} = \mathbf{p} \cdot \mathbf{x}_0 + \alpha \mathbf{p} \cdot \mathbf{y}_0. \tag{10}$$

From (6) we see that $\mathbf{p} \cdot \mathbf{y}_0 = -1$ and therefore by making α large, we can make P tend to $-\infty$. This shows that if (6) is consistent there is no lower bound on P as we wished to prove. The converse is left as an exercise.

When we have shown that there is a lower bound on P by the appeal to the Fundamental Theorem, finding that bound merely involves evaluation of all the penalty values for the complete set of basic feasible solutions; from (II) the smallest of these is the minimum penalty. Notice that the theorem makes no statement about the uniqueness of a basic optimal solution. It is quite possible that there are several basic optimal solutions, each achieving the same value for P. If that happens, it follows that there are other, non-basic optimal solutions too, because a convex linear combination of those vectors will still be an optimal solution but, as the reader can easily confirm, it need not be basic.

We should stress at this point that the above procedure is *not* the basis of a practical computational scheme. The number of operations grows astronomically with the size of the system if an exhaustive search is performed. Until very recently the standard numerical schemes were variants of the *Simplex algorithm*, which is discussed in all the texts on LP. This algorithm, while having excellent behavior in most cases (it requires only slightly more arithmetic operations than are needed to solve a least-squares problem of the same size), can be forced by certain exceptional constraint sets to execute a number of steps proportional to $\exp N$; this means there are quite small problems that cannot be solved by the Simplex algorithm in a practical amount of time. A completely different numerical approach has recently been proved to be free of this defect and may be faster in general; this is Karmarkar's method. Both algorithms seek a local minimum in the penalty functional by "going downhill," relying on the convexity of P and the constraint set for their assured convergence to the true minimum. In Simplex the successive trial points lie on the boundary of the constraint set, while in Karmarkar's method they are inside the constraints and the variables of the original problem are transformed by a nonlinear mapping. It is to be expected that the new algorithm will replace Simplex in the standard computer libraries over the next few years. For a readable account of Karmarkar's method see Strang (1986).

The standard LP problem appears at first to be a very rigid thing: the unknown model is nonnegative, and it must obey a number of linear equality constraints. In fact the constraint set can be generalized to include quite general inequalities as well as unrestricted model vectors. First, suppose that some of the components of $\mathbf{x}$ are permitted to be of either sign (or zero). By reordering if necessary let these be the first L: $x_1, x_2, \cdots x_L$. Now introduce for each x_j of this set a pair of dummy components $y_j \geq 0$ and $z_j \geq 0$ and let $x_j = y_j - z_j$, eliminating x_j from the linear equality constraints (1). Since every real number can be written as the difference of two nonnegative numbers, and the new system is still in the form of (1) we can transform any number of variables of unrestricted sign to the standard nonnegative case. Naturally, $y_j - z_j$ must be substituted in the penalty function in place of x_j.

Next suppose instead of only equality constraints we would like one or more constraints in the form:

$$\mathbf{c}_k \cdot \mathbf{x} \le q_k \qquad (11)$$

where $\mathbf{c}_k$ is a known vector. This is converted into an equality by adding for every such inequality a *slack variable*, a new unknown $s_k \ge 0$. Then the above inequality becomes an equality:

$$\mathbf{c}_k \cdot \mathbf{x} + s_k = q_k . \qquad (12)$$

Thus inequalities can be made to look like equality constraints at the expense of introducing additional variables that do not appear in the penalty function, or more properly, are associated with zero elements in modified penalty vector.

With these two simple devices we can reduce to standard form any collection of positive or general real unknowns constrained by any combination of equality or inequality constraints. The reader should understand, however, that properly written computer algorithms do not waste the storage and execution time associated with a naive implementation of slack variables; nonetheless the principle is important, as we shall shortly see.

As we stated at the outset we shall apply the finite-dimensional form of LP as an approximation for our problems in function spaces. We imagine that the dimension of the model space becomes very large. As a first illustration of this approach consider a linear forward problem in one spatial variable of the familiar kind: we make exact measurements of N linear functionals of the unknown model

$$d_j = \int_a^b g_j(\xi)\, m(\xi)\, d\xi, \ j = 1, 2, 3, \ \cdots \ N \qquad (13)$$

where it will be necessary to assume that the representers g_j are at least continuous functions. Suppose that in addition to (13) we know from physical considerations that $m(\xi) \ge 0$ throughout the interval $[a, b]$. We wish to obtain the smallest values of some predictor

$$P[m] = \int_a^b \tilde{g}(\xi)\, m(\xi)\, d\xi \qquad (14)$$

exactly as in section 4.03; notice how in place of a norm, positivity

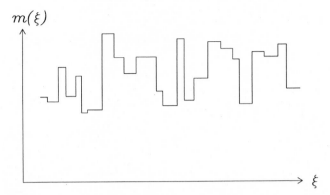

$m(\xi)$

Figure 4.06a: The approximation of a model function $m(\xi)$ in a step-function basis. Notice it is not necessary to assume the subdivisions must be equal in length.

supplies the extra information about m that permits a bounded outcome. How shall we approximate the integrals here? When we sought smooth solutions it was sensible to use numerical integration schemes that took advantage of the regularity of the integrand; smoothness is not a property of the solution for the present problem. Let us represent the model in a step-function basis—the model is piecewise constant with value m_k when $\xi_{k-1} \leq \xi < \xi_k$ and $a = \xi_0 < \xi_1 \cdots \xi_{M-1} < \xi_M = b$. See figure 4.06a. Then (13) and (14) become

$$d_j = \sum_{k=1}^{M} G_{jk} m_k, \ j = 1, 2, 3, \cdots N \tag{15}$$

$$P = \sum_{k=1}^{M} p_k m_k \tag{16}$$

where

$$G_{jk} = \int_{\xi_{k-1}}^{\xi_k} g_j(\xi) \, d\xi, j = 1, 2, 3, \cdots N, \ k = 1, 2, 3 \cdots M \tag{17}$$

and similar integrals for p_k. When the condition

$$m_k \geq 0 \tag{18}$$

is combined with (15) minimization of (16) is the standard LP problem. Thus the problem of consistency of (13) with $m(\xi) \geq 0$ is solved, at least to some numerical approximation, and of course we can obtain an upper bound on the optimal functional. Suppose

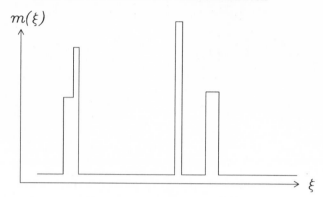

$m(\xi)$

ξ

Figure 4.06b: The approximation to the optimal solution minimizing $P[m]$ in (14) by a basic solution in the step-function basis.

that, as M grows, all earlier subdivision points are included in subsequent discretizations, and the subdivision is carried out so that the longest subinterval tends to zero. Then the estimate for the smallest P obtained by LP can never increase. Since it cannot decrease below zero, the sequence of approximants to the optimal, that is smallest possible, P must converge. Does it converge to the true optimal result? The answer is yes, provided the functions g_j and p are reasonably smooth, say continuous. We cannot prove this with the mathematics we have developed so far; the reader should consult Luenberger (1969) for the material on weak* convergence. What does a typical basic optimal solution look like? Since $M-N$ components must vanish in m_k, this means at most N are nonzero, no matter how large M becomes. Since $\xi_k - \xi_{k-1}$ tends to zero with increasing M, the only way that (13) can be satisfied is if the size of the nonvanishing m_k grows indefinitely. Thus we see that a basic solution tends to a finite set of δ functions; see the sketch in figure 4.06b.

It may not be obvious from what we have seen so far how the LP approach can minimize a norm or how the norm could be used to constrain the model or data. Let us first discuss the minimization of the uniform norm of a model. Assuming observations in the form of (13) which are represented again by (15), we seek the model with the smallest value of $\|m\|_\infty$. In the approximation of the step-function basis

$$\|m\|_\infty = \max_{1 \le k \le M} |m_k| . \tag{19}$$

Let us introduce the unknown $n \ge 0$ together with the condition that none of the values of m_k can exceed it; to be in standard form this necessitates the introduction of M slack variables, s_k:

$$m_k + s_k - n = 0, \quad k = 1, 2, 3, \cdots M . \tag{20}$$

If the model $m(\xi)$ is permitted to be negative, the condition $m_k > -n$ must be included; normally this would be accomplished through the introduction of M additional slack variables after having split m_k into two nonnegative parts as we mentioned earlier. In this problem, however, it pays to transform to a new set of variables:

$$m_k' = m_k + n . \tag{21}$$

The requirement $m_k' \geq 0$ automatically enforces the condition that m_k be bounded below by $-n$. Now (20) becomes

$$m_k' + s_k - 2n = 0, \quad k = 1, 2, 3, \cdots M \tag{22}$$

and (15) should be written in terms of m_k'. Thus we have put the constraints into the form of a standard LP problem. What penalty function must be used to minimize $\|m\|_\infty$ under the restrictions of (15)? Assuming m may take both signs, we have in the standard form an unknown vector $\mathbf{x}$ in $\mathbb{R}^{2M+1}$, which we construct as follows: the first M components carry m_k', then M components from the slack variables s_k, and finally n, which is also unknown. To minimize the norm, the penalty vector should be simply:

$$\mathbf{p} = [\, 0, 0, 0, \cdots 0, 0, 1\,] . \tag{23}$$

In other words, only the component of the vector corresponding to n is penalized. Because (22) and the implicit constraint $m_k' \geq 0$ guarantee that $|m_k|$ can never exceed n, minimizing n gives us the solution to the norm minimization problem. If instead of seeking the smallest norm, a bound on $\|m\|_\infty$ is known (see the previous section, for example), n can be moved to the right side of (22) to apply the constraint.

It is instructive to see how the Fundamental Theorem enables us to discover the form of a norm-minimizing solution in this case; as always we discuss only the basic solutions. Although our unknown vector lives in $\mathbb{R}^{2M+1}$ the number of equality constraints is $M+N$, from equations (15) and (22). It follows from the definition of a basic solution that no more than $M+N$ components of $\mathbf{x}$ can be nonzero. Let us assume that the smallest model is not identically zero; then the norm $n > 0$. Then in (22) one or both of the numbers m_k' and s_k must be positive too. Let us define an *intermediate* component of m_k as one that is *not* at its upper limit n or at its lower limit $-n$; correspondingly, an intermediate m_k' satisfies the condition $0 < m_k' < 2n$. How many components can be intermediate? If m_j' is intermediate, and therefore nonzero, the associated slack variable s_j in (22) is also positive and so all three components are nonnegative in this particular constraint equation. If there are J intermediate components, we simply count the total number of nonnegative components in $\mathbf{x}$, and apply the Fundamental Theorem:

$$2J + M - J + 1 \le M + N \tag{24}$$

or

$$J \le N - 1. \tag{25}$$

As we let M tend to infinity, we find almost all the components of the basic solution vector are at the upper or lower limit, while no more than $N - 1$ can hover between the extremes. This agrees with the piece-wise constant model described by equation 4.05(20).

The next important matter is the accommodation of uncertainty in the data. We have seen how easily the uniform norm can be treated by LP. The statement that the predictions of the model lie strictly in some data interval $[d_j^-, d_j^+]$ is, in the notation of (15)

$$d_j^- \le \sum_{k=1}^{M} G_{jk} m_k \le d_j^+, \ j = 1, 2, 3, \ \cdots \ N. \tag{26}$$

The translation of these conditions to the standard LP problem is of course trivial with slack variables.

In section 3.01 we discussed the properties of a good measure of model misfit. As we stated there the uniform norm for misfit $\|\mathbf{X}\|_\infty$ (where $\mathbf{X}$ is the standardized random vector) is traditionally regarded as unsatisfactory for two reasons: first, when the error distribution is truly Gaussian, the variance of the uniform norm is large; second, in practice the deviations from a precise Gaussian error law often appear as large outliers, whose presence completely spoils statistical tests that rely upon the Gaussian assumption. Therefore, even though linear programming can easily handle the uniform norm of data misfit, we shall not pay this measure any further attention. The second argument for rejecting $\|\mathbf{X}\|_\infty$ is often used against $\|\mathbf{X}\|_2$ as well; instead the *robust* statistic $\|\mathbf{X}\|_1$ is frequently favored.

Somewhat surprisingly perhaps the 1-norm can also be treated in a linear program. As an illustration, we suppose that a linear functional of the model like (15) is to be minimized, and the criterion for acceptability of the model is that the discrepancy

$$\|\mathbf{X}\|_1 = |X_1| + |X_2| + |X_3| + \cdots |X_N| \tag{27}$$

shall be less than T, where X_j is the difference between the kth prediction of the model (a linear functional) and the observed datum. Once again we rely upon (15) for an adequate approximation of the forward solution. Because X_j may have either sign, for the linear program we write it as the difference of two nonnegative parts, Y_j and Z_j. This generates the following set of equality constraints:

$$(\sum_{k=1}^{M} G_{jk} m_k) - Y_j + Z_j = d_j, \quad j = 1, 2, 3, \cdots N . \tag{28}$$

We express the demand $\|\mathbf{X}\|_1 \leq T$ like this

$$s + \sum_{j=1}^{N} (Y_j + Z_j) = T \tag{29}$$

where s is a slack variable to account for the inequality. As written, these constraints are apparently insufficient to apply the 1-norm constraint: (29) is a correct 1-norm only when one member of every pair Y_j, Z_j vanishes, so that whenever X_j is positive, Z_j vanishes, and when X_j is negative, Y_j vanishes. The LP algorithm must be encouraged to make one of the components of each pair zero. One way to achieve this is to modify (16) thus

$$P[m] = \sum_{k=1}^{M} p_k m_k + \sum_{j=1}^{N} \varepsilon(Y_j + Z_j) \tag{30}$$

where ε is a small positive number. Clearly in (28) the constraint is indifferent to the addition of a constant to both members of a pair Y_j, Z_j, so there is a family of pairs parameterized by the constant. If we imagine that some particular value of $Y_j - Z_j$ is correct in each equation of (28), the additional terms in the penalty vector selects from the family the one with the smallest sum, and that evidently arises when one of the two members is zero. When that happens (29) constrains the 1-norm as we require. Provided ε is small enough, the penalty functional acts strongly on the first sum and will not cause a significant depression in the value of the second. In practice ε should be chosen so that the ratio of the magnitude of the two sums is a thousand to one or more. No such delicate balancing act is needed when the functional to be minimized is the 1-norm of the misfit, or of the model itself; the reader should look at those two straightforward cases.

The question of how to choose T in this case has already been touched upon in 3.01. The mean and variance of the distribution of $\|\mathbf{X}\|_1$ for a random Gaussian N-vector are readily computed, but neither the pdf nor the distribution function are easily calculated and they are not widely available in the standard tables; therefore in appendix B a table is provided for this purpose.

Should we wish to persist in using the ordinary Euclidean norm for the purposes of misfit measure then clearly LP is no longer applicable. The optimization problem of Quadratic Programming (QP) is almost as well developed as its LP counterpart, provided that complete generality is not required. Good numerical methods exist for a QP problem in the following form:

$$\min_{\mathbf{x} \geq 0} \|A \, \mathbf{x} - \mathbf{b}\|_2 \tag{31}$$

where $\mathbf{x} \in \mathbb{R}^M$, $\mathbf{b} \in \mathbb{R}^N$ and A is a matrix of the appropriate size. Notice that here we minimize a convex functional over a convex set, with the consequent gratifying absence of multiple local minima. (Not all quadratic penalty functions are convex, for example $x_1 x_2$.) A computer code for (31) that is widely available and, because of its use of QR factorization, numerically stable is the FORTRAN program NNLS (for Non-Negative Least Squares) published in Lawson and Hanson's book (1974). Vectors equivalent to the basic solutions of LP are found with the same property: no more than N components of $\mathbf{x}$ will be positive, while the remainder (if any) must vanish. Slack variables and decomposition of components into the difference of positive terms can obviously generalize the constraint set to some degree, but an algorithm that provides the solution to problem (31) is not as it stands suitable for our purposes: we wish to apply the Euclidean norm in a *constraint*, rather than have it appear in the penalty functional. We describe next how to adapt NNLS or an equivalent QP solver to our ends.

To be specific, suppose once more the linear forward problem is approximated by a system like (15) and we wish to minimize the uniform norm $\|\mathbf{m}\|_\infty$ and to simplify the situation for the moment we restrict ourselves to $\mathbf{m} \geq 0$. This is the ideal body problem except that the misfit condition will be in the form of a 2-norm, namely,

$$\|G \, \mathbf{m} - \mathbf{d}\|_2 \leq T \ . \tag{32}$$

The very first thing to be done is to verify that the value of misfit is possible for nonnegative $\mathbf{m}$; we do this by solving the QP problem of (31) with G substituted for A and $\mathbf{d}$ for $\mathbf{b}$ to confirm that the smallest misfit lies below T. To find the smallest model our strategy is to solve a minimization problem with a quadratic penalty and inequality constraints. Define the function

$$\mu[n] = \min_{\substack{\|\mathbf{m}\|_\infty \leq n \\ \mathbf{m} \geq 0}} \|G \, \mathbf{m} - \mathbf{d}\|_2 \tag{33}$$

whose numerical evaluation we shall discuss in a moment. We sweep through positive values of n, evaluating μ for each one. The following familiar argument shows that μ cannot increase with increasing n. (The reasoning is the same as that used for generating depth rules from ideal bodies.) The vector $\mathbf{m}$ providing the solution to (33) at n satisfies the constraints for any other $n' > n$ and so we know at the very least that $\mu[n'] \leq \mu[n]$. As we sweep n upwards we generate a curve like the one in figure 4.06c that divides the plane into a feasible part above and to the right of the curve and nonfeasible below and to the left. Suppose we find n_T, the value of the model norm for which $\mu[n_T] = T$; then exactly as in section 4.05, n_T must be the minimum uniform norm associated with the specified misfit T, because all other models with this misfit have larger norms. It is not difficult to make the argument valid for the situation with discontinuities in μ or exactly flat portions of the boundary curve; in practice μ is found to be a smoothly decreasing function of n. Thus we have solved the original norm minimization problem if we can solve the QP problem (33), which is not in the standard form.

The stability of NNLS allows us to use a weighting method, along the lines just given for applying the 1-norm misfit constraint, or, to recall a much earlier problem, as in the weighted solution for the seminorm-minimizing vector of 2.08. Let us introduce a slack variable $s_k \geq 0$ for each component to apply the bound on the norm:

$$m_k + s_k = n, \ \ k = 1, 2, 3, \ \cdots \ M \ . \tag{34}$$

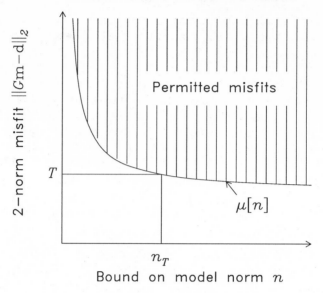

Figure 4.06c: Rough sketch of the plane of the bound on the model norm and the associated minimum 2-norm misfit. The data misfit of any model obeying the constraints in (33) must lie above the curve.

Obviously we have now doubled the dimension of the model space since there are exactly as many slack variables as original unknowns. To apply an equality condition like (34) we put it into the quadratic penalty with a heavy weight: the new functional for minimization is

$$\left[w \, \|\mathbf{m} + \mathbf{s} - n \, \mathbf{i}\|_2^2 + \|G \, \mathbf{m} - \mathbf{d}\|_2^2 \right]^{\frac{1}{2}} \tag{35}$$

where $\mathbf{i} \in \mathbb{R}^M$ is a vector of ones and w is a large real number. Equivalently the matrix $A \in M(N \times 2M)$ in (31) has become

$$A = \begin{bmatrix} w^{\frac{1}{2}}I & w^{\frac{1}{2}}I \\ G & O \end{bmatrix} \tag{36}$$

where $O \in M(N \times M)$ is a matrix of zeros and $I \in M(M \times M)$ is the unit matrix. The vector $\mathbf{b} \in \mathbb{R}^{2M}$ is now

$$\mathbf{b} = \begin{bmatrix} w^{\frac{1}{2}}n \, \mathbf{i} \\ \mathbf{d} \end{bmatrix} . \tag{37}$$

When the optimization program minimizes the functional (35) it pays most attention to the heavily weighted term and then uses whatever freedom remains in the system to minimize the misfit portion. Unless w is infinite, (34) cannot be exactly enforced but experience indicates that the behavior is identical to that seen with the simpler weighting solution discussed in section 2.08: it is possible by suitable choice of w to obtain compliance with (34) almost to the full accuracy of the computer.

For inverse problems of moderate dimension ($M > 100$ say) the wastefulness of the slack variables in the weighting approach becomes excessive, particularly as a sequence of optimizations must be performed as n is varied. It is not difficult to apply the standard theory used for NNLS (see Lawson and Hanson, 1974) to write a program for direct solution of the following minimization problem:

$$\min_{\mathbf{l} \leq \mathbf{x} \leq \mathbf{u}} \|A\,\mathbf{x} - \mathbf{b}\|_2 \tag{38}$$

where $\mathbf{l}$ and $\mathbf{u}$ are specified vectors of lower and upper bounds on $\mathbf{x}$. Philip Stark and I have written such a program called BVLS for Bounded Variable Least Squares (Stark and Parker, 1994). Aside from the economies achieved by directly applying the upper bound without the use of slack variables, the program gains important advantages by being able to start from a user-supplied guess for the solution, usually generated in an earlier call to BVLS. Since a parameter sweep requires repeated optimizations of systems that differ only slightly from each other, BVLS properly used can be very efficient in these circumstances.

The final problem we mention is the construction of bounds on a linear functional of the model subject to the 2-norm data misfit criterion and other constraints such as model positivity or a known upper bound on the uniform norm of the model. The approach we have just developed works in much the same way: the 2-norm misfit is used as the penalty functional and the linear functional is included as a constraint which is swept through real values in search of the desired misfit. The weighting scheme can also be used here to apply the linear constraint. The situation is slightly more complicated now because the curve of μ against the varying parameter exhibits a minimum and the linear functional will possess a maximum and a minimum value. For proofs of these remarks in a particular case the reader should consult Stark and Parker (1987).

The next section will provide illustrations of most of the ideas discussed here in the context of a new inverse problem and a very familiar one.

Exercises

4.06(i) Consider the problem of interpolation of a real function of a single variable discussed in section 2.07; instead of minimizing the 2-norm of d^2f/dx^2 apply LP theory to minimize the uniform norm of the second derivative. Show that the interpolating curve consists of connected pieces of a polynomial as in the case of cubic splines. What is the degree of the polynomial? Are the knots of the interpolation identical to those of cubic spline interpolation?

4.06(ii) Astronomical and satellite observations of the planet Mars show that the mean density $\bar{\rho}$ is 3933 kg m^{-3} and the ratio of moment of inertia C to Ma^2 is 0.365. Recall from exercise 4.04(i) that we may write

$$d_1 = \tfrac{1}{3}\bar{\rho} = \int_0^a \rho(r)\, r^2 \, dr/a^3$$

$$d_2 = \frac{\bar{\rho}C}{2Ma^2} = \int_0^a \rho(r)\, r^4 \, dr/a^5$$

where $\rho(r)$ is the density at radius r.
Suppose it is reasonable to assume the density can only increase with depth and that the surface density is 2700 kg m^{-3}. Show how these assumptions and the two data can be used to calculate an upper and a lower bound on the density at the center of Mars. Explain how such bounds can be obtained for other depths also. If the assumption about density gradient is dropped, can anything be concluded? Show how bounds on pressure might be found. Are these calculations significantly different? If so explain why.
Hint: Integrate by parts and write the density as the integral of its gradient.

4.07 Examples using LP and QP

Our first illustration will be an ideal body calculation, more elaborate than any of those in section 4.05. The following problem is patterned after one described by Grant and West in their famous text (1965): a gravity profile is measured across a glacier with the intent of estimating the thickness of the ice. In this setting it is permissible to approximate the glacier and the valley that it partly fills by an infinitely long system, thus allowing us to work

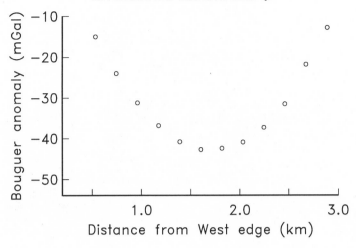

Figure 4.07a: Terrain-corrected gravity anomaly values at twelve equally spaced stations across a glacier.

in a cross-sectional plane. To be useful the original observations must be carefully corrected in several ways, for example, to remove the gravitational attraction of the surrounding mountains. We shall assume this has been done properly; the data shown in figure 4.07a represent the gravitational deficit from the replacement of dense rock (density, $\rho = 2700\,\mathrm{kg\,m^{-3}}$) by ice ($\rho = 1000$ $\mathrm{kg\,m^{-3}}$) measured at twelve equally spaced stations across the glacier. It is therefore convenient to think of the disturbance to the gravity as being caused by a body of negative density contrast $\Delta\rho = -1700\,\mathrm{kg\,m^{-3}}$, which naturally produces a negative gravity

Table 4.07A: Gravity Anomaly Values

Distance from edge (m)	Anomaly (mGal)	Distance from edge (m)	Anomaly (mGal)
535	−15.0	1819	−42.4
749	−24.0	2033	−40.9
963	−31.2	2247	−37.3
1177	−36.8	2461	−31.5
1391	−40.8	2675	−21.8
1605	−42.7	2889	−12.8

Observation sites

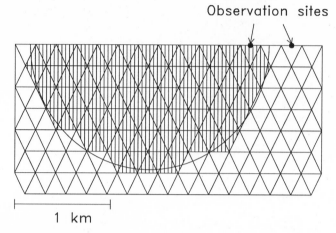

1 km

Figure 4.07b: Tessellation into triangles used for the LP approximate solution of the ideal body problem for least density based upon two data. The shaded triangles are those containing material. Also shown is the semicircular surface of the true ideal body associated with the same data.

anomaly. The solution to the forward problem was part of exercise 4.05(ii):

$$d_j = \Delta g\,(\mathbf{r}_j) = \int_D \frac{2G\,\hat{\mathbf{z}} \cdot (\mathbf{r}_j - \mathbf{s})}{|\,\mathbf{r}_j - \mathbf{s}\,|^2}\,\Delta\rho(\mathbf{s})\,d^2\mathbf{s},\, j = 1, 2, 3,\, \cdots\, N\,. \quad (1)$$

To obtain a finite-dimensional approximation like equation 4.06(15) we subdivide the x-z plane into elementary triangles within which the density is constant. The contribution from the kth unit-density triangle at the jth observer position forms the collection of matrix elements G_{jk} in 4.06(15). In this simple problem that integral can be expressed in elementary functions—it is a special case of the formula for a polygonal prism given in any book covering gravity interpretation in geophysics, such as Grant and West (1965). (The attraction of an arbitrary tetrahedron, needed for the 3-dimensional problem, can also be expressed in elementary terms, although the result is impressively complicated.) Having chosen a suitable level of subdivision we are ready to apply linear programming to make some deductions from the gravity data.

First we perform the calculation of the smallest uniform norm of density contrast, the first ideal body problem with the

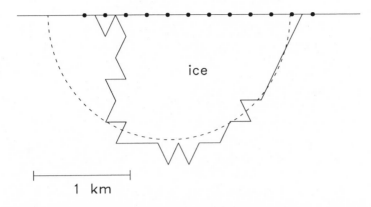

Figure 4.07c: Outline of the LP approximation of the minimum density contrast model satisfying all twelve gravity data. Also shown dashed is the semicircle of the two-data ideal body.

constraint that the contrast be negative. In this case it is an entirely academic exercise because the surface of the anomalous mass, the glacier, is exposed and the density of the surrounding rock is known too. Our object here is to see how good the finite-dimensional version of the exact theory is. Taking only two of the gravity data, the tenth and the twelfth in table 4.07A, we find the ideal body is a hemicylinder in exact analogy to the hemisphere of the full 3-dimensional solution; the bound obtained upon plugging in the numbers is $\|\Delta\rho\|_\infty \geq 1330 \, kg \, m^{-3}$. With the tessellation shown in figure 4.07b the solution of the LP problem for the bound yields $1339 \, kg \, m^{-3}$. This is excellent agreement. The approximate ideal body and the actual one are both shown in the figure: the shaded triangles contain nonzero density and, as we predicted in the previous section, all but two have the bounding density contrast. The approximation of the shape could not be better within the limitations of the discretization.

The two gravity data were chosen for this demonstration because they yielded the greatest bound among all pairs drawn from the table. How much greater is the bound when all the data are used together? The LP calculation proceeds without difficulty (in contrast to an expansion of the ideal body computations to twelve data) to yield $\|\Delta\rho\|_\infty \geq 1389 \, kg \, m^{-3}$. Thus the complete set does little to improve the bound obtained from a single data pair. Figure 4.07c shows the outline of the more complex minimum-contrast body in this case.

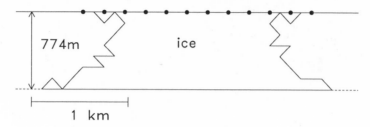

Figure 4.07d: Outline of the thinnest body of density contrast $-1700\,\mathrm{kg\,m^{-3}}$ exactly satisfying the anomaly data in table 4.07A.

As is usually the case in real life, the density of the anomaly causing mass is actually known quite well here; the density bound provides us with no new information. We are more interested in deducing constraints on the shape or spatial extent of the object; the same problem arises in other geophysical contexts, such as ore prospecting and in tracing the depth to the Moho. In the present example the depth to the top of the ice mass is not in question as it might be in some situations; but the depth to the base of the glacier is a significant unknown. We can ask, What is the shallowest possible structure compatible with the data and a density contrast of $-1700\,\mathrm{kg\,m^{-3}}$? We imagine finding the ideal body confined to lie between a horizontal plane at some depth and the surface; then the plane is raised until the density of the ideal body agrees with the known value. This gives the body with the minimum extent in depth by the argument of section 4.05. Carrying through the calculations (which required very modest computer time—less than 10 seconds on Sun Workstation) we discover the body shown in figure 4.07d and the thickness bound of 774 m. Thus we have conclusively demonstrated that the valley must be deeper than 774 m at some point. The reader might be curious to learn the upper limit of ice thickness in the valley: unfortunately, nothing very useful can be said. This is because thickness profiles like $h(x) = h_0/|x|$ have finite gravitational attractions at the surface, but extend to infinite depths; we shall discuss this in greater detail in section 5.07. Finally for this example, let us consider the minimization of another linear functional when the density contrast is set at the known value: we minimize the functional

$$L_1[\Delta\rho] = \int \Delta\rho(\mathbf{s})\, d^2\mathbf{s} .$$ (2)

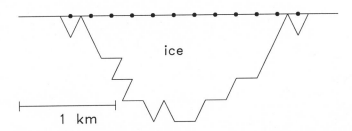

Figure 4.07e: Outline of the body with density contrast $-1700\,\mathrm{kg\,m^{-3}}$ exactly satisfying the anomaly data and possessing the least cross-sectional area.

Since we know the constraints cause the function $\Delta\rho(\mathbf{s})$ to be either zero or $\Delta\rho_0$, this model has the least area of cross-section. The details of the approximate LP problem are left for the reader. We find the minimum area to be $1.51\,\mathrm{km^2}$, with the discretization of figure 4.07b. We would expect the resulting model to be a very compact solution and indeed, as figure 4.07e shows, the valley wall is fairly plausible in shape, aside from two small indentations.

To obtain more information it will be necessary to assume more about the structure. In this problem one could reasonably rule out overhangs like those in figures 4.07d and 4.07e, or even very steep valley walls. To apply such constraints requires a reformulation of the forward problem that makes it nonlinear; we must wait until the next chapter for this development, although we shall discuss only the question of model construction.

Each of the models constructed so far by LP has satisfied the data values precisely. Rather than repeating all the calculations including appropriate uncertainties let us move on to another inverse problem with uncertain observations. We reconsider the seismic dissipation problem. In section 4.02 we concluded from an inspection of the resolving functions in figure 4.02b that the measurements could at best provide only crude averages of the attenuation in the upper mantle and lower mantle. The discussion in the following section should have convinced the reader that even this modest claim is unsupportable unless we inject addition constraints on the kinds of functions permitted. A suitable property of q mentioned several times already is positivity. We ask first, What is the positive attenuation function q fitting the estimates of Q_l with the smallest maximum value? From the

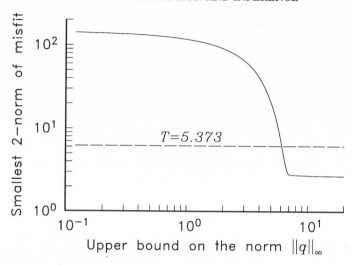

Figure 4.07f: Smallest 2-norm of misfit of positive attenuation models subject to the constraint that $\|q\|_\infty \leq n$; see equation (4). We have chosen the tolerance T so that 95 percent of the time the true data vector under random Gaussian perturbation will have a misfit less than T.

answer to this question we determine a value of the attenuation function that must be achieved somewhere in the mantle. The calculation falls squarely within the province of BVLS as defined in 4.06(38), provided we retain the weighted 2-norm for our misfit measure. Recall from 2.08 that we write the solution to the linear forward problem as

$$G\,W\,\mathbf{q} = \mathbf{d} \tag{3}$$

but here W corresponds to an integration scheme like the trapezoidal rule which is not upset by discontinuous functions. To find the upper bound on $\|\mathbf{q}\|_\infty$ we solve

$$\min_{0 \leq \mathbf{q} \leq \mathbf{u}} \|\Sigma^{-1}G\,W\,\mathbf{q} - \Sigma^{-1}\mathbf{d}\|_2 \tag{4}$$

where the upper bound vector $\mathbf{u}$ consists of

$$\mathbf{u} = n\,(1, 1, 1, \cdots 1)\,. \tag{5}$$

In these expressions we have had to include the diagonal matrix Σ of standard errors, because the uncertainties of the data are not identical. As discussed in the previous section we sweep through positive values of the parameter n. When the 2-norm matches

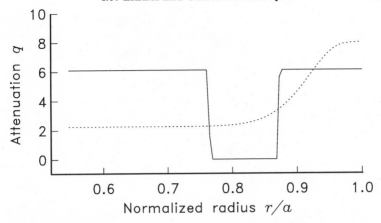

Figure 4.07g: Approximation to the minimum uniform-norm, positive solution with misfit $T = 5.373$. The dashed curve is the minimum seminorm model discovered in section 3.06.

some desired misfit T, n is the smallest value of $\|q\|_\infty$ for any model with that misfit. In figure 4.07f we plot the results of this calculation. We set the misfit tolerance $T = 5.373$, which from the standard χ^2 tables corresponds to a probability of 0.95 that the true vector of 18 data is contained within the ball of uncertainty. Notice that as n becomes very large, the misfit does not tend to zero: this implies there is no positive model fitting the given data exactly. Earlier we had suggested for problems of this kind that the minimum misfit model be found (with NNLS for example) to be sure solutions of some kind do exist, but in this case we already possess positive solutions with reasonable misfit levels from the calculations of section 3.06.

We find with $T = 5.373$ that $\|q\|_\infty \geq 6.198$. Note how steeply the curve falls for misfits between 3 to 50. This has the gratifying consequence that the minimum norm solutions are quite insensitive to the precise choice of misfit tolerance T: if the probability value is raised from 0.95 to 0.99, the corresponding bound on the norm changes to 6.110. We conclude that in order to fit the data, an attenuation of more than 6.198 must occur somewhere in the mantle. The minimum norm model, shown in figure 4.07g, is perhaps somewhat unexpected: it exhibits the high attenuation in the upper mantle that we have become used to in earlier solutions, but the high values in the lower mantle have not been seen before in our models. We can already guess that the behavior below $r/a < 0.76$ is not well constrained by the data.

To extract some information about the *shape* of the function q we appeal to the notion of resolution. The length scales discussed in 4.02 are not rigorous averaging intervals, but we can use them as a rough guide. Here we repeat the analysis of section 4.02 except that instead of a bounded model norm, positivity supplies the essential constraint to keep the average values bounded. Recall that the idea is to find upper and lower bounds on the average value of q in intervals of our choosing; the corresponding predictor is

$$p[q] = \frac{1}{r_2 - r_1} \int_{r_1}^{r_2} q(r)\, dr .$$

(6)

To apply the 2-norm misfit constraint together with the positivity condition, we use NNLS including a heavily weighted row in the matrix as sketched in the final paragraphs of the previous section; we minimize over nonnegative $\mathbf{q}$:

$$w\,(\mathbf{p}\cdot\mathbf{q} - P)^2 + \| \Sigma^{-1} G\, W\, \mathbf{q} - \Sigma^{-1}\mathbf{d} \|_2^2$$

$$= \left\| \begin{bmatrix} w^{1/2}\mathbf{p} \\ \Sigma^{-1} G\, W \end{bmatrix} \mathbf{q} - \begin{bmatrix} w^{1/2}P \\ \Sigma^{-1}\mathbf{d} \end{bmatrix} \right\|_2^2 .$$

(7)

The right side is the form required by NNLS, equation 4.06(31). By choosing a large value of w we obtain almost exact satisfaction of the equation

$$\mathbf{p} \cdot \mathbf{q} = P$$

(8)

which is the finite-dimensional approximation of $p[q] = P$. In the calculations about to be described, when w was chosen to be 10^6, the error in matching (8) was typically a part in 10^4 or less. We vary P over positive values until the minimum 2-norm of misfit to the data agrees with the stipulated tolerance T. The misfit curve as the average P is varied always exhibits a minimum, corresponding to the average of the best-fitting model. If the tolerance level intersects the curve twice, these points are the lower and upper bounds on $p[q]$: this is the case with the solid curve in figure 4.07h where we seek the average q over the interval $(0.9, 1)$ in normalized units. We find that $p[q]$ must lie between 3.55 and 16.8. If we move the interval of interest down to $(0.8, 0.9)$ we obtain the lightly dashed curve: now there is only

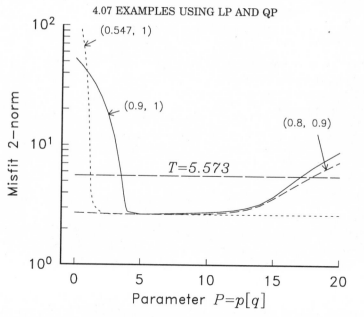

Figure 4.07h: Minimum misfit as a function of P, the averaging functional $p[q]$ for three intervals of averaging as indicated.

one intersection with the desired misfit at 17.6; thus the average q in the interval may be zero, which means of course that then q itself could vanish on the entire interval, but $p[q]$ must be less than 17.6. The third curve (long dashes) is the one corresponding to the problem of finding a bound on the average throughout the mantle. The lower bound on the q throughout the mantle is 1.17; the upper limit cannot be shown on the graph because it is enormous: 95,200. We can confirm that virtually nothing is known from these data about the lower mantle: when we seek bounds on the average over the interval (0.547, 0.9), the only constraint is that $p[q]$ must be less than 122,000. These calculations are based on an impoverished set of data with almost no sensitivity to the lower mantle and large error bars. Nonetheless, we have shown that rigorous conclusions can be obtained. With a reasonably dense and accurate set of observations we anticipate much stronger inferences will be possible. A seismic example of this kind can be found in the paper of Stark et al. (1986).

The magnetic anomaly problem does not benefit greatly from the techniques we have been using because the magnetization may take either sign. A completely different approach based upon statistical assumptions is appropriate, and that is the subject of the next section.

Exercise

4.07(i) In the calculations for the glacial valley an exact fit was
demanded to the gravity anomaly data. Consider the problems solved
when we require only that the 2-norm misfit be less than T. Show
how to use NNLS and BVLS with weights to solve the problems of
finding minimum sup norm of density, maximum depth of burial, subject to a bound on density and minimum cross section.

4.08 A Statistical Theory

In this chapter we have looked for additional assumptions that
could be brought together with the observations so that definite
inferences might be drawn about model properties. Statistical
theory has played a minor role, having appeared in chapter 3 only
to justify an appropriate value of the misfit tolerance. As many
readers no doubt are aware, there is a school of inverse theory in
which statistics lie at the heart of the entire enterprise; this
approach is laid out in great detail by Tarantola (1987). It would
be redundant to repeat that exposition here, but statistical
assumptions about the model present a powerful means of introducing subtle constraints in a way quite different from any we
have seen so far. We shall describe only one kind of statistical
technique and link it closely with an example. A brief inspection
is certainly in order.

   The fundamental assumption of the statistical theories is
that the model we seek is but a single realization of a random process whose general behavior we know more or less completely.
This extra knowledge, when combined with the observations, goes
a long way toward reducing uncertainty in estimates of desired
properties, which for the sake of consistency we take to be
bounded linear functionals of the model. A question that must be
faced is whether the representation of the model by a random process is a sensible idea from a physical or philosophical standpoint.
Ironically, we can say at the outset that any statistical characterization of the Earth can never be declared incorrect; its worst
defect is that of being wildly improbable!

   One's response depends on how one feels about a fundamental question in the foundation of statistics: whether the idea of
repetition necessarily plays any part in identifying a statistical
description with a physical system or not: at one end of the spectrum, Bayesian theorists like Jeffreys (1983) maintain that probability can be defined without the need for an actual or even

potential repetition of an observation. Most scientists, including the author, feel happier if a supposedly random process is capable of repeated realizations in time or space so that there is some real or potential population over which averages and other operations can be carried out. Thus the present-day magnetic field of the Earth can be plausibly identified as the current realization of a time-varying field with certain statistical properties (Cox, 1981; Constable and Parker, 1988a). One can, in imagination at least, observe a feature of the field again and again at different times and thereby generate a population whose statistics could be defined. A similar, but less compelling, justification could be given for properties like the profile of temperature in the Earth's mantle on the grounds that on scales of tens of millions of years, tectonic activity and convective overturn continually perturb the temperature. Spatial resampling provides a more convincing framework than this in many cases: for example, a crustal seismic p-wave profile might be considered to be just one of a population taken at a large number of different sites. The example we examine later treats a particular marine traverse as just one of a population of such profiles throughout the surrounding marine province. This discussion suggests that it is always possible, by stretching a point, to come up with a statistical context for any model, though a prudent person would be unlikely to place much credence in conclusions that depend directly on the more artificial constructs.

If we adopt a statistical description, we shall need two things for it: a mathematical model and the numerical parameters that enter in its definition. There are formidable mathematical difficulties in defining a perfectly general statistical definition of a random function: one is the technical question of the apparent need to define a "volume" in a space of functions; another is the complexity in the description of an arbitrary random function. To illustrate the latter point, imagine a statistical model of the density within the Earth. At each point we need to know the probability density function of the random variable, which could vary from place to place. But knowledge of the pdf at each point is far from a complete description, because it would be most artificial to demand that densities at a collection of points be independent of each other; therefore we must specify the conditional probabilities, for example, of density at pairs of points, so that when the density at point A is fixed, it will be known how the statistics of the value at B are modified. Or the similar question for sets of three points, and four, and so on. The amount of necessary specification in a general situation is almost inconceivable. To make practical progress, vast simplification is unavoidable. A most efficient scythe to cut through the proliferation is the idea of *stationarity*: the statistical properties of a stationary random function are identical at each point.

In the case of the global magnetic field, stationarity in time is an acceptable approximation, in suitably short geological interval, say about 10 million years, but over long spans the variable reversal rates might be a serious embarrassment. We shall appeal to this notion in the numerical example. If we think of functions that vary over the surface of the Earth, like topographic height, of magnetic field intensity, stationarity in space looks like a workable hypothesis, provided we remain within a single geological province. For vertical profiles of limited extent, we might be able to argue for stationarity, but not, I believe, with any conviction if the profile spans the mantle, or straddles any of the major discontinuities.

For simplicity's sake we consider models depending on only a single independent variable, traditionally identified with time, although in most of the applications in inverse theory a spatial coordinate would be more appropriate. For a case worked out on a plane, see Caress and Parker (1989). In geodesy the theory is carried out for a spherical surface and called *least-squares collocation* (Moritz, 1980). The literature on stationary processes of a single variable (time-series analysis) is huge; for reference the book by Priestley (1981) is recommended, and there the reader will find the proofs of the numerous assertions about to be made. For more on stationary processes in space see Vanmarcke (1983). We give a brief sketch of essential results.

A real *stochastic process*, which we denote by $X(x)$, is a random function; we imagine an infinite series of experiments each one of which produces a new realization of the function. When averaging or expectation is carried out, it is over the different realizations. We concentrate on *stationary processes* in which the statistical properties of the process are independent of the position x. In the geophysical literature it is common to assume in addition that the underlying pdf of the process is Gaussian, although this is inessential. In a stationary process the mean value of the function is obviously constant, and there is no loss of generality in taking that value to be zero. Thus

$$E[X(x)] = 0 . \tag{1}$$

Next we consider the covariance between samples of the process taken at two points; when their means are zero, the covariance is the expectation of the product of two random variables, as in 3.01(35). We define the *autocovariance function* by

$$R_X(y) = E[X(x+y)X(x)] . \tag{2}$$

Because of the assumption of stationarity, the function R_X does not depend on the value of x that is chosen on the right side. Clearly the autocovariance is telling us something about how smooth the stationary process is: for example, if R_X is a broad, slowly decaying function it means that there is no abrupt variation in space. A measure of the total variability of the process is the process variance which is obviously

$$\text{var}[X] = R_X(0) = \sigma_X^2 \tag{3}$$

and it probably will come as no surprise to the reader that σ_X can furnish a norm for the process. The property of stationarity requires $R_X(x)$ to be even in x.

Another function, carrying the same information about X as R_X is the *power spectral density*, or PSD: it is simply the Fourier transform of R_X

$$S_X(\lambda) = \int_{-\infty}^{\infty} R_X(x) e^{-2\pi i \lambda x} \, dx . \tag{4}$$

The evenness of R_X shows us that S_X is real and even; less obvious is the fact that S_X is never negative. Another useful property, derived from (3) and the inverse Fourier transform of (4) is

$$\text{var}[X] = \int_{-\infty}^{\infty} S_X(\lambda) \, d\lambda . \tag{5}$$

The importance of the PSD is that it tells us how much variability there is in X as a function of wavenumber (or, in the case of time series, frequency): if we imagine passing the X through a filter that lets through signals only in the narrow wavenumber interval $(\lambda, \lambda + d\lambda)$ the variance of the output is $S_X(\lambda) d\lambda$. Thus if the PSD has a single strong peak at a particular wavenumber λ_0, the corresponding stationary process resembles a sinusoidally fluctuating signal with period $1/\lambda_0$; the length scale over which the resemblance remains faithful is related to the width of the peak in the obvious way. More generally and less loosely, when a new stochastic process Y is generated from X by convolution with the function g (recall 2.04(8)):

$$Y = g * X \tag{6}$$

the corresponding PSD for Y is given by

$$S_Y(\lambda) = |\hat{g}(\lambda)|^2 S_X(\lambda) \tag{7}$$

where $\hat{g}$ is the Fourier transform of g. We now have enough background to begin an illustration. It should be noted that the generalizations of these results to stochastic processes on a plane (where x is replaced by $\mathbf{x} \in \mathbb{R}^2$) is fairly straightforward.

We take up once more the example of the magnetic anomaly data, abandoned in 4.03 for want of a suitable Hilbert space. We assert that the crustal magnetization can be thought of as a realization of a stationary stochastic process M. The reversals of the Earth's main field when viewed over a suitable interval appear to have just this property (Cox, 1981). Field reversals seem to be events with a roughly Poisson distribution in time, which means that the chance of one occurring in any small interval dt is proportional to dt but independent of the actual time itself. This is an approximation, but if we consider only the last 10 million years, say, it is a quite good model. Let us assert that the creation of marine crust at the ridge proceedes at a stately and quite invariant velocity u_0. Furthermore, we idealize the impression of the magnetic field into the magnetic crust by the production of a sequence of sections with constant magnetization in the normal and then reversed direction. In this way we have generated a stochastic process in space for the crustal magnetization. The autocovariance function of this process is easily shown to be (Jenkins and Watts, chapter 5, 1969)

$$R_M(x) = M_0^2 \, e^{-2\nu|x|/u_0} \tag{8}$$

where ν is the mean rate of reversal. Applying (4) we find the PSD

$$S_M(\lambda) = \frac{\nu u_0 M_0^2}{\nu^2 + \pi^2 \lambda^2 u_0^2} \tag{9}$$

where M_0 is the amplitude of the magnetization. These two functions are shown in figure 4.08a; the process by which we arrive at numbers for the various constants will be discussed later. The familiar linear connection between the observed magnetic anom-

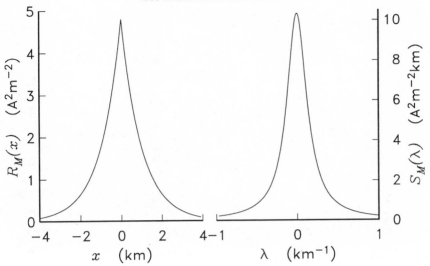

Figure 4.08a: Autocovariance function R_M and corresponding PSD S_M for the crustal magnetization stochastic process.

aly and the magnetization can be written as the convolution:

$$D = g * M \tag{10}$$

and from 2.04(10)

$$g(x) = -\frac{\mu_0}{2\pi} \frac{x^2 - h^2}{(x^2 + h^2)^2} . \tag{11}$$

In (10) we view the anomaly profile D as another stationary process.

    Unlike the other worked examples of this chapter, here we finally attempt to discover the actual value of the magnetization rather than some other linear functional of the solution; the power of the statistical approach makes this possible and also allows us to provide an estimate of the accuracy. The plan of attack is to seek a linear combination of data values with which to approximate the magnetization function at a specified point. It can be shown that if the pdf of M is Gaussian the linear combination is optimal in the sense of having the smallest variance among unbiased estimators; here, if we are to take our stochastic model seriously, the Gaussian assumption is inappropriate and it follows that a nonlinear estimator might do better. To approximate the model at the point $x = 0$ we write

$$M(0) = \sum_{j=1}^{N} w_j \, d_j + E \tag{12}$$

$$= \sum_{j=1}^{N} w_j \, (g * M)(x_j) + E \tag{13}$$

$$= \sum_{j=1}^{N} w_j \, (g_j * M)(0) + E \tag{14}$$

where E is the error in the approximation and we have introduced N new convolution functions, displaced appropriately:

$$g_j(x) = g(x - x_j) . \tag{15}$$

We now imagine moving the point of estimation around, but keeping the observation locations in the same relative locations; because of stationarity, optimal weights for (14) would be optimal in the more general situation:

$$M(x) = \sum_{j=1}^{N} w_j \, (g_j * M)(x) + E(x) . \tag{16}$$

The purpose of this almost trivial change is that it creates a relationship among a set of stochastic processes. The construction of a good model involves the choice of suitable coefficients w_j that make the error E small in some sense. We rearrange to focus on the error term; at this point it is clear we have an equation for one stationary process, E on the left, in terms of another:

$$E = M - \sum_{j=1}^{N} w_j \, (g_j * M) \tag{17}$$

$$= (\delta - \sum_{j=1}^{N} w_j \, g_j) * M \tag{18}$$

since the delta function is the unit element for the operation of convolution. A natural size for the error in this setting is its variance. So we compute the value of the variance and then examine how to make this as small as possible by varying w_j. Before doing this we note that the position of the origin, where we are estimating M, is a convenience, and we can focus on different points by simply displacing the positions x_j as necessary.

First we find the PSD of E with (6) and (7):

$$S_E(\lambda) = |1 - \sum_{j=1}^{N} w_j \, \hat{g}_j(\lambda)|^2 \, S_M(\lambda) . \tag{19}$$

To obtain (19) we have used the formal result that the Fourier transform of a delta function is unity. For the variance of E we turn to (5):

$$V = \text{var}[\,E\,] = \int_{-\infty}^{\infty} |\,1 - \sum_{j=1}^{N} w_j \hat{g}_j(\lambda)|^2 \, S_M(\lambda) \, d\lambda \qquad (20)$$

$$= \|\, 1 - \sum_{j=1}^{N} w_j \hat{g}_j \,\|^2 \qquad (21)$$

where we have introduced the complex Hilbert space S associated with the inner product

$$(f, g) = \int_{-\infty}^{\infty} f(\lambda) g(\lambda)^* \, S_M(\lambda) \, d\lambda \qquad (22)$$

and the corresponding norm. Equation (21) shows that minimization of V is equivalent to finding the best approximation under the norm of S to the function 1 in the subspace spanned by the elements $\hat{g}_j$. Because unity is the Fourier transform of a delta function at the origin, we can reformulate the problem as one of seeking the best approximation to that delta function built from the basis g_j and using the norm of an isometric S' as our measure; see exercise 4.08(i). This is exactly the same as the Backus-Gilbert approach for finding functional estimates described in section 4.02, namely, finding an approximation of the functional of evaluation by the representers: (12) is just a rewritten form of 4.02(9) and (21) is equivalent to 4.02(19). Of course the choice of norm is motivated here in a novel way.

The variance V computed in (21) is the one associated with the stochastic nature of the model and is a measure of the fundamental inability of linear combination of the data to predict the model at any point. But the treatment of chapter 3 also allowed the measurements themselves to be erroneous, presumably because of magnetospheric signals and other sources of magnetic field. Such errors would be independent of the prediction error, and by assertion they are independent of each other. The variance of each datum is the same and here it will be called σ_d^2; from (12) we see this contributes an additional term

$$\text{var}[\sum_{j=1}^{N} w_j d_j] = \sum_{j=1}^{N} w_j^2 \sigma_d^2 . \qquad (23)$$

Thus the overall variance, which we minimize by choosing w_j, is given by

$$V = \sum_{j=1}^{N} w_j{}^2 \sigma_d^2 + \| 1 - \sum_{j=1}^{N} w_j \hat{g}_j \|^2 . \tag{24}$$

To solve the new minimization problem we can apply the algebra given in 4.02 for the problem of Backus-Gilbert resolution with noise. We omit intermediate details, which are straightforward, and cut straight to the answer: translating to the natural vector and matrix notation, to determine the optimal weight vector $\mathbf{w} \in \mathbb{R}^N$, we must solve the linear system

$$(\Sigma^2 + \Gamma) \mathbf{w} = \mathbf{u} \tag{25}$$

where $\Sigma = \sigma_d I$ and $I \in M(N \times N)$ is the unit matrix, Γ is the Gram matrix with entries

$$\Gamma_{jk} = (\hat{g}_j, \hat{g}_k) \tag{26}$$

$$= \int_{-\infty}^{\infty} \mu_0^2 \pi^2 \lambda^2 e^{-4\pi h |\lambda|} e^{-2\pi i (x_j - x_k)\lambda} S_M(\lambda) \, d\lambda \tag{27}$$

and the vector $\mathbf{u} \in \mathbb{R}^N$ has components

$$u_j = (1, \hat{g}_j) \tag{28}$$

$$= \int_{-\infty}^{\infty} \mu_0 \pi |\lambda| e^{-2\pi h |\lambda|} e^{-2\pi i x_j \lambda} S_M(\lambda) \, d\lambda . \tag{29}$$

Recall the expression for the Fourier transform $\hat{g}(\lambda)$ was given in 2.04(29), from which we easily obtain

$$\hat{g}_j(\lambda) = \mu_0 \pi |\lambda| e^{-2\pi (h |\lambda| - i x_j \lambda)} . \tag{30}$$

Notice in (25) there is no trade-off parameter or Lagrange multiplier; the relative importance of the two matrix terms on the left is set by the relative sizes of the data errors and the intrinsic variance of the stochastic process which represents the model. Having found the coefficients w_j by solving (25), we can find the model estimate with (12) and its uncertainty by calculating the minimum variance, which turns out to be:

$$V_{\min} = \| 1 \|^2 - \sum_{j=1}^{N} w_j (1, \hat{g}_j) \tag{31}$$

$$= M_0^2 - \sum_{j=1}^{N} w_j u_j . \tag{32}$$

It is the ability to provide an uncertainty like (32) that distinguishes the statistical approach from the model construction efforts of chapter 3. The reliability of that uncertainty depends on the credibility of the long string of statistical assumptions that have gone into obtaining this result. Furthermore, to proceed with a practical calculation, we must come up with a plausible PSD.

We return now to (9), the equation for S_M, which we justified on the basis of an idealized model of the generation of crustal magnetization; now we need numbers for the parameters in its definition. We could argue that in a marine magnetic survey, a geophysicist would know very well from the water depth and bathymetric characteristics that the local marine crust was young and hence a suitable value of v, the average reversal rate, could be chosen: according to the magnetic time scale of Harland et al. (1989), $v = 4.64 \times 10^{-6}$ yr^{-1} over the last 10 Myr. It is more arguable that a spreading velocity u_0 could be determined without detailed analysis of the magnetic anomalies—we are evidently in danger of circularity in our reasoning. Nevertheless, we shall supply the "correct" value, $u_0 = 10$ mm yr^{-1}. The amplitude of magnetization M_0 is difficult to prescribe without using the magnetic anomaly data themselves. We can estimate the variance of the observed values in the familiar way, by summing squares; we find

$$\text{var}[D] + \sigma_d^2 = 25{,}921 \, \text{nT}^2 . \tag{33}$$

The reader may object that here we have averaged in space to find var[D], when according to the idea of a stochastic process, we ought to have averaged over different realizations. In practice it is often impossible to obtain a population of realizations, and then spatial (or time) averaging over the available realization is the only option. (In this example there is in fact a natural population of realizations: other magnetic profiles observed at different places in the region.) If the stochastic process satisfies a certain property, *ergodicity* (Priestley, 1981), it can shown that spatial averaging is a legitimate alternative, provided one averages widely enough. The key to success is that the samples span a broad enough interval that correlations among them are not important: the correlation scale of the magnetic anomaly is about $u_0/v \approx 2.2$ km which is small compared with the 30-km interval of the data set.

To find M_0 we can compare the estimate in (33) with the prediction given by the power spectrum; combining (5), (7), and

(10) together with the definition of the inner product on S, we can easily show that

$$\text{var}[D] = \|\hat{g}_j\|^2 \tag{34}$$

$$= \int_{-\infty}^{\infty} \mu_0^2 \pi^2 \lambda^2 e^{-4\pi h |\lambda|} S_M(\lambda)\, d\lambda \tag{35}$$

$$= M_0^2 \int_0^{\infty} \frac{2\pi^2 \mu_0^2 \nu u_0}{\nu^2 + \pi^2 u_0^2 \lambda^2} \lambda^2 e^{-4\pi h \lambda}\, d\lambda. \tag{36}$$

We shall need to evaluate a lot of integrals like (36) for the Gram matrix; we defer the explanation for a moment. Having substituted values for ν, u_0, and h (which has been 2 km throughout the book), we can calculate the integral in (36): it is $5{,}406\,\text{nT}^2/\text{A}^2\,\text{m}^{-2}$. Since $\sigma_d = 10\,\text{nT}$, by comparing (33) with (36) we find that $M_0 = 2.19\,\text{A}\,\text{m}^{-1}$. Gratifyingly, this value is perfectly consistent with the magnitudes of magnetization found for the models in chapters 2 and 3.

All that remains to be done, now that a PSD for the model process is in hand, is to compute the Gram matrix and solve (25) for a series of points along the real axis. The reader will have no difficulty in seeing that after (9) has been inserted, the integrals in (27), (29), and (36) can be written in terms of the functions

$$f_k(a, b, c) = \text{Re} \int_0^{\infty} \frac{\lambda^k}{c^2 + \lambda^2}\, e^{-(a - ib)\lambda}\, d\lambda \tag{37}$$

where $a, b, c \in \mathbb{R}$, $a, c > 0$ and $k = 1$ for (29) and $k = 2$ for (27) and (36). Some light algebra yields

$$f_1(a, b, c) = \text{Re}\left\{ \frac{1}{2}[e^{iz} E_1(iz) + e^{-iz} E_1(-iz)] \right\} \tag{38}$$

$$f_2(a, b, c) = \text{Re}\left\{ \frac{c}{z} - \frac{ic}{2}[e^{iz} E_1(iz) - e^{-iz} E_1(-iz)] \right\} \tag{39}$$

where $z = ac - ibc$ and E_1 is the exponential integral function (Abramowitz and Stegun, chapter 5, 1968):

$$E_1(z) = \int_z^{\infty} e^{-t}\, \frac{dt}{t} \tag{40}$$

and the path of the integral runs parallel to the real axis. Abramowitz and Stegun list convenient Taylor series or continued

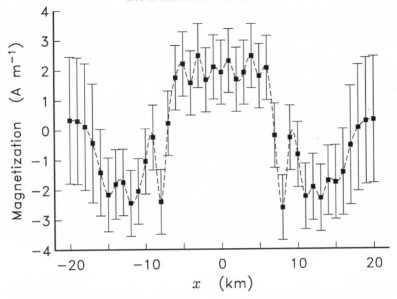

Figure 4.08b: Magnetization model estimates and associated uncertainties based on the stochastic model for magnetization with the autocovariance function shown in figure 4.08a.

fraction expansions for numerical calculations of E_1, although one needs to be careful in (39) to avoid loss of accuracy from cancellation when $|z|$ is large.

Finally we are in a position to compute all the parts of (25) and to solve for the weight vector **w**. This can be done for a series of points along the profile, and at each we can find the minimum variance estimate for M, via (12) and the corresponding uncertainty from (32). This we do at a set of points spaced by 1 km; the results are plotted in figure 4.08b. Strictly, the dashed curve is not a magnetization model but just a set of disjoint estimates of magnetization; for example, there is no reason why the function obtained by joining the points should generate a magnetic anomaly fitting the data within the observational errors. In this it is like the functional estimates of Backus-Gilbert theory, but in this case there is no resolving width—these are point estimates of the model. In the figure we see the oscillations in the magnetization, indicators of normal and reversed blocks of crust familiar from chapters 2 and 3; but notice the decline in magnetization amplitude away from the site of anomaly observation. Here is more evidence that the union of the point estimates does not constitute a true magnetization model, consistent with the original assumptions: these demand that the magnetization be statistically stationary, and therefore of similar amplitude everywhere on the real line. Notice how the uncertainties (given by the $V_{\min}^{\frac{1}{2}}$) grow as the estimation site moves away from the interval where there are observations: we may deduce from (32) that the uncertainty can never exceed M_0 or $2.19\,\mathrm{A\,m^{-1}}$. It is obvious from the figure that the estimates are not uncorrelated; one can easily compute the covariance between any pair and this is left as an exercise, 4.08(ii). Smaller error bars would result if we set up the theory to discover average magnetization within an interval, another exercise. The reason why the "model" in figure 4.08b looks rather rough compared with those in section 3.04 is that the covariance is derived from a model process permitting step discontinuities, and this is reflected in the PSD which falls off only as λ^{-2} as λ grows.

The attraction of the statistical approach is that it is able to do something other methods normally have great difficulty accomplishing: it provides a concrete value for the solution at each point and an accompanying uncertainty. Yet in achieving the goal of this chapter, the discovery of trustworthy information about the Earth, the statistical program leaves much to be desired. We must assume the unknown function is an example of a random process with a number of very special and exactly known properties. Not only must the norm of the model be known, but underlying smoothness as described by the covariance in space must be specified. In other geophysical settings it will often be hard to defend sweeping assumptions or to place much confidence in the numbers entering the statistical framework. While the assumptions and parameters are reasonably easy to justify in the magnetic anomaly problem, if one were to attempt to solve the seismic dissipation problem of 2.08 in a similar way, could one take seriously the stationarity of q, and where would one look for its autocovariance function?

Exercises

4.08(i) Write an inner product acting on f and g rather than on $\hat{f}$ and $\hat{g}$ and giving the same result. Express (22) as

$$(\hat{f}, \hat{g}) = \int_{-\infty}^{\infty} [\hat{f}(\lambda) S_M(\lambda)^{\frac{1}{2}}][\hat{g}(\lambda) S_M(\lambda)^{\frac{1}{2}}]^* \; d\lambda$$

and apply Parseval's Theorem to obtain the inner product of the space S'

$$(\hat{f}, \hat{g}) = \int_{-\infty}^{\infty} [f*s][g*s]\,dx = (f,g)'.$$

Show that with the magnetic anomaly PSD in (9)

$$s(x) = \frac{2M_0}{\pi}\sqrt{\frac{\nu}{u_0}}K_0\left[\frac{2\nu x}{u_0}\right]$$

where K_0 is the modified Bessel function of the second kind.

4.08(ii) Show that in general the Gram matrix is the covariance matrix of the data when measurement noise is ignored. Hence show that the components of the random vector

$$L^{-1}\mathbf{D}$$

are uncorrelated, where $\mathbf{D}$ is a random sample of the data vector and L is the Cholesky factor of Γ. Further, if the pdf of the model process is assumed to be Gaussian, show that the random variable

$$\mathbf{D}\cdot\Gamma^{-1}\mathbf{D}$$

is distributed as χ_N^2. Test the sample used in the magnetic anomaly example (listed below) to see if χ^2 is in an acceptable interval.

x_j (km)	d_j (nT)	x_j (km)	d_j (nT)	x_j (km)	d_j (nT)
−15	−175.20	−4	179.52	7	−34.74
−14	−185.61	−3	185.60	8	−158.59
−13	−194.74	−2	156.64	9	−115.69
−12	−218.23	−1	157.24	10	−121.50
−11	−187.71	0	155.84	11	−182.81
−10	−131.03	1	163.99	12	−205.81
−9	−108.89	2	154.81	13	−208.38
−8	−138.44	3	176.88	14	−180.26
−7	−8.84	4	208.54	15	−155.15
−6	127.49	5	193.84		
−5	190.90	6	128.84		

The model estimate is (16) with E set to zero. Find an expression for the covariance between the estimates of M based on this formula at two points x_1 and x_2; why is this different from $R_M(x_2-x_1)$?

4.08(iii) Investigate the question of estimating from the data d_j linear functionals of M other than the magnetization at a point. Give an explicit expression for an estimate of the average magnetization over an interval of x and find the associated uncertainty. If you have computer access, calculate the uncertainty of magnetization averaged over the central 10 km and compare this number with a typical error bar in that interval in figure 4.08b.

CHAPTER 5

NONLINEAR PROBLEMS

5.01 Some Familiar Questions

Most geophysical forward problems are nonlinear, and so by our definition the corresponding inverse problems are nonlinear also. But the concerns of linear inverse theory remain important: we should address the questions of existence, uniqueness, and construction of solutions, and finally of inference. Moreover, the fact the problems are nonlinear, does not mean we need abandon the setting of *linear* vector spaces; indeed, Hilbert spaces will be convenient for most of this chapter. From now on we cannot simplify the relationship between observation and model to be expressible as an inner product; instead we have to accept the more general functional given by 2.02(2)

$$d_j = F_j[m], \; j = 1, 2, 3, \; \cdots \; N \; . \tag{1}$$

Concrete illustrations follow shortly. Failure of linearity would seem to imply that none of the techniques we have developed can be applied to the class of problems. In a strict sense this is true, but as we shall see we will be able to build a useful approximate theory based on the linear theory of the earlier chapters.

The first question we tackled in 2.03 was that of the existence of solutions. We showed there that, aside from the technicality of linear dependence of the representers, it is *always* possible to find a model in accord with the linear version of (1), even if exact agreement between measurement and theory is required. In practice, the question turns out to be somewhat more complicated because we found later, in section 3.04, for example, that unrealistically rough or energetic models might be generated. Or if positivity conditions are imposed, these might make it impossible to satisfy observation. Nonetheless, in every case we have been able, through regularization or the use of linear or quadratic programming, to answer the question, "Do solutions exist?" when the forward problem is linear. For nonlinear problems the situation is quite different: there is in general no theory for answering this question.

The next issue is that of uniqueness: if it is granted that there are solutions, is there only one, or is there a family of acceptable models? Here the answer is the same as in the linear case. When we confine ourselves to a finite number of measurements that place constraints on a parameter distribution, which is a function, nonuniqueness is inevitable. If idealized measurements are proposed, in the shape of a complete function, free of error, uniqueness can sometimes be proved, but we shall devote no space to this kind of analytical problem.

Given the lack of uniqueness, the task of model construction entails the making of a choice: which particular model should be selected from the set of possible answers? Once more we follow the principle explored in the linear case: by means of optimization we seek the simplest, least exciting kind of solution. But, because the existence question is open, we are in the dark as to whether it is possible to find a solution of any kind. So construction carries with it the additional burden of the existence problem. This would be unimportant if the nonlinear construction techniques were wholly reliable, as they are for linear systems, but this is not the case. In fact the approach is to minimize an appropriate functional, whose domain is in principle an infinite-dimensional space, without the property of convexity. If the search turns up a satisfactory model, the existence question is obviously answered in the affirmative; if it does not, we cannot say whether this is because of the lack of a solution or the inadequacy of the search.

When we come to the question of making valid inferences, the reader will probably anticipate that again there is no satisfactory theory for the nonlinear case. We can put together an approximation to the Backus-Gilbert notion of resolution, but as we have already commented, even for linear problems this theory is more qualitative than quantitative.

The fact that there is a unifying framework for the linear theory means that linear problems are all fairly similar in some fundamental manner. In contrast, we find nonlinear problems are too diverse for there to be a rigorous general technique applicable to every one. Special methods, tailored for a few individual nonlinear problems have been discovered to provide complete answers to the questions of existence, construction, and even inference, and we shall look at some of these at the end of the chapter. The only general approach we offer concerns model construction, and this is based upon an approximation which turns the original problem into a linear one. But first, we describe two simple problems which will be our working examples to illustrate the approximate theory.

5.02 Two Examples

In this section we give the solution to the forward problem for two simple geophysical systems, one involving gravity data, the other electromagnetic measurements; both are nonlinear problems. Perhaps surprisingly, we begin by discussing a version of the gravity inverse problem introduced in 4.07, where it appeared as a linear problem. We make an apparently minor modification so as to avoid the possibility of re-entrant boundaries like those seen in figure 4.07d: we assert that the gravity anomaly is caused by the variation in the function $h(x)$, the thickness of a surface layer of ice whose density contrast with the basement is $\Delta\rho$. Because h is a single-valued real function of x, the horizontal coordinate, overhangs are forbidden. It is assumed the upper surface of the ice is the line $z = 0$; then we can integrate 4.07(1) to find the gravitational attraction observed at $(x_j, 0)$:

$$\Delta g(x_j) = \int_0^a dx \int_0^{h(x)} dz \; \frac{2G\,\Delta\rho\,z}{z^2 + (x_j - x)^2}$$

$$= \int_0^a G\,\Delta\rho \ln\left[\frac{(x_j - x)^2 + h(x)^2}{(x_j - x)^2}\right] dx \quad . \tag{1}$$

Here $\Delta\rho$ is constant, not a function of position as it was earlier; z is measured positive downwards.

Now the model is merely the shape of the boundary rather than the density contrast within a region. Considered as a functional over a space of functions (yet to be decided upon) (1) is evidently nonlinear in h. We need to restrict the domain of the functional to those functions with $h \geq 0$ since the derivation is incorrect for the case when the boundary breaks the surface of the glacier, a situation that is obviously improper in any case. What is an appropriate space for such models? Since L_2 has the simplest inner product, it might be a good idea to start there. Is $\Delta g_j[h]$ defined for every $h \in L_2$, or must further restrictions be placed on the domain in addition to positivity of h? This is an almost trivial question for linear functionals but it is quite difficult to answer even in this, the simplest of examples. The proof that Δg_j exists for every $h \in L_2(0, a)$ is left to the reader in exercise 5.01(i), with plenty of hints. Of course, the difficulties arise from the fact that elements of L_2 are not point-wise bounded, which really makes this space too big for geophysical purposes, something we have already discussed at length in chapter 2. We continue to use the numerical data tabulated in table 4.07A. Here we have an example of a nonlinear functional, which solves the forward problem and is a simple formula; next we describe a much more complicated, but not uncommon kind of forward functional, in which the model predictions are obtained by solving a differential equation.

The magnetic field of the Earth varies on time scales ranging from milliseconds to hundreds of millions of years. Roughly speaking, variations that take place over periods shorter than a year have their cause outside the Earth while those fluctuating more slowly than this have their origin in the core. The most notable exception to this rule is the eleven-yearly variation associated with the solar cycle. Externally generated magnetic fluctuations (due to solar-terrestrial electromagnetic and plasma interactions —see Parkinson, 1983) induce electric currents in the ground and, as one might expect, the intensity of those currents is related to the electrical conductivity of the Earth: this is the basis of the *magnetotelluric method* for probing the electrical structure of the Earth, invented by Caignard (1953). One must record simultaneous time series of the horizontal components of the electric and magnetic fields at a point on the surface of the Earth.

We make a number of drastic simplifications to reduce the problem to its simplest terms; even so, the resulting system is far more complex than the gravity problem just described. We shall not attempt to justify all the simplifications here, but we enumerate them in order of decreasing credibility:

(a) the displacement current term dropped from Maxwell's equations;

(b) the curvature of the Earth can be ignored;

(c) the constitutive relation between current density and electric field is the linear Ohm's law:

$$\mathbf{j} = \sigma \mathbf{E} \qquad\qquad (2)$$

with a scalar conductivity σ;

(d) effects of electrical polarization and magnetic permeability are ignored;

(e) in the atmosphere above the Earth, the magnetic field is uniform and horizontal in the $\hat{x}$ direction;

(f) the conductivity varies only with depth, z.

The observations recorded at the surface $z = 0$ are time series of the electric and magnetic fields. For theoretical purposes we imagine an experiment in which the magnetic source field could be controlled to be of unit amplitude in the $\hat{x}$ direction and set into pure sinusoidal oscillation at a number of frequencies $f_1, f_2, f_3, \cdots f_N$. In this experiment we would measure the corresponding electric field, which, because of the linearity of Maxwell's equations and Ohm's law, would also vary sinusoidally, with an amplitude and phase that depends only on the conductivity in the ground. It can be shown that the magnetic field throughout the system is in the $\hat{x}$ direction, and the electric field is at right angles to it. Thus we can introduce two complex scalar functions B and E of z via:

$$\mathbf{B}(x,y,z,t) = \hat{x} B(z) e^{2\pi i f t} \tag{3}$$

$$\mathbf{E}(x,y,z,t) = \hat{y} E(z) e^{2\pi i f t} . \tag{4}$$

Of course such an experiment cannot be performed, but that is unnecessary because we can obtain the same information by statistical analysis of the time series measured under normal conditions. The natural magnetic field has a broad Fourier spectrum and so we may consider it to be composed of a sum of sinusoidal variations; similarly with the electric field. Naively all we need do to obtain the response of the Earth to a unit magnetic field at a specific frequency f is to divide the complex Fourier transforms of the two observed series at that frequency:

$$Z(f) = \frac{\hat{E}_y(f)}{\hat{B}_x(f)} = \frac{E(0)}{B(0)} . \tag{5}$$

Dividing Fourier transforms to calculate Z is not a very satisfactory method of processing the observations, because of the poor bias and variance properties of the estimate in the presence of noise signals. There are much better techniques but it would take us too far afield to discuss them (see Jones et al., 1989); suffice it to say that Z can be estimated from the experimental time series. Our next concern is to predict $Z(f)$ from a model conductivity profile, $\sigma(z)$.

The Maxwell equations corresponding to Faraday's and Ampère's (without the displacement-current term) laws are

$$\nabla \times \mathbf{E} = -\frac{\partial \mathbf{B}}{\partial t}; \qquad \nabla \times \mathbf{B} = \mu_0 \mathbf{j} . \tag{6}$$

Putting (2), (3), and (4) into (6) we find after a little algebra that

$$\frac{dE}{dz} = -2\pi i f B(z) \tag{7}$$

$$\frac{d^2E}{dz^2} = 2\pi i f \, \mu_0 \sigma(z) E(z) \, . \tag{8}$$

Equation (8) gives us the differential equation for the electric-field scalar within the conductor ($z \geq 0$, since z positive is downward); we can find the magnetic field from (7) and the current density from (2). Now we need some boundary conditions.

There are two kinds of solution to (8): one increases exponentially with z, the other decays. The physics of the problem is that magnetic-field fluctuations generated outside the atmosphere cause eddy currents to flow in the Earth and these cannot increase indefinitely with depth. Therefore we may reject the increasing solutions; one way to supply a boundary condition is to demand $E \rightarrow 0$ as $z \rightarrow \infty$. It is often more convenient to solve (8) over a finite interval, rather than allowing infinite depth. So we propose that the system ends at $z = H$. This may seem to be an additional approximation, but if H is chosen to be very large (say a parsec) it is certainly less significant than (b) in the list of approximations. Then the boundary condition at the bottom becomes

$$E(H) = 0 \, . \tag{9}$$

An interpretation of (9) is that at $z = H$ there is a conductor with infinite conductivity. At the surface of the Earth, the magnetic field just above the boundary and just below are equal since current sheets are absent: this supplies a boundary condition at $z = 0$: from (7)

$$E'(0) = \left. \frac{dE}{dz} \right|_{z=0} = 2\pi i f B(0) \, . \tag{10}$$

Thus if (8) is solved under the boundary conditions (9) and (10) we can predict the complex electric-field scalar $E(0)$ that would be observed at frequency f when the conductivity profile σ is given. Hence the forward problem has been solved. Some minor rearrangement brings our solution into a more conventional form. For each frequency f, the complex *magnetotelluric response* is defined as

$$c(f) = -\frac{E(0)}{E'(0)} \tag{11}$$

$$= \frac{E(0)}{2\pi i f B(0)} = \frac{Z(f)}{2\pi i f} \, . \tag{12}$$

It will be noticed from (11) that c does not depend on the amplitude of the magnetic field, because (8) is a homogeneous equation and changing the size of $B(0)$ in (10) scales the solution by a constant factor, which cancels in the definition of c. Therefore, for each frequency f, the response c is a nonlinear functional of the conductivity profile. Equation (12) shows how c is connected to observation.

The physical meaning of c is not very apparent at this point. To understand what is going on, let us consider the simplest physical system, a uniform halfspace in which the conductivity is σ_0 for all $z \geq 0$. Then (8) is easy to solve: the general solution is

$$E(z) = \alpha e^{+kz} + \beta e^{-kz} \tag{13}$$

where

$$k = (2\pi i f \mu_0 \sigma_0)^{1/2} = (1+i)(\pi f \mu_0 \sigma_0)^{1/2}. \tag{14}$$

We must suppress the exponentially growing term and so:

$$E(z) = \beta e^{-(1+i)z/z_0} \tag{15}$$

where z_0 is the characteristic length scale for decay of the electric field, the *skin depth*, given by $z_0 = (\pi f \mu_0 \sigma_0)^{-1/2}$; notice the effective depth of penetration decreases like $f^{-1/2}$. From (7) we see that the magnetic field dies away with depth in the same way. Substituting (15) into (11) we find the response at frequency f of a uniform halfspace is

$$c(f) = \frac{1-i}{(4\pi i f \mu_0 \sigma_0)^{1/2}} = \frac{z_0(1-i)}{2}. \tag{16}$$

Thus the magnitude of the complex response, which has the physical dimensions of length, is about 0.7 times the skin depth in a uniform conductor; in a nonuniform conductor, we might expect $|c|$ to be a rough indicator of the average depth at which the electric current is flowing; see exercise 5.02(ii). For finite thickness models it is easy to see from the differential equation that $c(f) \to H$ as $f \to 0$.

Equation (16) suggests another way of expressing experimental results. Instead of calculating the magnetotelluric response c from the experimental data, we find the *apparent resistivity* thus:

$$\rho_a(f) = \frac{\mu_0 |Z(f)|^2}{2\pi f} = 2\pi \mu_0 f \, |c(f)|^2. \tag{17}$$

Table 5.02A: Magnetotelluric Data Set

Period (s)	ρ_a (Ωm)	Uncertainty std. dev.	Φ degrees	Uncertainty std. dev.
463.7	21.64	6.19	55.58	8.19
233.1	20.42	2.59	52.31	3.63
112.6	25.43	2.31	52.02	2.61
58.56	31.74	1.62	49.12	1.46
29.76	36.55	1.61	41.26	1.26
14.65	31.28	0.80	32.22	0.73
7.452	21.74	1.38	25.80	1.82
3.477	12.93	0.66	25.36	1.47
1.791	8.15	0.43	33.77	1.52
0.9290	8.16	0.60	49.53	2.10
0.4610	9.14	0.66	57.78	2.08
0.2123	15.87	0.86	62.65	1.56
0.1066	23.08	1.14	61.78	1.41
0.05159	33.42	1.46	61.47	1.25
0.02661	34.26	1.34	55.58	1.12
0.01202	50.82	2.84	53.43	1.60
0.006059	67.33	4.76	49.64	2.02

When the solution for the uniform halfspace is inserted into this expression we find that $\rho_a = 1/\sigma_0$, the resistivity of the material, independently of the frequency. Again, we might hope that if the conductivity profile is relatively smooth, (17) may indicate the ground resistivity and even its behavior with depth; so just by plotting ρ_a against period, we get a crude idea of the one-dimensional solution. This idea has intuitive appeal, and geophysicists prefer to report their magnetotelluric results as ρ_a and the phase of Z, rather than as complex c or Z. Alternative ways of expressing the experimental results (ρ_a instead of c) or different parameterizations of the model (resistivity ρ instead of σ) are options we have not encountered in our study of linear inverse problems. The only transformations we could make in the linear theory were linear ones (for example, in 3.03)—anything else would make the problem nonlinear, which would constitute a foolish complication; now the problem is already nonlinear and a nonlinear mapping does not necessarily complicate matters and it may even simplify them.

Table 5.02A lists a magnetotelluric data set, which will be used later to illustrate the theory. The numbers are based on a

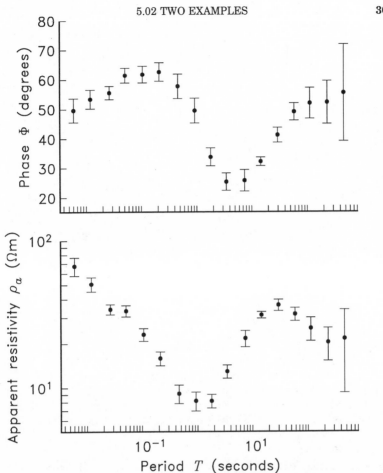

Figure 5.02a: The magnetotelluric data set, displayed as apparent resistivity and phase versus period. Note the enormous span of time scales, covering nearly five decades. To make them more easily visible, the error bars are drawn as plus-and-minus two standard deviations rather than the usual one.

study reported by Young et al. (1988) performed by four research groups in Oregon; for the purposes of illustration the author cooked up a composite data set for station 5 of this study, a site in the Cascade Mountains roughly 100 km east of Eugene. Values measured independently by the four groups have been averaged and the uncertainties estimated from the internal consistency of the observations. The table is presented in the manner conven-

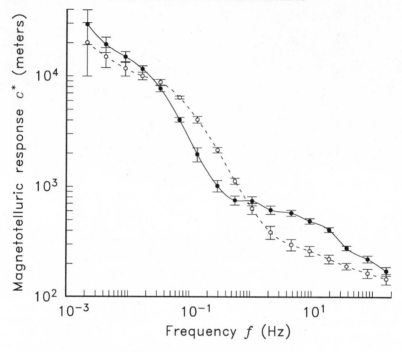

Figure 5.02b: The magnetotelluric data set, plotted as real and imaginary parts of c^* against frequency. The solid line connects the real part and the dashed line the imaginary part of the response. The error bars, like those on figure 5.02a, are twice the normal size.

tional in exploration geophysics, with ρ_a and Φ listed as function of period $T = 1/f$. The quantity Φ is the phase when Z is written $|Z| e^{i\Phi}$.

Figure 5.02a shows the same information; note the logarithmic time axis. Unlike the frequency, the apparent resistivity does not vary over a very large range, and the phase lies within twenty degrees of 45°, the value for a halfspace; on the assumption that the uniform conductor model is a valid guide for interpretation, we would conjecture from the figure that the true ground conductivities lie somewhere between 0.012 and 0.12 $S\,m^{-1}$. (The SI unit of conductance is the siemens, and $1\,S = 1\,\Omega^{-1}$.) In figure 5.02b we plot the magnetotelluric response against frequency. Since the imaginary part is always negative (a fact we shall prove later) we must plot the complex conjugate c^* if we wish to use log axes, a virtual necessity given the range of values.

The values of c suggest that at the highest frequencies in the study, the magnetotelluric currents are flowing only a few hundred meters below the surface, while at the longest periods there are currents at depths of over twenty kilometers.

Considered as a (complex) functional of σ, the response is obviously nonlinear—one need only look at the results for a uniform halfspace to see this. Elementary physics dictates that only positive σ can occur in the Earth, so the domain of the functional is the set of positive functions σ. We can ask what an appropriate space of functions might be, again in anticipation of the need for an inner product. Equation (8) is a linear differential equation of a standard form; the standard texts (e.g., Coddington and Levinson, 1955) show that on the finite interval $(0, H)$ solutions for this kind of equation exist for $\sigma \in L_1$; on a finite interval, the family L_2 is a subspace of L_1 and so c might be defined for every element in the positive subset of L_2. The only doubt about this is the possibility that $E'(0)$ in (11) might vanish for some f. We shall show later that this cannot happen for any real f, but when the theory is extended to cover complex f there are points in the complex plane where $c(f)$ is not defined and these points have the deepest significance for the theory; see section 5.08.

Exercises

5.02(i) Prove that Δg is defined on all of $L_2(0, a)$ by showing $\Delta g[h] \le c \|h\|^{1/2}$. First relabel the coordinates so that the observer is at $x = 0$ and set $G\Delta\rho = 1$; then

$$\Delta g[h] = \int_0^a \ln[1 + (h(x)/x)^2]\, dx \ .$$

Prove for all $\xi \ge 0$ that $\ln(1+\xi) \le c_1 \xi^{1/4}$ for some sufficiently large constant c_1. (In fact, whenever $c_1 > 1.479$.) Combine this result, the definition of $\Delta g[h]$ and Hölder's inequality with $p = 1/4$ and $q = 3/4$ to obtain the required result. Extend the result to $L_2(-\infty, \infty)$.

5.02(ii) Show by integrating (8) appropriately that c is a kind of complex center of mass for the electric-current density:

$$c = \int_0^H j(z)z\, dz \div \int_0^H j(z)\, dz$$

where $j(z) = \hat{\mathbf{y}} \cdot \mathbf{j}$.

5.02(iii) In both of the examples, let the models be defined on an infinite interval, for the gravity problem on $(-\infty, \infty)$ and for the magnetotelluric problem $(0, \infty)$. In the MT problem let $\rho(z) = 1/\sigma(z)$ be the model variable. Now show that on the one-dimension linear subspace of models defined by m = constant, the forward problem is linear for both kinds of problem.

5.02(iv) Solutions of the differential equation (7) in terms of elementary functions are not easy to find. However, with Bessel functions, power-law conductivity profiles $\sigma(x) = \sigma_0(x/a)^\nu$ are amenable to treatment (Watson, section 4.3, 1966). If u obeys

$$\frac{d^2u}{dx^2} = -\beta^2\gamma^2 x^{2\beta-1} u(x)$$

then u is a linear combination of Bessel functions:

$$u(x) = Ax^{1/2}J_{1/(2\beta)}(\gamma x^\beta) + Bx^{1/2}Y_{1/(2\beta)}(\gamma x^\beta).$$

If you have access to a computer program for complex argument Bessel functions of real order, calculate and plot ρ_a, Φ and c for the following conductivity profiles, taking $\nu = -1/4, 1/2, 1, 2$:

(a) $\sigma(z) = z^\nu$, $0 \le z \le 1$
(b) $\sigma(z) = (1-z)^\nu$, $0 \le z \le 1$
(c) $\sigma(z) = z^\nu$, $0 \le z < \infty$.

Which of these profiles is in the class treated in this section? You will have to decide how to choose A and B on the basis of (9) or the condition $E \to 0$ at infinity. The classic result for the derivative of J_ν will prove useful:

$$J_\nu' = -J_{\nu+1} + \frac{\nu}{x}J_\nu$$

and similarly for Y_ν.

5.03 Functional Differentiation

In this section we extend the familiar idea of differentiation to cover the derivative of a functional on a Hilbert space. It will then be fairly self-evident how to differentiate functionals on arbitrary normed spaces. Looking ahead, we shall need to be able to differentiate nonlinear functionals in order to apply something like Newton's iterative method for solving the equations for constructing models.

We get straight to the point: let $m \in$ H, a Hilbert space, and let $F: X \subseteq$ H $\to$ IR be a functional on some domain X in H, then F is said to be *Fréchet-differentiable* at the element m when we can find D such that for all Δ with $\|\Delta\| < \varepsilon$

$$F[m + \Delta] = F[m] + (D, \Delta) + R[\Delta] \tag{1}$$

and

$$R[\Delta]/\|\Delta\| \to 0 \quad \text{as} \quad \|\Delta\| \to 0 . \tag{2}$$

In this definition $\Delta, D \in$ H and R is another functional on H. The element D is called the *Fréchet derivative* of F at m. As we shall see, it plays the role of the gradient in ordinary multivariate calculus. Another way of writing the definition, perhaps a little more attractively, is:

$$F[m + \Delta] = F[m] + (D, \Delta) + o \|\Delta\| . \tag{3}$$

Intuitively, the definition is saying that in the neighborhood of an element m where F is Fréchet-differentiable, the functional behaves like a constant plus a bounded linear functional, that is, like an inner product with a fixed element D. It turns out that there are several kinds of derivatives for functionals and this one makes the strongest demands on F in terms of requiring good behavior in the vicinity of the point. As (3) makes clear, to find D we must perform a perturbation analysis of the forward problem, discovering how small variations about a fixed model change the value of a particular measurement. The difficulty from a mathematical standpoint is often not in finding D, but in being certain the functional R vanishes quickly enough, as described by (2).

As a first, almost trivial illustration, consider a continuously differentiable function on E^N with the ordinary Euclidean norm and associated inner product. Then

$$(\mathbf{x}, \mathbf{y}) = \mathbf{x} \cdot \mathbf{y} = \sum_{i=1}^{N} x_i y_i . \tag{4}$$

If F on this space is Fréchet-differentiable, we have from (1)

$$F[\mathbf{m} + \Delta] = F[\mathbf{m}] + (\mathbf{D}, \Delta) + R \tag{5}$$

$$= F[\mathbf{m}] + \Delta \cdot \mathbf{D} + R \tag{6}$$

$$= F[\mathbf{m}] + \sum_{i=1}^{N} \Delta_i \frac{\partial F}{\partial m_i} + R \tag{7}$$

which follows from the Taylor series expansion of F. Thus we can identify the element $\mathbf{D}$ with the ordinary gradient at $\mathbf{m}$, namely, $\mathbf{D} = \nabla F$.

Now to a more general, but nonetheless simple, example: suppose F is a bounded linear functional on H. Then by the Riesz Representation Theorem:

$$F[m] = (g, m).$$ (8)

It follows that

$$F[m + \Delta] = (g, m + \Delta) = (g, m) + (g, \Delta)$$ (9)

$$= F[m] + (g, \Delta).$$ (10)

Evidently the remainder R is identically zero here, so (2) is trivially obeyed, and we see that the Fréchet derivative D is just g itself. Notice that in this case D does not vary with m, the point of evaluation.

These examples show us that there is nothing very frightening about Fréchet derivatives. There is a fairly obvious version of the Chain Rule. Let f be a real, continuously differentiable function and suppose we wish to find the Fréchet derivative of $f(F)$ when we know F is differentiable at m with derivative D_F. Let

$$\Phi[m] = f(F[m]).$$ (11)

Then the reader should have little difficulty showing the Fréchet derivative of Φ is

$$D_\Phi = f'(F[m])D_F.$$ (12)

We can use this result to find the derivative of the norm of H. First let $F[m] = \|m\|^2$. Then we calculate the Fréchet derivative as follows:

$$F[m + \Delta] - F[m] = (m + \Delta, m + \Delta) - (m, m)$$ (13)

$$= (m, m) - 2(m, \Delta) + (\Delta, \Delta) - (m, m)$$ (14)

$$= (2m, \Delta) + \|\Delta\|^2.$$ (15)

Since in this case $R/\|\Delta\| = \|\Delta\|$, condition (2) is evidently satisfied for every m and the Fréchet derivative of $\|m\|^2$ is $2m$. To calculate the derivative of $\|m\|$ we apply (12) with $f(x) = x^{\frac{1}{2}}$; note that f is *not* differentiable at $x = 0$. Writing

$$\Phi[m] = \|m\| = (F[m])^{\frac{1}{2}}$$ (16)

we have, after some light algebra

$$D_\Phi = m/\|m\|$$ (17)

provided $m \neq 0$. When $m = 0$ we may calculate explicitly:

$$\Phi[m + \Delta] - \Phi[m] = \|0 + \Delta\| - \|0\| = \|\Delta\| \tag{18}$$

and so $R / \|\Delta\| = 1$, which does not vanish in the limit as $\|\Delta\|$ tends to zero as (2) requires. This means that $\|m\|$ is not Fréchet-differentiable at the zero element of H, although it is elsewhere.

As we noted earlier, Fréchet differentiability demands a high degree of regularity of the functional at the point of evaluation and it may be hard to prove, or may be untrue that F really is Fréchet-differentiable. For almost all our purposes it is sufficient for F to possess another kind of differential called the *Gateaux derivative*. For each $h \in H$ (called the *increment*) as $t \in \mathbb{R}$ tends to zero, if we have

$$F[m + th] = F[m] + t\, G[m; h] + o(t) \tag{19}$$

then the functional G is the Gateaux derivative of F at m with increment h. An equivalent and explicit definition is

$$G[m; h] = \frac{d}{dt}F[m + th]\Big|_{t=0}. \tag{20}$$

Notice how the dependence of G on h is more complicated than in the case of the Fréchet derivative and that there is no requirement that G be a linear functional of h; for example, the Gateaux derivative may take on one form if h is confined to a certain subspace, and something else otherwise; see exercise 5.03(ii). It is shown in all the functional analysis books that if the Fréchet derivative exists for F at m, then the Gateaux derivative always exists, and for all increments h

$$G[m; h] = (D, h). \tag{21}$$

A local minimum of the functional F occurs at the element m_* if for some $\varepsilon > 0$ and every $\Delta \in H$ with $\|\Delta\| < \varepsilon$,

$$F[m_*] \leq F[m_* + \Delta]. \tag{22}$$

An important theorem in optimization theory says that if F is Gateaux-differentiable at m_*

$$G[m_*, h] = 0 \tag{23}$$

for all h. When F is Fréchet-differentiable, it follows from (21) that $D(m_*) = 0 \in H$. These useful results are the obvious analogs of the conditions for local minimum in ordinary multidimensional calculus. The general nonlinear theory for optimization is necessarily a local theory.

The reader will probably be wondering what all the fuss is about. The distinction between Fréchet and Gateaux differentiability is not entirely academic, even in geophysical circles. In a famous paper Backus and Gilbert (1967) gave what they thought was the Fréchet derivative of the frequency of free oscillation of the Earth as a function of density and seismic velocities. Nearly ten years later numerical experiments discovered that the derivative incorrectly predicted the changes in certain frequencies when the radius of the core was changed; this was shown by Woodhouse (1976) to be due to the omission of certain terms that arise only for discontinuous perturbations. We may interpret this by saying that those frequencies are not Fréchet-differentiable in L_2 as Backus and Gilbert had believed, since the same D is not valid for all L_2-perturbations Δ.

Exercise

5.03(i) Let $m, g_j \in H$, a Hilbert space and d_j be a set of real numbers, with $j = 1, 2, 3 \cdots N$. Find the Fréchet derivative of the functional

$$X^2[m] = \sum_{j=1}^{N} (d_j - (g_j, m))^2.$$

If the Fréchet derivative of the F_j at m is D_j find the Fréchet derivative of

$$X^2[m] = \sum_{j=1}^{N} (d_j - F_j[m])^2.$$

5.03(ii) Let F be a functional on a Hilbert space H and let $m, f, g \in H$ and $f \neq g$. Suppose F is defined by

$$F[m] = (g, m), \text{ if } (f, m) = 0$$

$$= (f, m), \text{ otherwise.}$$

Show this functional is Gateaux-differentiable at the origin, but not Fréchet-differentiable there.

5.04 Constructing Models

To begin we consider the situation in which we have been given N real data values which are exactly known; the solution to the forward problem is represented by a set of nonlinear functionals F_j over an appropriate Hilbert space. We are asked to find any model that will exactly fit the data; thus we wish to discover $m \in H$ such that

$$d_j = F_j[m], \ j = 1, 2, 3, \ \cdots \ N . \tag{1}$$

Such a request presupposes there are models capable of exactly reproducing the observations; in many cases, however, observational noise or theoretical simplifications (see 5.02, assumptions (a)–(f)) may make this demand unrealizable. For the moment we shall not address this difficulty, but when we come to the point of designing a practical algorithm, the issue cannot be neglected. Our strategy will be based upon iterative refinement of a guessed solution. We assume there is some solution, call it m_*, and we have somehow generated a guess model m_1. We further assume that each F_j is Fréchet-differentiable at m_1. So

$$d_j = F_j[m_*] = F_j[m_1 + m_* - m_1] \tag{2}$$

$$= F_j[m_1] + (D_j, m_* - m_1) + R_j . \tag{3}$$

Or, rearranging

$$(D_j, m_* - m_1) = d_j - F_j[m_1] - R_j . \tag{4}$$

Furthermore, suppose our guess model is so accurate that the remainder term in the differential formula is very small. We can use (4) as a basis for an iterative scheme: we drop the terms R_j and replace the true model m_* with m_2, our next approximation:

$$(D_j, m_2 - m_1) = d_j - F_j[m_1], \ j = 1, 2, 3, \ \cdots \ N . \tag{5}$$

With $\Delta = m_2 - m_1$, this becomes

$$(D_j, \Delta) = d_j - F_j[m_1], \ j = 1, 2, 3, \ \cdots \ N . \tag{6}$$

We may view (6) as a set of N equations for the unknown element Δ; everything on the right is known, and D_j must be evaluated at m_1, which is also known. Since m_1 is given, we can find the next approximation to the solution by $m_2 = m_1 + \Delta$. The set of equations is exactly in the form of our standard linear inverse problem for exact data, first encountered in chapter 2, but here the representers are the Fréchet derivatives. We know how to solve such systems, but we also know the solution to the linear system is not unique: which of the infinitely many possible solutions for Δ should we choose? The answer given by Backus and Gilbert

(1967) is plausible: we choose the smallest one, because in that way we keep as close to m_1 as possible and thereby have the greatest chance that R_j will be negligible, thus justifying the fundamental approximation in (6). An obvious iterative procedure suggests itself: if m_2 predicts the observations successfully, we stop; if not, we repeat the process using m_2 as the base model for another approximation m_3. The procedure is repeated until a satisfactory answer is obtained or the investigator's patience (or computer budget) is exhausted.

As described the process has an almost negligible chance of success. It is frequently the case that after one iteration of the scheme, we are further away from a solution than when we started, so applying the procedure again is a certain recipe for disaster. Indeed, it may be impossible to proceed at all because the first step may have taken the element outside the domain on which the forward functionals are defined; for example, it may demand negative conductivities in an electrical problem. Despite its pretensions to be a conservative process, the method is actually not cautious enough, at least in the early stages when the approximate model is far from a solution that fits the data. The following modification, called *step-length damping*, improves the situation considerably. Let us agree to measure discrepancy between the model predictions and the observations by the Euclidean norm. Then the squared misfit is

$$X^2[m] = \sum_{j=1}^{N} (d_j - F_j[m])^2 . \tag{7}$$

Instead of taking the next approximation to be $m_1 + \Delta$, we shall be more timid and step only partly along the way toward m_2 for a new model

$$m_\gamma = m_1 + \gamma\Delta \tag{8}$$

$$= (1-\gamma)m_1 + \gamma m_2 . \tag{9}$$

where $0 < \gamma < 1$. If we take the Fréchet differential of X^2 (this was exercise 5.03(i)) we can compute the squared misfit of the model m_γ and then calculate the change in misfit in going from m_1 to m_γ:

$$X^2[m_\gamma] - X^2[m_1] = \sum_{j=1}^{N} -2(d_j - F_j[m_1])(D_j, \gamma\Delta) + o\,\|\gamma\Delta\| \tag{10}$$

$$= -2\gamma \sum_{j=1}^{N} (d_j - F_j[m_1])^2 + o\,\|\gamma\Delta\| \tag{11}$$

$$= -2\gamma X^2[m_1] + o \, \|\gamma\Delta\| \qquad (12)$$

where we have appealed (6). Since $X^2[m_1]$ and γ are both positive, (12) shows us that by making γ small enough, we can
guarantee a better misfit at m_γ than at m_1. The existence of the
Fréchet derivative at m_1 means that in the neighborhood of this
model the functionals F_j are approximately linear and we can use
the locally linear behavior to find a direction in which the misfit
improves; if we travel too far in that direction (usually m_2 is too
far) the approximation fails badly enough that the misfit will be
larger than at the starting point.

Armed with this result we propose a modification of the
naive iterative scheme. We solve (6) for Δ as before and now
examine a set of m_γ given by (9) for a range of values of $\gamma > 0$, computing $X^2[m_\gamma]$, and searching for the value that makes the misfit
as small as possible; we are certain that the proper choice of γ will
yield a misfit smaller than that at m_1. To make further progress
one must repeat the whole process substituting the current best
solution for m_1. Since X^2 is bounded below by zero, the sequence
of decreasing misfits we obtain will converge to a limit, although
there is nothing to say the sequence of models converges to an
end point in the Hilbert space. Because exact data fitting is
artificial in the geophysical setting, one can imagine repeating the
iteration until the misfit is driven below the acceptable tolerance
level, or until it becomes clear the limit point lies above the
hoped-for accuracy.

We call this approach the *creeping* algorithm, because it cau-
tiously advances by the smallest step at each stage. Creeping is
still in use today, although it exhibits some serious drawbacks. It
is a Hilbert space relative of the classical method of *steepest des-
cents* in numerical optimization theory, which has a well-earned
reputation for slow convergence (see Luenberger, 1984 or Gill et
al., 1981). The reason for the slow convergence is essentially as
follows: at m_γ no further progress can be made by moving in
either direction along the line between m_1 and m_2; to descend,
the next move must be perpendicular to this line. Thus the progress of the method is traced by series of zig-zag line segments
with right-angle bends, which is obviously not the fastest way to
reach a destination.

More fundamentally, creeping lacks specificity. When a
solution to the nonlinear inverse problem does exist, there are

normally infinitely many others, just as with linear equations. If the creeping algorithm is used, the final result is an unpredictable function of the initial model. We find in the geophysical literature that the mysterious dependence on the initial model has been mistakenly attributed to the nonlinear nature of the problem rather than to its correct source, the creeping algorithm. The geophysicist must choose the kind of model he or she wants when solving an inverse problem and not leave that decision to an accidental feature of a computer program, or an arbitrary initial guess at the solution. Let us adapt the creeping algorithm to seek a particular solution.

Continuing briefly with the fiction that exact matching of the data is required, we return to (5) and rearrange the equation thus:

$$(D_j, m_2) = d_j - F_j[m] + (D_j, m_1). \tag{13}$$

Again we have a set of linear relations obtained by dropping the remainders R_j, but now we select the solution that minimizes the norm of m_2 rather than the perturbation Δ; obviously an appropriate seminorm could be minimized here equally well. Naturally m_2 might be quite far from m_1 and the neglect of the remainder terms would then be totally unjustified; thus step-length damping is vital here. The same theory for damping applies because the reasoning that led to (12) did not need to inquire which of the infinitely many solutions to (5) was chosen. Thus when (13) is solved for the smallest m_2 the algorithm is seeking a particular solution at each step within the local linear approximation. It can be shown that if this process converges to a model with zero misfit, the model is at a local minimum of the norm.

As we described in the introduction to this chapter, construction methods have to be pressed into service to answer the existence question, by discovering well-fitting models. The optimization problem of finding the best-fitting model over a Hilbert space cannot be solved by simple linearization because at every stage, the linear problem appears to have infinitely many exact solutions, while the true nonlinear system may have none at all. The descent methods just described have the apparent virtue of being capable of reducing the data misfit at every step and they have been applied in practical problems. But the author's experience is that they are not very effective in practice, something that will be apparent in 5.06, when we attack a modest-sized problem. An alternative commonly applied technique, which we shall not discuss in detail, is the nonlinear version of collocation: reduce the number of model parameters to less than the number of observational constraints; now the linearized version of the problem has a unique solution, and an iterative scheme can easily be constructed. The obvious danger is that by removing model parameters one does not leave enough freedom in the solution to satisfy the observations. A better alternative is through regularization, where we retain enough model freedom, but strive for model simplicity through a penalty on complexity. Paradoxically, we find controlling model complexity in this way is often the key to rapid improvement of misfit.

Let us jettison the unrealistic demand that the data values be matched exactly; we can always turn down the tolerance parameter that measures misfit if we want to recover that case. For simplicity's sake we treat the problem of finding the model with the smallest norm subject to some desired misfit level. This is the major question of chapter 3, but in the more general, nonlinear setting. It is probably a good idea for the reader to review section 3.02 before proceeding. A Lagrange multiplier μ is introduced to weight the misfit term:

$$U[m, \mu] = \|m\|^2 + \mu \sum_{j=1}^{N} (d_j - F_j[m])^2 / \sigma_j^2 . \tag{14}$$

For each positive real value of μ we see that U is a functional on H; suppose the minimum value of U over $m \in H$ for given value of μ is attained for the element $m_*(\mu)$. Then the theory of Lagrange multipliers (which applies virtually unchanged to Hilbert spaces; see Luenberger, 1969, chapter 7) says that if we can pick a $\mu > 0$, which we call μ_*, such that the second term in the sum matches the desired misfit, that is,

$$T^2 = \sum_{j=1}^{N} \frac{(d_j - F_j[m_*(\mu_*)])^2}{\sigma_j^2} \tag{15}$$

then the solution $m_*(\mu_*)$ is the one of smallest norm with misfit T. To avoid the messy notation let us agree to write m_0 for $m_*(\mu_*)$, a notation in accord with the one in chapter 3. Of course (15) is nothing more than equation 3.02(11) for arbitrary functionals instead of bounded linear ones. We would like to differentiate (15) to discover the location of the minimum, but this must be done over models m in Hilbert space rather than over the expansion coefficient α in the linear theory. This is where 5.03(23) or its Fréchet equivalent comes in. Applying the Chain Rule 5.03(12) we find the Fréchet derivative of (14) and set it to zero; we obtain a condition that must hold at m_0:

$$0 = 2m_0 + 2\mu_* \sum_{j=1}^{N} \frac{(d_j - F_j[m_0])D_j(m_0)}{\sigma_j^2} . \tag{16}$$

Rearranging, and noting that everything under the sum except the derivatives can be combined into a set of N unknown scalars, we find

$$m_0 = \sum_{j=1}^{N} c_j D_j(m_0) \tag{17}$$

where

$$c_j = -\mu_*(d_j - F_j[m_0])/\sigma_j^2 . \tag{18}$$

In other words, at the norm minimizer, the solution can be written as a linear combination of Fréchet derivatives. This was the key result of section 3.02 that unlocked the linear problem; here, since the derivatives D_j depend on the point of evaluation, m_0, (17) is an intractable nonlinear relationship. Furthermore, it is only a necessary condition, and might hold at any number of other elements of H besides the true norm minimizer, m_0. Life in the nonlinear world is tough.

Since (17) is not particularly helpful in the discovery of a solution, we can develop an approach along similar lines to the creeping method. At first we ignore the question of numerical efficiency. Suppose we have a trial value of Lagrange multiplier $\mu_1 > 0$, and our immediate objective is to minimize (14) starting from some initial guess model m_1. In our first attempt at an algorithm, we keep the Lagrange multiplier fixed while we seek $m_*(\mu_1)$ by a descent method. Then we move to another value of μ and repeat the process. In this way we can develop the decreasing curve of misfit X^2 considered as a function of Lagrange multiplier, a curve like the one shown for example in figure 3.04c, but for a nonlinear system; and as in the linear case, we seek the intercept of the curve with the line $X^2[m_*(\mu)] = T^2$. To minimize $U[m, \mu_1]$ over $m \in$ H we shall formulate a Hilbert space version of the Gauss-Newton method for functional minimization (see Gill et al., 1981). By the definition of Fréchet-differentiability at m_1, we know that

$$F_j[m] = F_j[m_1] + (D_j, m - m_1) + R_j \tag{19}$$

$$= F_j[m_1] - (D_j, m_1) + (D_j, m) + R_j \ . \tag{20}$$

Let us proceed as if the guess m_1 is close enough to the minimizing solution that dropping the terms R_j in (20) causes only a small error and then we may replace F in (14) with a *linearized* version, namely, (20) without R_j: we obtain a modified functional

$$\tilde{U}[m; \mu_1] = \|m\|^2$$

$$+ \mu_1 \sum_{j=1}^{N} \frac{(d_j - F_j[m_1] + (D_j, m_1) - (D_j, m))^2}{\sigma_j^2} \tag{21}$$

$$= \|m\|^2 + \mu_1 \sum_{j=1}^{N} \frac{(\tilde{d}_j - (D_j, m))^2}{\sigma_j^2} \tag{22}$$

where we have introduced the abbreviation

$$\tilde{d}_j = d_j - F_j[m_1] + (D_j, m_1) \ . \tag{23}$$

We know how to minimize $\tilde{U}$—it is the problem solved in 3.02 with the Fréchet derivatives D_j in place of the representers g_j, and furthermore, μ_1 is fixed. The minimizer of $\tilde{U}$ is

$$m_2 = \sum_{j=1}^{N} \alpha_j D_j \tag{24}$$

where the coefficients α_j are the components of the solution to the linear system

$$(\mu_1^{-1} \Sigma^2 + \Gamma)\boldsymbol{\alpha} = \tilde{\mathbf{d}} \tag{25}$$

and Γ is the Gram matrix of Fréchet derivatives. We know the $\boldsymbol{\alpha}$ that solves this system makes $\tilde{U}$ minimum, but not that it minimizes U, the functional we really want to minimize. Therefore it seems natural to appeal to step-length damping again, with a model m_γ that lies between m_1 and m_2 according to (9), choosing γ to make $U[m_\gamma, \mu_1]$ minimal. It is a straightforward repetition of the algebra required to find (12) to show that for some positive γ, the value of $U[m_\gamma, \mu_1]$ is smaller than at m_1.

Thus we have a systematic program to find the norm-minimizing solution satisfying (15): (a) for a fixed Lagrange multiplier μ_1 we minimize $U[m, \mu_1]$ with the Gauss-Newton scheme just described, employing step-length damping as necessary; (b) if, at the minimizer we find that the squared misfit $X^2[m_*(\mu_1)]$ exceeds the target value T^2, we repeat the process for a new value of the Lagrange multiplier $\mu_2 > \mu_1$ because misfit is a strictly decreasing function of Lagrange multiplier; conversely, we would

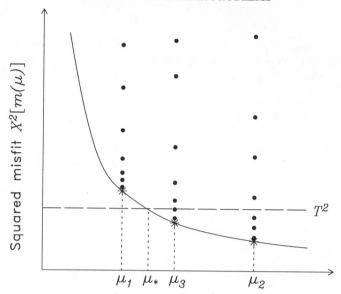

Figure 5.04a: Systematic iterative approach to the construction of smallest m. For a set of constant μ the functional $U[m, \mu]$ is minimized over m and the value of squared misfit calculated; misfits found during the iterative process are shown as dots, the limiting values by asterisks. The monotone decreasing curve joining the asterisks intersects the target value of T^2 when $\mu = \mu_*$.

select a smaller value for μ_2 if the data were over-fit; (c) the process is repeated until μ_* has been located. The procedure is illustrated in figure 5.04a.

Obviously there are a number of deficiencies in this recipe. We do not know if the minimum of U obtained by the Gauss-Newton method is really the smallest attainable value, or merely a local minimum in a functional with a multitude of potholes. Ignorance on this score is inescapable when we attempt a minimization without some theoretical warranty, like the one provided by convexity. Another difficulty is that we cannot know beforehand whether any value of μ can deliver a small enough misfit, since the observations and condition (15) may be inconsistent; this is the question of existence discussed in section 5.01. Nothing can be done about these fundamental issues without a deeper understanding of the mathematical problem, something usually requiring a radically different approach from iterative optimization (see, for example, section 5.08). On a pragmatic level, however, the scheme can be improved: it is clearly very wasteful from a numerical perspective. For example, it is unnecessary after having found $m_*(\mu_1)$ to return to m_1 as the initial guess with the new Lagrange multiplier μ_2; it is likely that starting at $m_*(\mu_1)$ would yield a more rapidly converging sequence, particularly if the Lagrange multipliers are close. The huge mathematical literature in optimization theory (e.g., Gill et al., 1981) has hardly been touched in the solution of geophysical inverse problems, and there are doubtless many untried approaches that would work wonders on the functional (14).

We conclude this section with a completely different scheme, a heuristic algorithm, lacking theoretical underpinnings, but one that has been found to be remarkably effective in practice. It was introduced in a paper (Constable et al., 1987) entitled "Occam's Inversion," after William of Occam (whose razor cut away unnecessary hypotheses) to emphasize the importance of seeking the simplest resistivity profile in electrical sounding studies, an exhortation surely unnecessary for the reader at this point. Nonetheless we call the empirical recipe *Occam's process* on the grounds that the scheme parsimoniously dispenses with an unnecessary search parameter, γ; the Lagrange multiplier μ serves both for exploration and to constrain the solution to a particular misfit level.

One of the evidently wasteful aspects of the systematic procedure just described is the necessity of minimizing U with high accuracy for values of the Lagrange multiplier other than m_*; the trouble is that μ_* is unknown. Nonetheless, the inefficiency of searching along lines of constant μ as shown in figure 5.04a might be avoided if μ were allowed to vary much earlier in the procedure, thus obviating the need to converge on intermediate solutions of no intrinsic interest. One way of moving "horizontally" in figure 5.04a would be to treat (22) as an exact equation and to *solve* for μ_1 by matching the misfit term to T^2 just as we did for the linear inverse problem in section 3.02. Suppose this is done; the model obtained from the expansion (24) would not in fact possess squared misfit T^2 because the misfit term in (22) is only an approximation to $X^2[m]$ of (7), not the real thing. We fix this flaw by computing a suite of models by varying μ_1 in (24) and (25) and selecting the one for which $X^2[m] = T^2$, that is, by matching the true misfit, rather than the approximation, to the target value. On the assumption that such a solution can be found, would this model, which has the correct misfit, be the solution of smallest norm? Not necessarily, because it is comprised of a linear combination of Fréchet derivatives *evaluated at* m_1, whereas the condition for optimality, equation (17), says it must be built from derivatives evaluated at the model element itself. All this immediately suggests that the process be repeated, forming $\tilde{U}$ in (22) on the basis of derivatives evaluated at the new solution. If the process converges, it is easy to see that (17) and (24) coincide and the necessary condition for optimality is achieved.

In practice, the assumption that a unique value of μ_1 in (25) exists allowing the true misfit criterion to be satisfied can fail. When positive values of μ_1 are swept out, the approximate misfit derived from linearization of F_j decreases monotonically as in the familiar linear theory; in contrast, the true misfit, X^2, usually exhibits a minimum, and in the early stages of the iteration the minimum lies above the target level T^2. Therefore the Occam process is modified to choose as the basis for the next iteration the model found at the misfit minimum, since a primary objective is to find a model in accord with the demands of the data, and misfit minimizer is the best available approximation in that regard. Repetition of the cycle normally produces rapid improvement in misfit at the minimum of the misfit curve, until the best misfit point falls below the desired tolerance. Of course this may never happen, either because it is simply impossible, or perhaps because the approximations suggesting the iterative procedure were never valid. We shall call the iterations designed to bring the misfit down to the target level, Phase 1 of the Occam process. In Phase 2 we choose from the family of solutions generated by varying μ_1 the one with the correct misfit. If, as is usually the case, the function X^2 crosses the level T^2 at more than one point, we must decide which one to choose. Unlike the misfit, the model norm must exhibit monotone behavior with varying μ_1, increasing with that parameter. Since we want the model of smallest norm, the required model is always the one associated with the smaller value of μ_1. Derivatives are computed at this model and the process is repeated until the model norm has converged. While this description may seem quite complex, an example will make things clearer; the next two sections are devoted to illustrations.

Occam's process is difficult to analyze theoretically because there is no guarantee that any member of the family of models produced by varying μ_1 will be close to m_1, the model at which the derivatives are evaluated. This makes the familiar perturbation methods of analysis impossible to use since they depend upon a model being in a neighborhood of m_1 in which the linear approximation is valid. As we shall see, there is no guarantee that Occam's process will always produce in its initial phase a model with smaller misfit than that of m_1, but experience with a variety of problems points to dramatic improvement in the early stages, and to much better performance than the conservative creeping approach.

5.05 The Gravity Profile

The first thing to be done before we can apply the theory of the previous section is to find a functional derivative for the forward problem in some appropriate space. This may look formidable, but in fact we can discover a derivative of some kind quite easily; verifying rigorously that it is a true Fréchet or Gateaux derivative can be much more difficult. The general idea is to write out an expression for

$$F_j[m+\Delta] - F_j[m] \tag{1}$$

and to perform some algebraic manipulations in order to isolate a part that looks like an inner product. From 5.03(3) we observe that an inner product is a candidate for the Fréchet derivative. Let us illustrate this carefree approach with the gravity profile problem of section 5.02.

For the gravity problem we take 5.02(1) and turn the algebraic crank:

$$\Delta g_j[h+\Delta] - \Delta g_j[h] = \int_0^a G\Delta\rho \ln\left[\frac{(x_j-x)^2 + (h(x)+\Delta(x))^2}{(x_j-x)^2}\right] dx$$

$$- \int_0^a G\Delta\rho \ln\left[\frac{(x_j-x)^2 + h(x)^2}{(x_j-x)^2}\right] dx \ . \tag{2}$$

Let us abbreviate the right side of (2) by dF_j; then

$$\frac{dF_j}{G\Delta\rho} = \int_0^a \ln\left[\frac{(x_j-x)^2 + (h(x)+\Delta(x))^2}{(x_j-x)^2 + h(x)^2}\right] dx \tag{3}$$

$$= \int_0^a \ln\left[1 + \frac{2h(x)\Delta(x)}{(x_j-x)^2 + h(x)^2} + \frac{\Delta(x)^2}{(x_j-x)^2 + h(x)^2}\right] dx \ . \tag{4}$$

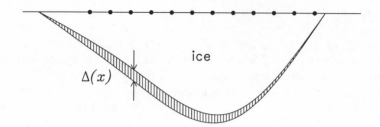

Figure 5.05a: Equation (5) describes the attraction of a thin additional layer, thickness $\Delta(x)$ at the base of the gravitating body, here a lens of ice.

At this point everything is exact. To define a functional derivative we must discover what happens when Δ is small in some sense. Obviously we would like to drop the term in Δ^2 and then linearize the log with the approximation $\ln(1+\xi) \approx \xi$. Suppose we do this, without worrying about the details for the moment: we find

$$\frac{dF_j}{G \Delta \rho} = \int_0^a \left[\frac{2h(x)}{(x_j - x)^2 + h(x)^2} \right] \Delta(x) \, dx + R_j[h, \Delta] \qquad (5)$$

$$= (D_j, \Delta) + R_j \qquad (6)$$

where we have used the inner product of $L_2(0, a)$. Obviously

$$D_j = \frac{2h(x)}{(x_j - x)^2 + h(x)^2} \qquad (7)$$

and from the definition 5.03(1)-(2), D_j is the Fréchet derivative of Δg_j on L_2 provided that R_j goes to zero rapidly enough when Δ tends to zero.

Before looking into this question, let us invest (5) with some physical meaning. In the gravity profile problem we can easily show that (5) represents the gravitational attraction of a thin strip of material, with thickness $\Delta(x)$ and with a shape given by $z = h(x)$ exactly as we would expect from simple perturbation theory; see figure 5.05a. The neglect of R_j is equivalent to condensing the material in the strip into surface distribution of matter on the boundary.

We can write out R_j explicitly:

$$R_j[h,\Delta] = \int\limits_0^a \left\{ \ln\left[\frac{(x_j - x)^2 + (h(x) + \Delta(x))^2}{(x_j - x)^2 + h(x)^2} \right] \right.$$
$$\left. - \frac{2h(x)\Delta(x)}{(x_j - x)^2 + h(x)^2} \right\} dx \ . \tag{8}$$

To show that D_j truly is the Fréchet derivative we must prove that for all $\Delta \in L_2(0,a)$ such that $\|\Delta\| \to 0$, $R_j/\|\Delta\| \to 0$. This is a tall order, and is certainly untrue unless some restrictions are placed on h. In fact a number of more limited results would serve as well as this one. Recall from 5.03(23) that the optimization conditions require merely Gateaux differentiability. This is somewhat easier to show as we shall shortly see. The mathematical difficulties in proving the differentiability stem directly from the fact that functions in L_2 are not pointwise bounded and so it simply is not true that as $\|\Delta\|$ shrinks to zero, the value of $\Delta(x)$ grows small for any particular x. This suggests one way out of our difficulties: we should work in a space of smoother functions, like W_2^1, in which all valid elements are bounded pointwise. This is not an onerous restriction: in the context of the gravity problem it corresponds with geological experience in that bottomless valleys are never encountered in real life. In the construction phase of the linear inverse problem we have seen how smooth solutions are desirable and they remain so now; it is therefore perfectly natural to adopt a space of better-behaved functions as the proper setting for building models.

These remarks notwithstanding, we shall now prove that the gravity functional is Gateaux-differentiable for a large class of depth profiles in L_2. The reader who is eager to get on with the numerical examples can skip the next few pages, but the author hopes there are some who would like to see how a proper derivation is carried out. Our approach is to apply 5.03(20). First we streamline the expression: clearly there is no loss in generality if we move the origin of coordinates and then set $x_j = 0$; we drop all the subscripts; we also let $G\Delta\rho = 1$. Now

$$F[h] = \int\limits_b^c \ln(1 + h(x)^2/x^2) \, dx \ . \tag{9}$$

Naive application of 5.03(20) immediately produces

$$G[h;\Delta] = \int\limits_b^c \frac{2h(x)\Delta(x)}{x^2 + h(x)^2} \, dx \tag{10}$$

which is just equation (5). The real problem comes in identifying
the conditions under which this result is valid.

We introduce some restrictions natural for the gravity problem: first of course $h, \Delta \in L_2(b,c)$. Recall from 5.02 that we
require all admissible profiles to be nonnegative; thus

$$h(x) \geq 0; \qquad h(x) + \Delta(x) \geq 0 . \tag{11}$$

In addition we shall assert that in some open neighborhood of
$x = 0$, $h(x) > 0$; this says that the ice layer has positive thickness
under the observer and it has the consequence that there is a
positive number c_1 such that for all $a \leq x \leq b$

$$h(x)^2 + x^2 \geq c_1 . \tag{12}$$

While this condition may seem artificial and could probably be
weakened, something like it is required, because otherwise we see
from (5) that D_j might lie outside L_2; see exercise 5.05(i). We
shall return to this point shortly.

To prove (10) we write the derivation thus:

$$G = \lim_{t \to 0} \int_b^c \frac{1}{t} \ln \left[\frac{x^2 + (h(x) + t\,\Delta(x))^2}{x^2 + h(x)^2} \right] dx . \tag{13}$$

We want to reverse the limit and the integral. According to the
Lebesgue Dominated Convergence Theorem (Korevaar, 1968), this
is permitted if we can find a function $\phi(x)$ independent of t and
larger in magnitude than the integrand, yet integrable on (b,c).
Let us write

$$\frac{x^2 + (h(x) + t\,\Delta(x))^2}{x^2 + h(x)^2} = 1 + \tau(x) \tag{14}$$

where

$$\tau(x) = \frac{2t\,\Delta(x)\,h(x) + t^2\Delta(x)^2}{x^2 + h(x)^2} . \tag{15}$$

Then our objective is to discover an integrable function $\phi(x)$ with
the property

$$\left| \frac{\ln(1 + \tau(x))}{t} \right| \leq \phi(x) \tag{16}$$

for all points in (b,c). To find ϕ we bound the integrand above
and below. Since the logarithm is a monotone function we may
bound its argument: Assume henceforth that $t > 0$. From (12) we

have

$$\tau(x) \leq \frac{2t \ |\Delta(x)| \ |h(x)| + t^2 \Delta(x)^2}{c_1} . \tag{17}$$

It is an elementary fact, shown by looking at the integral of $1/\xi$, that

$$\ln(1 + \xi) \leq \xi, \qquad \xi > -1 . \tag{18}$$

Therefore, by (17) and (18) the integrand of (13) satisfies

$$\frac{\ln(1 + \tau(x))}{t} \leq \frac{2 \ |\Delta(x)| \ |h(x)| + t \ \Delta(x)^2}{c_1} \tag{19}$$

$$\leq \frac{2 \ |\Delta(x)| \ |h(x)| + \Delta(x)^2}{c_1} . \tag{20}$$

where we can safely replace t in (19) by 1 in (20) because t is tending to zero.

Turning to the lower bound, we call upon (11) to obtain

$$h(x) \Delta(x) \geq -h(x)^2 . \tag{21}$$

Substituting this into (15) yields

$$\tau(x) \geq \frac{-2t \ h(x)^2 + t^2 \Delta(x)^2}{x^2 + h(x)^2} \geq \frac{-2t \ h(x)^2}{x^2 + h(x)^2} \geq -2t . \tag{22}$$

Thus returning to original integrand of (13)

$$\frac{\ln(1 + \tau(x))}{t} \geq \frac{\ln(1 - 2t)}{t} . \tag{23}$$

The expression on the right tends to -2 from below as t tends to zero; thus for a conservative bound we can use any smaller constant, say -3. Combining (20) and (23) we can now construct the function $\phi(x)$ of (16) as

$$\phi(x) = \max\left[3, \ \frac{2 \ |\Delta(x)| \ |h(x)| + \Delta(x)^2}{c_1}\right] . \tag{24}$$

It remains to be shown that $\phi(x)$ is integrable. Clearly if each of the entries of the maximum is integrable, ϕ is. Equally obviously the first entry is integrable. Finally we appeal to the fact that Δ and h are in L_2: it follows from Schwarz's inequality and the definition of the norm that

$$\int_b^c [2 \ |\Delta(x)| \ |h(x)| + \Delta(x)^2] \ dx \leq 2(\|\Delta\| \ \|h\|)^{\frac{1}{2}} + \|\Delta\|^2 . \tag{25}$$

Hence when h and Δ are valid elements of L_2 the function ϕ is integrable and the reversal of the limit and integration in (13) is allowed. Therefore (10), or equivalently (5), is a valid Gateaux differential for the problem for all increments $\Delta \in L_2$ and a wide class of profiles h.

The restriction (12) was needed to make the proof go through but, as noted earlier, there is a more fundamental need to restrict the class of elements: the gravity functional is simply not Gateaux-differentiable at every element in its domain. Exercise 5.05(i) gives a seemingly innocuous example of a function h for which the functional Δg is not Gateaux-differentiable (and therefore not Fréchet-differentiable either). This is entirely analogous to the situation in which a function (like $|1-x^2|$) is not differentiable for every x. As we shall observe shortly, this is not just a technicality, for if the solution we seek is an element for which one or more of the forward functionals fails to be differentiable, we may have great difficulty in discovering that solution. Moreover, if we encounter elements during the iteration where Δg_j is nondifferentiable, moving away from them in the proper direction toward a solution will be problematic because our whole strategy for improvement depends on the validity of a local linear approximation.

Let us now proceed to numerical illustrations based on the glacier traverse introduced in section 4.07 using the gravity anomaly values in table 4.07A. In our calculations we shall take $\Delta \rho$ to be $-1700 \, \mathrm{kg \, m^{-3}}$, a value for which exact solutions to the inverse problem are known to exist, albeit with overhangs like the one shown in figure 4.07d. Initially, we shall demand an exact match between the predictions of the model and the measured gravity anomaly values although this situation is rare in real geophysical problems. To generate an approximation suitable for numerical work we rewrite 5.02(1)

$$\frac{\Delta g_j}{G \Delta \rho} = \int_0^a \ln \left[(x - x_j)^2 + h(x)^2 \right] dx$$

$$- 2 \left[(a - x_j)(\ln |x_j - a| - 1) + x_j \ln x_j - x_j \right] \qquad (26)$$

where we have performed the singular integral for the denominator within the logarithm analytically because it can be evaluated exactly and is in any case awkward to treat accurately by numerical means. We specify that the valley floor outcrops at $x = 0$ and $x = a = 3.42$ km. As in section 2.08 on seismic dissipation, we approximate the integral in (26) by a weighted sum over samples of the integrand, chosen to be equally spaced at 51 points, including the ends of the interval. The choice of quadrature rule depends on the expected smoothness of the solution—Simpson's rule would be acceptable if we could depend upon continuous second derivatives, but to begin with we want to look at some rough models and so the primitive trapezoidal rule will be better matched to the problem because it remains accurate for functions with discontinuous derivatives. Iterative techniques require an initial guess from which to begin. One might be tempted to set out from the smallest model $h_1 = 0$, but that is precisely where the gravity functionals all fail to be differentiable and it would be a most unfortunate launching point. Instead, we shall start at a bland model, the constant function $h_1 = 500$ m. To measure misfit between the observed gravity and the prediction of the model we use the norm:

$$X[h] = [\sum_j (d_j - \Delta g_j[h])^2]^{1/2} \tag{27}$$

rather than the square of the norm, because then the misfit has familiar units of milligals. The misfit of our starting guess, h_1, is 32.52 mGal.

Our first exercise will be to attempt the original scheme of Backus and Gilbert, working in $L_2(0, a)$ or in a finite-dimensional equivalent. Converting 5.04(6) to the finite-dimensional approximation, we seek the shortest length vector that solves the (underdetermined) linear system

$$D \Delta = \mathbf{d'} \tag{28}$$

where $\Delta \in E^L$ is the perturbation to the starting model vector that we seek, $D \in M(N \times L)$ is a matrix whose rows are L samples of the Gateaux derivative (7), evaluated at h_1 and weighted by the quadrature weights; the components of $\mathbf{d'} \in E^N$ are

$$d_j' = d_j - \Delta g_j[h_1], \quad j = 1, 2, 3, \cdots N. \tag{29}$$

Remember in our numerical realization of the problem, $N = 12$ and $L = 51$. It was asserted at the beginning of the previous section that the plain solution of this linear system is unlikely to be successful and indeed that is the case here: the first perturbation Δ oscillates between ± 4000 m, changing sign 10 times. Having immediately stepped out of the domain of Δg (owing to negative values of h), the new model apparently cannot have its misfit taken. The reader may have noticed that the model appears in

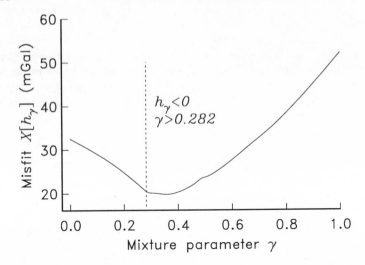

Figure 5.05b: Misfit of the linear combination model h_γ of (30) as γ varies; note that before the expected minimum in the curve is reached, the function runs out of the strict domain of Δg.

the forward functionals only as h^2 and that suggests we could simply treat sign changes in h as irrelevant to the evaluation of Δg; we could replace h by $|h|$. If we do this, we find at the naive Backus-Gilbert solution $X[h_2]=53.67$ mGal, more than 1.5 times larger than the misfit of the initial guess. We conclude that h_1 is too far from an exact solution for the unmodified method to work properly. We must resort to step-length damping of the Backus-Gilbert scheme, which we called creeping. The variation of $X[h_\gamma]$ is shown in figure 5.05b for the linear combination profile

$$h_\gamma = (1-\gamma)h_1 + \gamma h_2 . \tag{30}$$

as a function of the parameter γ. We proved in section 5.04 that the misfit must initially decrease and it does; but before a significant improvement in the misfit is attained, the model acquires negative values. Clearly we must face the question of the positivity of the model:

As we noted, the simplest policy is to ignore the sign of h and evaluate the model predictions from the magnitude of the model function as we described in the previous paragraph. Other options for enforcing positivity constraints in nonlinear problems will be presented in the second example; for now we will keep everything as simple as we can. As we see in figure 5.05b, further improvement is obtained when we use model magnitudes,

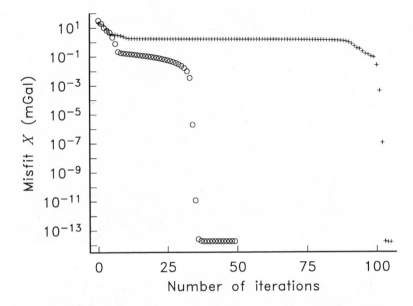

Figure 5.05c: Performance of the damped Backus-Gilbert method (creeping) with iteration number; each point represents the result of minimization over the parameter γ. Circles illustrate progress of the solution starting at a constant-depth model with $h = 500$ m; plus signs correspond to a starting model with sine-squared profile.

and an optimum model is discovered at $\gamma = 0.340$ with a misfit of 19.72 mGal. Encouraged by the progress, we follow the prescription given at the beginning of section 5.04, reevaluate the derivative, and repeat. The circles in figure 5.05c demonstrate how X decreases as the cycle is repeated; we plot the value of the minimum misfit found after each sweep through γ. Initially, progress in reducing the misfit is made rapidly, but then after 8 iterations the rate levels off; around iteration number 32 the current solution approaches close enough to an exact solution that the linear approximation becomes very good and X drops precipitously. By iteration 37, the iteration has converged, and the remaining mismatch between the predictions of the model and the observed data is due to computer round-off noise, which obviously cannot be reduced. When the misfit converges to zero, equations (28) and (29) show that Δ, the step from one model to the next, is reduced to zero also, and the solution has converged to a specific element of the model space. The rate of convergence is, as

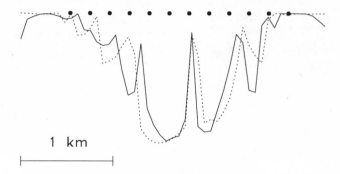

Figure 5.05d: Two exact solutions of the gravity inverse problem
for the shape of the valley: the solid curve is obtained by creeping
in L_2 starting at $h_1 = 500$ m; the dashed model was found starting
from the model $h(x) = \sin(\pi x / a)^2$. The sites of the gravity observations are plotted as the solid dots.

expected, highly dependent on the initial guess. Shown in figure
5.05c by plus signs is the misfit record starting at a somewhat
better initial guess, where h is in the shape of a sine wave that
spans the valley. Despite the lower initial X of 20.65 mGal, the
improvement levels off sooner than in our first experiment and
the plateau lasts much longer; the final turndown is not reached
until iteration number 99. In a third trial, the initial valley is set
at a uniform 1000 m; now exact matching is attained after 360
cycles of the creeping process.

   We have already discussed the fact that not only does convergence rate depend on the initial function, but in Backus-
Gilbert creeping, so does the solution arrived at. The final model
initiated at the flat-bottomed, 500-meter deep valley appears as
the solid curve in figure 5.05d; it is an unlikely candidate for the
shape of the floor of a glacial valley, which can be partly attributed to the fact that we are working in L_2 without any attempt at
regularization. The second solution in this figure is the one found
when we started at the sine-wave profile. It is somewhat
different, but not in any easily comprehensible way.

   Next we examine a solution in which a definite model is
sought, one of smallest seminorm. The undulations of the models
in the figure remind us how unsuitable L_2 can be, and so instead
of seeking the smallest model in this space we shall move on to
W_2^1 where the slope is penalized. It is time to select a target
misfit rather than attempting to achieve the geophysically

unnatural goal of an exact fit. The proposed misfit is $X[h]=0.5$ mGal which is equivalent to RMS discrepancy of 0.144 mGal. In the present problem this is merely an arbitrary value not dictated by specific experimental information but consistent with the accuracy of typical field measurements. We return to 5.04(14); adapted to the current problem, and with the seminorm of RMS slope, the new minimization problem is

$$U[h,\mu] = \left\| \frac{dh}{dx} \right\|^2 + \mu \sum_{j=1}^{N} (d_j - \Delta g_j[h])^2 \tag{31}$$

where the norm is the L_2-norm. The minimization is carried out for a choice of Lagrange multiplier $\mu > 0$ such that $X[h]=0.5$, where X is defined in (27) and the square of which is the second term in (31). To solve this problem we apply the Occam process which we summarize here. We postulate a model h_1 close to the desired solution, and argue as we did for 5.04(22) that the functional U may be approximated by replacing the forward functionals Δg_j with linear approximations, thereby obtaining

$$\tilde{U}[h,\mu] = \left\| \frac{dh}{dx} \right\|^2 + \mu \sum_{j=1}^{N} (\tilde{d}_j - (D_j, h))^2 \tag{32}$$

where

$$\tilde{d}_j = d_j - \Delta g_j[h_1] + (D_j, h_1). \tag{33}$$

Here and in (32) the derivative D_j is evaluated at h_1. The finite-dimensional version of (32) reads

$$\tilde{U} = \| \hat{R} \mathbf{h} \|^2 + \mu \| \tilde{\mathbf{d}} - D \mathbf{h} \|^2 \tag{34}$$

where $\hat{R} \in M(L \times L)$ is a regularizing matrix: for the problem corresponding to (32) we set $\hat{R}$ proportional to ∂, the first difference operator defined in 3.05(7), choosing a scaling to make $\|\hat{R}\mathbf{h}\|$ the RMS slope. If the misfit term in (34) were exact rather than a linear approximation, we would vary μ, minimizing $\tilde{U}$, in order to seek a solution whose misfit equals the target value. In Occam's process we minimize $\tilde{U}$ for a series of values of $\mu > 0$ by one of the methods discussed in section 3.05 (I invariably use equation 3.05(18) for stability), generating a suite of models h_μ from the same derivative matrix D but we examine the corresponding true misfits $X[h_\mu]$, not the approximations. Initially all the misfits are too large, and so we proceed with Phase 1 in which we select the model with the smallest misfit, and repeat

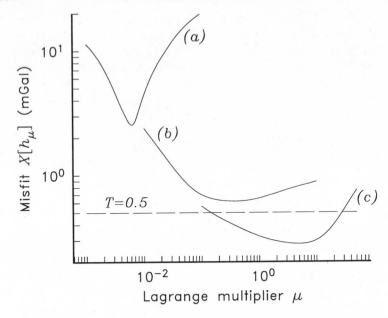

Figure 5.05e: Phase 1 of the Occam process. Misfits for three families of models: (a) solutions h_μ minimizing $\bar{U}$ in (34) with D evaluated at the starting guess, $h = 500$ m; (b) D evaluated at the minimizer of curve (a), where $X = 2.54$ mGal; (c) D evaluated at the minimizer of curve (b), where $X = 0.618$ mGal.

the process, recalculating D at that model. The misfit is minimized over the new family of solutions h_μ until the agreement with observation is satisfactory or improvement of the misfit ceases. When the minimum misfit of a family is below the target level (assuming that this happens) Phase 2 can be begun. Now we choose the solution with correct misfit and the smaller value of μ, compute derivatives at this model, and repeat until the seminorm has converged to its minimum value.

In the first numerical experiment we start from the constant model $h = 500$ m; we achieve the desired misfit of 0.5 mGal in only three cycles of Phase 1, illustrated in figure 5.05e. We needed seven such minimizations to reach this misfit by creeping, starting at the same initial guess. When the starting model had a constant depth of 1000 m, creeping went through 225 iterations to attain the same misfit; initiated at this profile, Occam's process again took three cycles of Phase 1 to reduce the misfit below the 0.5 mGal level.

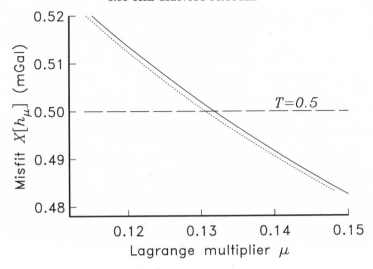

Figure 5.05f: Phase 2 of Occam's process: Solid curve is (c) of figure 5.05e showing intersection with target misfit in detail. The dashed curve represents the family of models based on the solution from curve (c) with misfit of 0.5 mGal. Further iterations lie almost exactly on top of this curve.

Phase 2 concentrates on the value of μ in the family where the target misfit of 0.5 mGal is achieved; on curve (c) in figure 5.05e this is found to be $\mu=0.1319$; at this point the model roughness, or RMS slope, $\|\hat{R}\mathbf{h}\|=0.7876$. We recompute the derivative matrix D for the corresponding h_μ and sweep through values of μ again, seeking the Lagrange multiplier giving $X=0.5$ mGal; now it is $\mu=0.1304$ and the roughness is reduced very slightly to 0.7871. We see in figure 5.05f that the two curves are very close and a further repetition yields an almost identical curve, intercept, and roughness value: Phase 2 has converged. The minimum roughness solution appears as the solid curve in figure 5.05g; it is very smooth, in contrast to those shown in figure 5.05d. This comparison is not strictly fair since the rougher models fit the measurements essentially exactly. But if we examine the solution at step 7 of the creeping calculation, with misfit 0.39 mGal, we find a very irregular model, equally rough in the measure we are using as the final solution found by creeping. Notice that the valley floor reaches the surface at $x=0$ and $x=a$; this is because the constraint that $h=0$ at the end points has been added. In practice this was done by modifying $\hat{R}$; the reader might like to think

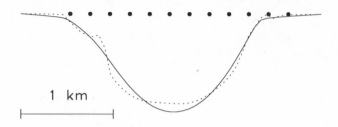

Figure 5.05g: Solid line: approximation to the solution in W_2^1 with smallest RMS slope and misfit $X[h] = 0.5$ mGal. This model is also constrained to outcrop at $x = 0$ and $x = a$. Dashed line: smallest solution in L_2 subject to the same misfit.

about this and alternative approaches to the constraint question. Inserting these conditions in L_2 would be pointless, because arbitrary jump discontinuities are permitted there, but in W_2^1 the constraint introduces useful additional information into the system.

   The author performed a number of other computer runs on this problem. Several different starting guesses were used: the sine-squared profile, with initial misfit of 12.56 mGal took the same number of cycles to reach the target misfit and generated the same model as shown in figure 5.05g. Another starting point was the exactly fitting model found by creeping shown as a solid line in figure 5.05d; this resulted in the same smooth solution. Here, however, the first sweep through μ generated a family of models all of which agreed with the data less well than the initial guess. Phase 1 with the target misfit of 0.5 was complete in two cycles, but Phase 2 required five cycles before reaching a stable solution. Thus with Occam's process it does not necessarily pay to initiate the solution at a model fitting the data well if that model is much rougher than the final solution. This experience also shows that a sweep through the family h_μ may not produce any solution in as good accord with the observations as the model used to generate the derivative matrix D. If that happens it may be necessary to revert to a slower but guaranteed descent method.

   In another experiment, the regularizing matrix $\hat{R}$ in (34) was set to the unit matrix, so that ordinary L_2 size was penalized instead of RMS slope. Surprisingly (to the author anyway) Occam's method happily reduced the misfit to 0.475 mGal in only four steps starting at the usual constant half-kilometer guess. Iterating for the smallest model with the standard 0.5 misfit produced the dashed model in figure 5.05g.

Exercises

5.05(i) Suppose the equation for the layer thickness is given by

$$h(x) = |x - a/2|, \quad 0 \le x \le a$$

and gravity is measured at $x = a/2$. Show by evaluating 5.03(19) that the corresponding gravity functional is not Gateaux differentiable with the increment $\Delta(x) = 1$. Notice this model violates condition (11). Is there something unusual about the integral in (4) for this h?

Suppose $h(x) = 0$ for all x on the interval $(0, a)$. Show the gravity functional is not Gateaux differentiable for any increment.

(Beware: h is the *increment* in 5.03(19) but it is the *model* in (2) and in this question!)

5.05(ii) Figure 5.05g shows the smallest solutions in two Hilbert spaces. Consider the question of finding the smallest model under the 1-norm. Linear perturbation theory and the results of section 4.06 seem to suggest it would consist of a series of delta-function spikes. But recall figure 4.07e: if we ignore the minor overhangs on the left, is this not an approximation to the minimum 1-norm solution? Resolve the apparent contradiction.

5.06 The Magnetotelluric Problem I

As in the previous section we need to calculate a functional derivative of the measurement functional, in this case the complex admittance response c. In essence we must apply linear perturbation theory to see what changes appear in c when a small change is made to the electrical conductivity profile. This example is less straightforward because there is no simple expression, like the integral 5.02(1), relating a measurement c to the model σ; they are related through the solution of a differential equation.

We work at a fixed real frequency f. It is convenient to solve the differential equation 5.02(8) for the complex electric field scalar

$$\frac{d^2 E}{dz^2} = 2\pi i f \, \mu_0 \sigma(z) E(z) \tag{1}$$

subject to the boundary conditions

$$E(H) = 0 \quad \text{and} \quad E'(0) = -1. \tag{2}$$

A difficulty arises if (1) has no solution under this two-point boundary system: that can happen only if f is an eigenfrequency of the system under the condition $E'(0) = 0$; we shall show in 5.08

that the eigenfrequencies are all imaginary, while f is real. From
5.02(11) the admittance is simply $c = E(0)$. Now consider two
conductivity profiles on the interval $(0,H)$ called σ_1 and σ_2; associated with each are electric fields E_1 and E_2, along with
corresponding admittances $c_1 = c[\sigma_1]$ and $c_2 = c[\sigma_2]$. Our task is to
calculate $c_2 - c_1$ for a small change $\sigma_2 - \sigma_1$. We do this by finding
and solving the differential equation for the change in the electric
field.

By the definitions

$$\frac{d^2E_1}{dz^2} = 2\pi i f \mu_0 \sigma_1(z)E_1(z) \tag{3}$$

$$\frac{d^2E_2}{dz^2} = 2\pi i f \mu_0 \sigma_2(z)E_2(z). \tag{4}$$

We define ΔE as $E_2 - E_1$ and $\Delta\sigma = \sigma_2 - \sigma_1$. Then subtracting (3)
from (4) and rearranging we find

$$\frac{d^2\Delta E}{dz^2} - 2\pi i f \mu_0 \sigma_1(z)\Delta E(z) = 2\pi i f \mu_0 E_2(z)\Delta\sigma(z) \tag{5}$$

under boundary conditions

$$\Delta E(H) = 0 \quad \text{and} \quad \Delta E'(0) = 0. \tag{6}$$

Obviously the solution of (5) at the surface is the perturbation we
seek; in other words,

$$\Delta E(0) = c_2 - c_1 = c[\sigma_1 + \Delta\sigma] - c[\sigma_1]. \tag{7}$$

Since (5) is inhomogeneous the simplest approach is to apply
Green's function:

$$\Delta E(z) = \int_0^H 2\pi i f \mu_0 E_2(y)\Delta\sigma(y)G_1(z,y)\, dy \tag{8}$$

with $G_1(z,y)$ arising from the solution of the inhomogeneous form
of (3), which explains the subscript one. It is shown in any text
on ordinary differential equations (e.g., Lanczos, 1961) that

$$G_1(z,y) = \begin{cases} U(z)V(y), & 0 \le z \le y \le H \\ V(z)U(y), & 0 \le y \le z \le H \end{cases} \tag{9}$$

when U and V are any two linearly independent solutions of (3)
satisfying $U'(0) = 0$ and $V(H) = 0$ and with normalized Wronskian

$$U'V - UV' = -1. \tag{10}$$

An obvious candidate for V is E_1; if we are interested in $\Delta E(0)$, then according to (9) we only need to know $U(0)$. But by (2) and (10), $U(0) = -1$. Therefore we find, combining (9) and (8) at $z = 0$

$$G_1(0,y) = -E_1(y) . \tag{11}$$

Notice that, by virtue of the relationship $c = E(0)$, (11) gives a result most useful to us later on:

$$c[\sigma_1] = -G_1(0,0) . \tag{12}$$

To return to the main issue, let us insert (11) into (8) to obtain:

$$\Delta E(0) = -\int_0^H 2\pi i f \mu_0 E_1(y) E_2(y) \Delta\sigma(y) \, dy \tag{13}$$

$$= -\int_0^H 2\pi i f \mu_0 E_1(y)^2 \Delta\sigma(y) \, dy + R \tag{14}$$

where

$$R = -\int_0^H 2\pi i f \mu_0 E_2(y) \Delta E(y) \Delta\sigma(y) \, dy . \tag{15}$$

Intuitively, we expect R in (15) to be second-order in $\|\Delta\sigma\|$, and so (14), which is a linear functional of $\Delta\sigma$, may be the Fréchet or Gateaux derivative for c. It is perfectly possible, though somewhat mysterious, to obtain (13) directly from (3) and (4), without any appeal to Green's function; see exercise 5.06(iii).

The following proof that R is small enough for (14) to be a genuine Fréchet differential is due to McBain (1986). Let us use $L_2(0,H)$ as the Hilbert space, which gives (14) an obvious interpretation as an inner product. The general idea of the proof is to show that ΔE is the same order of size in the 2-norm as $\Delta\sigma$ and then use this in (15). First we need the fact that Green's function $G_1(z,y)$ is bounded on the square $0 \le z \le H$, $0 \le y \le H$. This follows from (9), which states that G_1 is a combination of two solutions of the ordinary differential equation (3) which has bounded solutions for positive σ_1 in L_2 and real frequency f; Green's function would not exist at eigenfrequencies as we noted earlier, but a real f cannot be an eigenfrequency. We assert that

$$|G_1(z,y)| \le K, \quad 0 \le z \le H, \ 0 \le y \le H \tag{16}$$

where the real constant K obviously depends on σ_1 and f but obviously not on σ_2. Apply this bound to (8) and then use

Schwarz's inequality:

$$|\Delta E(z)| \le 2\pi\mu_0 f K \int_0^H |E_2(y)| \, |\Delta\sigma(y)| \, dy \tag{17}$$

$$\le 2\pi\mu_0 f K \|E_2\| \, \|\Delta\sigma\| \, . \tag{18}$$

We now square and integrate (17) to find an inequality for $\|\Delta E\|$:

$$\|\Delta E\|^2 = \int_0^H |\Delta E(z)|^2 \, dz \tag{19}$$

$$\le 4\pi^2\mu_0^2 f^2 K^2 H \, \|E_2\|^2 \, \|\Delta\sigma\|^2 = 4\pi^2\mu_0^2 f^2 K^2 H \, \|\Delta\sigma\|^2 \|E_1 + \Delta E\|^2 \tag{20}$$

$$\le 4\pi^2\mu_0^2 f^2 K^2 H \, \|\Delta\sigma\|^2 (\|E_1\|^2 + 2\,\|E_1\| \, \|\Delta E\| + \|\Delta E\|^2) \, . \tag{21}$$

Let us introduce the abbreviation:

$$\varepsilon = 2\pi\mu_0 K H^{1/2} \|\Delta\sigma\| \, . \tag{22}$$

With this we can rearrange (21) to read:

$$\|\Delta E\|^2(1-\varepsilon^2) - 2\varepsilon^2\|E_1\| \, \|\Delta E\| - \varepsilon^2\|E_1\|^2 \le 0 \tag{23}$$

$$(\|\Delta E\| - r_-)(\|\Delta E\| - r_+) \le 0 \, . \tag{24}$$

Since r_-, the lower root, is negative it places no constraint on $\|\Delta E\|$; thus we learn from (24) that

$$\|\Delta E\| \le r_+ = \frac{\varepsilon\|E_1\|}{1-\varepsilon} \, . \tag{25}$$

When $\|\Delta\sigma\|$ is smaller than $1/(4\pi\mu_0 K H^{1/2})$ we have from the definition (22) that $\varepsilon < 1/2$; since $\|\Delta\sigma\|$ is tending to zero, it can be chosen to be smaller than any fixed amount and therefore we can arrange $\varepsilon < 1/2$ in the denominator of (25) and so by (22) we have

$$\|\Delta E\| \le 4\pi\mu_0 K H^{1/2}\|\Delta\sigma\| \, . \tag{26}$$

Armed with this result we can return to (15), the expression for the remainder. First we insert (18) and then apply Schwarz's inequality once more:

$$|R| \le 4\pi^2\mu_0^2 f^2 K \, \|E_2\| \, \|\Delta\sigma\| \int_0^H |E_2(y)| \, |\Delta\sigma(y)| \, dy \tag{27}$$

$$\le 4\pi^2\mu_0^2 f^2 K \, \|E_2\|^2 \, \|\Delta\sigma\|^2 \, . \tag{28}$$

Writing $E_2 = E_1 + \Delta E$ and applying the triangle inequality

$$|R| \leq 4\pi^2 \mu_0^2 f^2 K \, \|\Delta\sigma\|^2 (\|E_1\|^2 + 2\|\Delta E\| \, \|E_1\| + \|\Delta E\|^2) \, . \tag{29}$$

It is now obvious from (29) and (26) that as $\|\Delta\sigma\| \to 0$

$$\frac{R}{\|\Delta\sigma\|} \to 0 \tag{30}$$

and so we have shown (14) is a true Fréchet derivative for c when $\Delta\sigma \in L_2$.

It is self-evident that before one can begin an inverse problem, one should know how to solve the forward problem. Equations (1) and (2) state the mathematical task: solve the differential equation at each measured frequency for the given conductivity profile σ. But how shall we do this numerically, for an arbitrary function? First, we must represent the model in a finite-dimensional way, suited to the information available. In the gravity profile, equal spacing of the sample of the valley depth was quite natural. Here, if we inspect figure 5.02b, we discover a range of length scales covering 100 m to 30 km. If this span is to be covered in an equally spaced sampling in depth we will need more than 1000 sample points in z. This seems to be extravagant, when we understand that the solution in the deeper part of the model will be varying very slowly, since the wavelength of the penetrating waves is comparable to the depth of penetration. These considerations suggest an uneven sample spacing: one in which the sample spacing is proportional to depth, namely a geometrical increase in spacing. In principle we can now apply one of the standard numerical equation solvers, like Runge-Kutta integration (see, e.g., Press et al., 1992) to (2). But the conventional approach in the literature, which we adopt, relies on a different idea: the conductivity profile is taken to be a set of constant-conductivity layers (of variable thickness) instead of a smooth curve sampled at various points. Then (1) can be solved exactly within each layer. Suppose there are L layers with interfaces at $z_l, l = 0, 1, \cdots L$ and $0 = z_0 < z_1 < z_2 < \cdots < z_L = H$. Then if $z_{l-1} \leq z < z_l$, the conductivity $\sigma(z) = \sigma_l$. From (1) it is easy to see that

$$E(z) = A_l e^{k_l z} + B_l e^{-k_l z}, \quad z_{l-1} \leq z < z_l \tag{31}$$

where $k_l = (1+i)(\pi f \mu_0 \sigma_l)^{1/2}$. The boundary conditions of continuity of E and E' are applied at the interfaces; then the complex electric field need be computed only at each interface. In fact one can carry the admittance itself through the system starting at $z = H$, where we set E and c to zero, working up to the surface:

$$C_{l-1} = \frac{\tanh k_l (z_l - z_{l-1}) + k_l \, C_l}{k_l (1 + k_l \, C_l \tanh k_l (z_l - z_{l-1}))}, \quad l = L, L-1, L-2, \cdots 1 \quad (32)$$

where $C_l = -E(z_l)/E'(z_l)$ and obviously $C_0 = c$, the admittance of the model. The calculation must be performed for every frequency at which c has been measured. The advantage of this approach over solving the differential equations numerically is that we find an exact solution for c. This would not be so important except for the possibility of finding an exact functional derivative corresponding to this solution, which we describe in a moment. We shall soon see that the matrices of derivatives can become very ill-conditioned so that tiny errors in the derivatives lead to serious errors in computed quantities; it is very important to avoid such errors as far as possible. Of course we sacrifice the ability to represent exactly smooth models, or any of the singular models that exist in L_2. We note that in the gravity problem of the previous section, the derivative obtained by numerical integration of the Gateaux formula is also an exact derivative of the approximate problem, provided the same integration scheme is used for the integrals.

The Fréchet derivative for the system of layers based on (14) can obviously be written as a sum, one term from each layer; because the conductivity is constant in each layer we have

$$c[\sigma + \Delta\sigma] - c[\sigma] = -\sum_{l=1}^{L} \int_{z_{l-1}}^{z_l} 2\pi i f \, \mu_0 E(y)^2 \Delta\sigma(y) \, dy \quad (33)$$

$$= -\sum_{l=1}^{L} \Delta\sigma_l \int_{z_{l-1}}^{z_l} 2\pi i f \, \mu_0 E(y)^2 \, dy \quad (34)$$

$$= -\sum_{l=1}^{L} \Delta\sigma_l \, D_l \, . \quad (35)$$

Since the form of $E(z)$ in each layer is known (it is (31)) the integral in (34) can be performed exactly, permitting an exact derivative to be found. We omit the details which are straightforward.

Throughout the theoretical treatment we worked in a complex Hilbert space, because the observations are complex functionals of the model. But since we allow only real conductivity profiles, for the numerical work it is simpler to revert to a real inner product space in which the real and imaginary parts of c at any frequency are just separate real measurements. This does not mean giving up the convenience of the complex calculations for c or the rows of the Fréchet derivative matrix; the shift to real arithmetic is made only in the matrix computations of the iterative inversion. As well as being real, σ must be constrained to be positive too. In the first example the forward functional fortuitously ignored the sign of the solution and we could continue without taking any special steps to enforce the condition. The most obvious way to maintain a positive solution is to apply the constraint $\sigma_l \geq 0$ on each element and to use quadratic programming to minimize the approximate objective functions at each iterative step, for example 5.04(22). An alternative, more traditional approach for this particular problem is suggested in a very natural way by the observation that in the Earth the property of electrical conductivity varies over an enormous range: sea water has a conductivity of $4.4 \ \mathrm{S\,m^{-1}}$ and dry olivine $10^{-6}\,\mathrm{S\,m^{-1}}$. This makes it almost essential in displaying results of electrical sounding to consider the logarithm of σ. Because this is already a non-linear problem, we do not increase its complexity by choosing the model to be $m = \ln \sigma$. Now there is no restriction on the sign of m, while the material property σ remains positive. If we focus our attention on strictly positive models in W_2^1 for example, the continuity of the log function keeps the admittance function Fréchet differentiable. In any case, in the context of our finite-dimensional representation, we can proceed as though we were taking a conventional gradient (see 5.03(7), for example): then we find in place of (34):

$$c\,[m+\Delta m\,] - c\,[m\,] = -\sum_{l=1}^{L} \Delta m_l \int_{z_{l-1}}^{z_l} 2\pi i f\, \mu_0 E\,(y\,)^2\, e^{m_l}\, dy \quad (36)$$

where $m_l = \ln \sigma_l$.

We can now turn to the magnetotelluric data described in 5.02. To implement the numerical scheme we must choose a value for H, the depth to the bottom of the system. We have seen in exercise 5.02(ii) that $|c\,|$, which has units of length, is related to the depth at which currents flow; figure 5.02b shows the largest c to be about 30 km and so we shall use a value of $H = 60$ km. The number of layers in our numerical illustrations is 75, ranging in thickness from 10 m at the top to 4.8 km at the bottom. In the plots of the conductivity profiles, the individual layers are not shown; instead the functions appear as continuous curves. In our first example, we were fairly confident of the existence of solutions exactly matching the gravity data, because of our calculations in chapter 4 on this problem. In the present case we do not know if a one-dimensional conductivity model can fit the observations to a tolerance consistent with the uncertainties, let alone exactly. So the first job for an interpreter of the data is to see how well he or she can satisfy the data with the predictions of a model. Later in 5.08 we shall see that the magnetotelluric inverse problem is one of the few nonlinear problems for which this question can be answered. For now, we will apply our numerical techniques. There are 17 frequencies in table 5.02A; this means that there are 34 real data, counting real and imaginary parts of c_j separately. We select the familiar weighted 2-norm to be the measure of misfit:

$$X[m] = \left[\sum_{j=1}^{N} \frac{|c[m,f_j] - c_j|^2}{\varepsilon_j^2} \right]^{1/2} . \tag{37}$$

Here we have assumed the uncertainties in the real and imaginary parts of the admittance at each frequency are equal; the notation $c[m,f_j]$ means the admittance of the model m at frequency f_j. The target misfit level T for this norm is 5.788 according to 3.02(21) when we choose the expected value of the norm.

We begin with the allegedly foolproof creeping technique, in the expectation that with enough iterations we can drive the misfit to a satisfactory level. A starting profile is needed: from figure 5.02a we guess that a uniform conductor with $\sigma = 0.1\,\mathrm{S\,m^{-1}}$ might be a reasonably good initial model. For this conductivity function $X = 70.56$. The first cycle of the creeping algorithm with step length damping produces an improvement of less than 10^{-13} in X! At this rate we could run the program until the sun becomes a red giant and still not be close to a good fit. Evidently the gradient provides a valid description for the misfit functional in the tiniest neighborhood of the starting model. The smallest m in the linear regime possesses a 2-norm of around 10^8, and m is the log of conductivity! This is an indication of the minute condition number of the Gram matrix. Such behavior naturally suggests there is a bug in the program. To check, I computed the

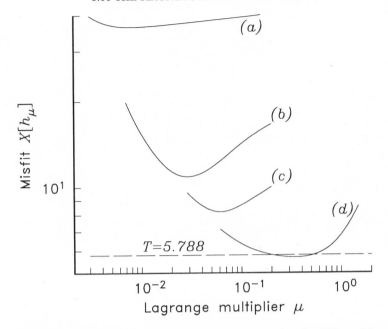

Figure 5.06a: Phase 1 of the Occam process for the MT problem. Misfit curves for four families of models (a)–(d): derivatives are calculated from the minimum misfit model of the preceding curve. The misfit of the starting model is $X = 70.56$.

admittances of a uniform conductor with $\sigma = 0.1\,\mathrm{S\,m^{-1}}$ and applied creeping to these responses. Initialized at a model with $\sigma = 0.10001\,\mathrm{S\,m^{-1}}$ one cycle of creeping yielded a much bigger improvement: 6×10^{-7}. Next I reduced the original system to one consisting of a subset of data at three frequencies drawn from the table—the highest, lowest, and one in the middle. Now the method reduces the misfit to 0.005 in five cycles. But the performance deteriorates rapidly as the number of frequencies is increased. This evidence tends to confirm the program is correct and that therefore the creeping algorithm is very inefficient. One possible repair might be as follows: find a solution from three frequencies, add a fourth, and start the iteration at the previous solution; repeat until all the frequencies are included. But rather than attempting to revitalize the moribund algorithm, we turn to Occam's process.

Phase 1 of the Occam process is illustrated in figure 5.06a, with a uniform $0.1\,\mathrm{S\,m^{-1}}$ conductor as the initial guess. The

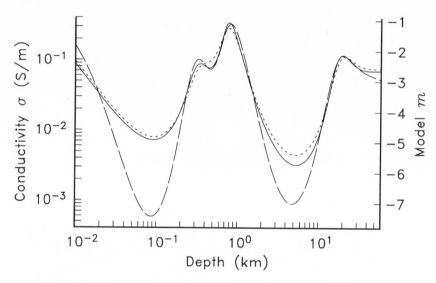

Figure 5.06b: Regularized solutions of the MT problem. Solid line: smoothest solution in the sense of norm of first difference of the log of conductivity with misfit $X = 5.788$. Light dashed line: misfit increased to 6.971, corresponding to 95 percent confidence. Long dashed line: solution regularized by the second difference in the log with the smaller misfit.

roughening matrix $\hat{R}$ used in this run is the simple first difference operator. Because of the increase in layer thickness with depth, this has the effect of introducing a weight function rising with depth into the 2-norm of the derivative in the equivalent continuous problem; the regularization is applied to m, the log of conductivity, not to σ itself. As the figure shows, Occam's process is incomparably better at improving the misfit than the conservative creeping method. After four cycles of Phase 1 the minimum X falls below the target value of 5.788. The success of the iterative construction method answers the question of existence in the affirmative, at the level of the mean value of the norm of misfit. How much more accurately can the data be matched? If we just continue Phase 1 it soon becomes apparent that the minimum misfit never drops below 4.89, but unfortunately nothing we have done so far allows us to conclude that this is the true minimum misfit. In fact we shall see later that $X_{\min} = 4.602$ so that in this case Occam's process has come gratifyingly close to the smallest misfit.

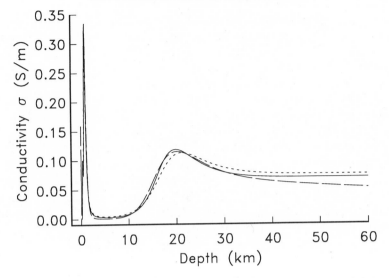

Figure 5.06c: Same as figure 5.06b without log scales. Notice how difficult it is to see the details near the surface.

To find the smoothest model we apply Phase 2, which converges in three steps to the solution shown in figure 5.06b. Also shown in the figure as a lightly dashed line is the result of fitting within a 95 percent confidence level, corresponding to $T = 6.971$. This change in misfit tolerance has no significant effect on the solution. The third solution, drawn with long dashes, is the result of changing the regularizing matrix $\hat{R}$ to the second-difference operator, the square of the original matrix so that we look for solutions in W_2^2. In regions of higher conductivity the solutions agree remarkably well, but when σ is small, the new solution is an order of magnitude less conducting in some places. We might conclude from this that while regions of low conductivity are detected in the solutions, the actual value of σ there is poorly determined.

Despite the emphasis on solutions with small derivatives imposed by the regularization, the profile is quite rugged, suggesting that such a structure is truly needed in order to satisfy the data. If the models are drawn as in figure 5.06c, without the benefit of log scales, near the surface we see a very thin, highly conductive layer which heavy regularization does not remove. A plausible geophysical interpretation of the solution is that there is a conducting sedimentary layer near the surface, which lies over nonconducting crust down to about 15 km, where the commonly observed mid-crustal conductor is seen. The origin of this high conductivity is controversial; one explanation is the presence of carbon.

How much confidence can we place in this conductivity profile? If we leave aside the inadequacies of the mathematical model (issues (a)–(f) in 5.02)), this is the question of inference that occupied us in chapter 4. A common approach is to borrow the idea of resolution from the linear theory; the assumption is made that the entire range of models fitting the observations can be treated as though they were sufficiently close to one another to allow a complete description of all the solutions in terms of linear perturbations. We shall briefly examine resolution in the next section. Yet even for a linear problem, resolving power calculations do not offer a rigorous way of drawing conclusions about the set of solutions. The alternative of maximizing and minimizing a suitable functional discussed in 4.03 works equally well in principle on nonlinear problems. But there is no general theory allowing us to be certain the minimum or maximum we compute is truly the smallest or largest value and not merely a local extremum. In the author's opinion this is a serious gap, although interesting results have been obtained. The interested reader is referred to the work of Oldenburg (1983) and Weidelt (1985) for applications to the magnetotelluric problem. In section 5.08 we return once again to the MT problem and show how the existence question can be solved in a more satisfactory way than by numerical iteration.

Exercises

5.06(i) For the simplest system $\sigma =$ constant, verify the validity of the Fréchet derivative for c, in the case of a small constant perturbation. Does the formula still work when the original model is $\sigma = 0$? Compare with exercise 5.05(i).

Perform a similar check for some of the profiles in exercise 5.02(iv); not for the faint-hearted.

5.06(ii) Consider again a simple uniformly conducting layer, say $\sigma = 0.1 \, \mathrm{S \, m^{-1}}$, of thickness 50 km; calculate the Fréchet derivatives at the frequencies in table 5.02A, and plot them.

Notice how the conductivity changes at greater depths almost exclusively affect the lower frequency admittances. If the linear approximation were correct and R were always negligible, (14) would imply that a large enough increase in σ at a great depth (say, 49 km in this system) could cause an arbitrarily large change in c at any frequency. In fact this is not true: show numerically that the largest possible change in c that can be produced by increasing σ uniformly below 49 km, is about 2 percent.

5.06(iii) Prove (13) without invoking Green's function. Consider the quantity

$$E_2 \frac{d^2 E_1}{dz^2} - E_1 \frac{d^2 E_2}{dz^2}$$

integrated on $(0, H)$. Integrate by parts and use the boundary conditions to show, on the one hand, it equals $\Delta E(0)$; use (3) and (4) to show, on the other hand, it is the integral on the right of (13).

5.07 Resolution in Nonlinear Problems

We would like to have some way of assessing the reliability of our models. The idea of the resolution of a set of data was invented by Backus and Gilbert specifically for a nonlinear inverse problem, the one based on the set of frequencies of free oscillation of the Earth (see 2.08), although they realized it must be an approximation. The recipe we shall apply is straightforward: after a satisfactory solution has been found, one assumes the accuracy of the linearization

$$F_j[m + \Delta m] = F_j[m] + (D_j, \Delta m) \tag{1}$$

for describing all small perturbations about that solution. We compute the consequences in the data set of a delta-function perturbation to the model (always in the approximation (1), for otherwise, as we shall see, strange things can happen), and invert the perturbation in the data, using the parameters that were involved during the original solution. The resultant functions are the (linearized) images of delta-function perturbations and their widths are an indication of smearing in detail introduced by the inversion process. Let us carry out this prescription for the gravity problem.

We have computed the resolving functions for two base models. Consider first smoothest solution in W_2^1 shown solid in figure 5.05g and again in figure 5.07a. Linearizing about this solution we obtain the solid lines in figure 5.07b; here the vertical dashed lines indicate the site of the delta-function perturbation, the point on which a perfect solution would center a delta

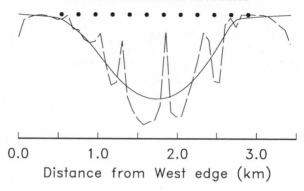

0.0   1.0   2.0   3.0
Distance from West edge (km)

Figure 5.07a: Two solutions of the gravity data inverse problem. Solid line: the smooth solution obtained by Occam's process in section 5.05; dashed line: the solution obtained by creeping.

function. It is obvious the results are far from perfect. In general better resolution is seen at the edges of the model where the valley is shallow. We can say that detail on scales of about 500–700 meters could be discerned at the best locations. Near the valley center the scale of resolution climbs to 1500 m.

A possible source of a concern is that the shape of these functions may depend strongly on the solution used as the center for linearization. Perhaps the variations are so large as to render this kind of interpretation worthless. Using the same Lagrange multiplier, the computations have been repeated for the model obtained by creeping, the ragged solution shown dashed in figure 5.07a. The corresponding curves are also drawn dashed in figure 5.07b. Despite the great differences in the models, the resolving functions are actually quite similar in a qualitative way, which is all that can be asked for. The biggest departure of the curves shown is the one centered at $x = 1900$ m, where the rougher model predicts a somewhat greater resolving length than the smooth one. This picture of the resolution abilities of the data warns us to be skeptical of the oscillations exhibited by the rough model in figure 5.07a, since the length scale of these undulations is less than the resolving length and, of course, that is quite proper since much smoother models do exist.

Is there any point in computing the response of the non-linear system to a delta-function input, rather than keeping to the linearized model for all the resolving calculations? In this case we find a surprising answer. Consider adding to a profile $h(x) > 0$ a positive perturbation:

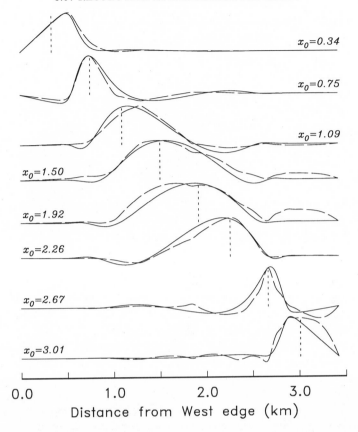

Figure 5.07b: Resolving functions of the inverse gravity problem corresponding to the two solutions shown in figure 5.07a. The vertical dashed line shows where the delta-function approximation should be centered. All curves have been normalized to have the same maximum amplitude for ease of comparison.

$$D(x) = \begin{cases} 1/\varepsilon, & |x - x_0| \le \varepsilon/2 \\ 0, & |x - x_0| > \varepsilon/2 \end{cases} \tag{2}$$

and then letting ε tend to zero. We calculate the gravitational change from 5.02(3):

$$\frac{dF_j}{G \Delta\rho} = \int_{x_0-\varepsilon/2}^{x_0+\varepsilon/2} \ln\left[\frac{(x-x_j)^2 + (h(x)+1/\varepsilon)^2}{(x-x_j)^2 + h(x)^2}\right] dx \tag{3}$$

$$= \int_{x_0-\varepsilon/2}^{x_0+\varepsilon/2} \ln \varepsilon^{-2} dx + \int_{x_0-\varepsilon/2}^{x_0+\varepsilon/2} \ln\left[\frac{(1+h(x)\varepsilon)^2 + \varepsilon^2(x-x_j)^2}{(x-x_j)^2 + h(x)^2}\right] dx . \tag{4}$$

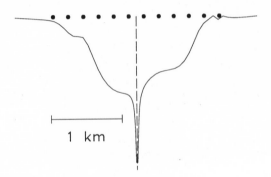

Figure 5.07c: A singular solution to the inverse gravity problem: the valley depth has a $1/|x - c|$ singularity with $c = 1.71$ km, but the profile satisfies the data since $X < 0.5$ mGal.

The second term tends to zero as ε becomes small because the integrand tends to a constant while the interval of integration shrinks; the first term reduces to $-2\varepsilon \ln \varepsilon$, which also goes to zero in the limit. Thus, without approximation, the delta-function perturbation has no effect on the gravity at all.

One of the appeals of the resolution analysis is the underlying assumption that certain averages are universal properties of the data set. In the linear problem smoothing a solution by integrating against a Backus-Gilbert resolution function for a point produces the same result no matter which solution is smoothed, so that such averages are universal properties. For nonlinear problems we hope the same idea continues to hold approximately: that the values of appropriate averages will remain, if not exactly constant, at least within a finite interval for the whole class of models fitting the data. This is certainly true for some linear problems with positivity constraints (recall the attenuation problem in 4.07) and for some nonlinear problems (see Oldenburg, 1983), but it does not hold generally. In particular we can show for our gravity problem, the data by themselves provide no upper bound on any weighted average of the depth function h. To do so we consider the singular depth profile

$$h_0(x) = \frac{b^2}{|x - c|}, \quad 0 \le x \le a, \quad 0 < c < a, \quad x \ne c . \tag{5}$$

It is easy to see that this "bottomless trough" possesses a finite gravity anomaly at the surface on account of the mollifying effects of the log in the forward formula. But if w is a positive weight function, none of the integrals

$$\int_0^a w(x) h_0(x)\, dx \tag{6}$$

exists, because the growth at $x = c$ is too strong. Now suppose a valley profile of the form $h(x) = h_0(x) + m(x)$, with m a smooth function could be found satisfying the gravity data; then any average like (6) of that solution would be undefined or could be made arbitrarily large by clipping the model at some very great depth. Needless to say, the existence of such solutions is not speculation. In the first place, one could reduce b in (5) so as to make the gravitational effect as small as desired, yet this would still leave an infinite integral in (6) as long as b is nonzero. More concretely, the author modified the construction program to find perturbations m away from a particular h_0, seeking smooth solutions in the sense of a small 2-norm of the first derivative in m. Occam's process found models with no difficulty, and one is displayed in figure 5.07c.

We report briefly the resolution analysis of the magnetotelluric problem. In figure 5.07d we treat m, the log of conductivity as the model, and calculate the resolving functions based on the smoothest model of W_2^1, shown at the top of the figure. The diagram shows that resolving scale increases roughly exponentially with depth, as we might expect. The high-frequency admittances allow us to distinguish considerable detail in the shallow part of the column, but because the energy at high frequencies (and short wavelength) is so strongly attenuated, the deeper parts are probed only with long-wavelength fields. Despite this, the resolving functions are very small outside a region of influence surrounding the target depth. This says that variations in m near the surface are not confused with structure at depth and conversely. Note a worsening of resolution in the regions of low conductivity. However, lack of resolving power is not enough to explain the great differences in m between acceptable solutions shown in figure 5.06b; apparently such differences are not within the linear realm. In fact we shall see in the next section that σ could vanish exactly over large depth intervals, so that $\ln \sigma$ may be arbitrarily large and negative. This represents another breakdown of the linear resolution approximation.

These few numerical experiments illustrate how resolution calculations can give a qualitative impression of the amount of

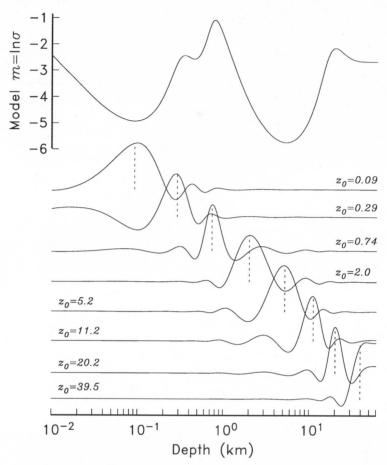

Figure 5.07d: Resolution functions for the magnetotelluric inverse problem, taking $m = \ln\sigma$ as the model. The functions have been normalized to be of unit amplitude.

trustworthy detail in a solution just as they do in linear systems. But such an analysis is far from rigorous; in particular, the notion that integrals of the model may be well constrained, even though fine-scale structure is not, turns out to be false in both of the cases we have studied. It may be true that one-sided bounds are well defined: for example the cross-sectional area of the valley is bounded below by gravity measurements as we saw in section 4.07, but the data cannot provide an upper bound because of the existence of solutions like the one shown in figure 5.07c.

5.08 The Magnetotelluric Problem II

For our finale we return to the magnetotelluric inverse problem. This is one of the rare instances of a nonlinear problem for which the questions of existence and construction of solutions can be settled in a definite way without any of the uncertainties surrounding iterative model building. This section comprises a somewhat simplified reworking of three papers (Parker, 1980, 1982; Parker and Whaler, 1981); the approach we take is intended to make the ideas as clear as possible, but to do so we will occasionally have to sacrifice mathematical rigor or skip some of the details.

To begin we set about understanding the properties of the forward problem in greater depth. In sections 5.02 and 5.06 the calculation of the admittance c was merely a recipe: find the ratio of the solution to a second-order ordinary differential equation and its derivative. The recipe provides little mathematical insight about how c might vary with frequency for a fixed model. By broadening the definition of the frequency f to become complex, and by looking at the normal modes of the system, neither of which has at first sight any connection with the geophysical problem, we discover a remarkable fact: c can be written as a *linear* functional of a function called the spectral function, which is directly, but nonlinearly, related to conductivity. Because of this we can break the solution into two stages: the first treats a typical linear inverse problem (with positivity constraints), the second provides a mapping between the spectral function and σ. As a result of having a means of deciding if a particular data set is compatible with the forward modeling assumptions, we can answer another question: How deep can we "see" in the Earth with a particular set of observations?

Although it is difficult to realize experimentally, there is nothing difficult about *calculating* how an electrically inhomogeneous layer would respond to an excitation with complex frequency: one simply solves the differential equation 5.02(8) with complex f on the right side. In this way we can imagine associating with each point in the complex f plane a complex value of c. It will probably come as no surprise that c is a complex analytic function of f. To show this, and much more, we invoke Sturm-Liouville eigenvalue theory. Consider the eigenvalue problem of 5.02(8) which we write:

$$-\frac{d^2u}{dz^2} = \lambda\mu_0\sigma(z)u(z) \tag{1}$$

subject to boundary conditions

$$u(H) = 0, \quad \text{and} \quad u'(0) = 0 .\tag{2}$$

We seek values of λ for which (1) and (2) can be satisfied ignoring the trivial $v(z)=0$. Obviously, for any such λ there is a corresponding frequency $f = i\lambda/2\pi$. In addition to being strictly positive, we shall suppose for the moment that σ is in some class of well-behaved functions (it is sufficient that $\sigma \in C[0, H]$).

We summarize the classical Sturm-Liouville theory (see, e.g., Coddington and Levinson, 1955): The differential equation on $(0, H)$:

$$\frac{d}{dz}(p(z)\frac{dv}{dz}) + q(z)v(z) = \lambda v(z)\tag{3}$$

where p is of one sign on $(0, H)$ and (3) is subject to boundary conditions

$$\alpha_1 v(0) + \alpha_2 v'(0) = \beta_1 v(H) + \beta_2 v'(H) = 0\tag{4}$$

where at least one of the constants α_1, α_2 is nonzero and similarly for β_1, β_2, constitutes a *self-adjoint system* (the differential operator equivalent of a symmetric matrix). We simply state these properties proved in every book on ordinary differential equations:
(a) There exist infinitely many distinct, real values of λ for which (3) possesses a nontrivial solution. These are the *eigenvalues*:

$$\lambda = \lambda_1, \lambda_2, \lambda_3, \cdots\tag{5}$$

and either λ_n or $-\lambda_n$ increases without bound.
(b) The solutions of (3) corresponding to λ_n are $v_n(z)$ and are called *eigenfunctions*; these functions are mutually orthogonal:

$$(v_m, v_n) = 0, \quad m \neq n .\tag{6}$$

If we adopt the inner product and norm of $L_2(0, H)$, we can scale each one to be of unit norm: $\|v_n\| = 1$.
(c) The eigenfunctions are *complete* in the sense that every continuous function on $(0, H)$ can be written as a convergent infinite expansion:

$$f(z) = \sum_{n=1}^{\infty} \alpha_n u_n(z)\tag{6}$$

where

$$\alpha_n = (f, v_n) .\tag{7}$$

We would like to assert these properties for (1), but as it stands our equation is not in the proper form. A simple transformation remedies the problem: suppose $v = \sqrt{\sigma}u$ in (1); then some light algebra shows (3) and (4) are satisfied when $p = 1/\sigma$ and $q = 3/4\sigma^3$. It follows that the eigenfunctions u_n of (1) are orthogonal, *but with the weight σ:*

$$\int_0^H u_m(z)\,u_n(z)\,\sigma(z)\,dz = 0, \quad m \neq n \tag{8}$$

and that there functions are also complete on $(0, H)$. For convenience we normalize so that

$$\int_0^H u_n(z)^2 \mu_0\,\sigma(z)\,dz = 1. \tag{9}$$

We can easily show that the eigenvalues λ_n are all positive. When λ is the eigenvalue λ_n multiply both sides of (1) by u_n and integrate:

$$\int_0^H -u_n(z)\,\frac{d^2u_n}{dz^2}\,dz = \lambda_n \mu_0 \int_0^H u_n(z)^2\,dz \tag{10}$$

$$= [-u_n(z)\,u_n{}'(z)]_{z=0}^{H} + \int_0^H u_n{}'(z)^2\,dz \tag{11}$$

where we integrated the left side by parts. But the first term in (11) vanishes on account of the boundary conditions (2), and so λ_n must be positive.

The eigenfrequencies $f_n = i\lambda_n/2\pi$ are all purely imaginary, and because $u_n{}'(0) = 0$, it is at those frequencies that the admittance c is not defined. These are also the resonant frequencies of the system, since near the eigenfrequencies in the complex plane, $|c|$ becomes arbitrarily large. A different physical model can be helpful: consider a thin string with linear mass density $\rho(z)$ under tension Θ; it is clamped at one end, while the other is free to move perpendicularly to the tension force. Such a string will have frequencies of free oscillation given by $(\lambda_n/2\pi)^{1/2}$ if we replace $\sigma(z)$ by $\rho(z)/\Theta$.

We are going to find an expression for c in terms of the eigenfunctions u_n. To do so we return to Green's function for 5.06(5). Recall that we found 5.06(12)

$$c = -G(0,0) \tag{12}$$

where Green's function $G(z,y)$ formally solves

$$\frac{d^2 G}{dz^2} - 2\pi i f \mu_0 \sigma(z) G(z,y) = \delta(z-y). \tag{13}$$

Since the eigenfunctions u_n are complete, G must possess an expansion like (6) in terms of them. Again formally, we just substitute the expansion into (13); we make no effort to be rigorous here but assume that interchanging differentiation and summation is allowed:

$$\sum_{n=1}^{\infty} [\alpha_n \frac{d^2 u_n}{dz^2} - 2\pi i f \mu_0 \sigma(z) u_n(z)] = \delta(z-y). \tag{14}$$

Now substitute from (1) and tidy up:

$$\sum_{n=1}^{\infty} \alpha_n u_n(x) \mu_0 \sigma(x) [\lambda_n + 2\pi i f] = -\delta(z-y). \tag{15}$$

Next, multiply both sides by u_m, integrate over the interval, and appeal to (8) and (9); then rearranging we find:

$$\alpha_n = -\frac{u_n(y)}{\lambda_n + 2\pi i f} \tag{16}$$

and so

$$G(z,y) = -\sum_{n=1}^{\infty} \frac{u_n(z) u_n(y)}{\lambda_n + 2\pi i f}. \tag{17}$$

And thus with (12) we arrive at our destination:

$$c(f) = \sum_{n=1}^{\infty} \frac{u_n(0)^2}{\lambda_n + 2\pi i f}. \tag{18}$$

This formula shows us that $c(f)$ is an analytic function, whose only singularities are simple poles on the imaginary axis associated with the eigenfrequencies of (1). Let us take the real and imaginary parts of (18), and let f be complex:

$$c(f) = c(p+iq)$$

$$= \sum_{n=1}^{\infty} \frac{(\lambda_n - 2\pi q) u_n(0)^2}{(\lambda_n - 2\pi q)^2 + 4\pi^2 p^2} - i \sum_{n=1}^{\infty} \frac{2\pi p u_n(0)^2}{(\lambda_n - 2\pi q)^2 + 4\pi^2 p^2}. \tag{19}$$

Thus on the positive real frequency axis (where we can measure it) c has a positive real part and a negative imaginary part as we asserted earlier; also we can see that the real part decreases

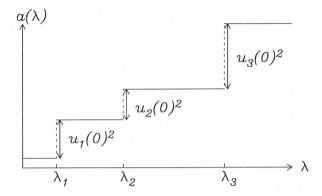

Figure 5.08a: The spectral function $a(\lambda)$ for equation (18), when the terms in the series are represented by a series of positive jumps at the eigenvalues.

monotonically. More generally, $\mathrm{Im}\, c < 0$ for the whole right half-plane where $\mathrm{Re}\, f = p > 0$; this will be useful later. Weidelt (1972) derives a host of interesting properties of the admittance from (18).

Equation (18) can be expressed in a different way. We allow λ to become a continuous variable. This would be necessary if we were treating an infinitely thick layer, because it is then possible that the points of an entire interval of the real line (even the whole real line) can be eigenvalues; see exercise 5.08(i). Then we must replace the sum by an integral and allow the index on u_n to become continuous too: this is best done by using λ itself. We write

$$c(f) = \int_0^\infty \frac{da(\lambda)}{\lambda + 2\pi i f} . \tag{20}$$

This is a Stieltjes integral over the nondecreasing *spectral function* $a(\lambda)$. To represent (18) the spectral function is constant, except for the values $\lambda = \lambda_1, \lambda_2, \lambda_3, \cdots$, where it exhibits a discontinuity, a simple jump of $u_n(0)^2$; see figure 5.08a. Thus the single spectral function contains the information about the two sequences $\{\lambda_n\}, \{u_n(0)^2\}$. Equation (20) can be considered to be *representation* of the allowable admittances c corresponding to the magnetotelluric problem. The representation holds for a much wider class of conductivity models than the continuous, strictly positive σ on $(0, H)$ we have considered. For example, it remains valid when we let H be infinite and work in a conducting halfspace; then the function a may increase continuously over some interval of λ, implying that numbers in a continuous range are eigenvalues. See Weidelt (1972) for a treatment of this case. The relation also holds when σ is allowed to vanish in part of the profile, something completely inadmissible in our derivation. Furthermore, it accommodates some badly behaved functions, including delta functions in conductivity but to show this the derivation must be completely recast, and centers on a representation theorem for analytic functions in the complex plane (Parker, 1980).

We turn now to the inverse problem. It is necessary to match N complex measurements $c_1, c_2, \cdots c_N$ made at the frequencies $f_1, f_2, \cdots f_N$ with the predictions of a model. Since every one-dimensional model of interest possesses a spectral function representation, the existence of a function a in (20) is a necessary condition for the existence of a conductivity profile. For an exact fit we demand

$$c_j = c(f_j) = \int_0^\infty \frac{da(\lambda)}{\lambda + 2\pi i f_j}, \quad j = 1, 2, 3, \cdots N. \tag{21}$$

which is a linear forward problem for the spectral function a with the constraint that a does not decrease—effectively $da/d\lambda \geq 0$. We have discussed this class of problems in section 4.06, where we applied linear and quadratic programming. The general idea is to approximate the integral by a sum, essentially reversing the step from (18) to (20). We sample the positive real axis at L points, $\lambda = \Lambda_1 < \Lambda_2 < \cdots < \Lambda_L$ and at these we allow positive jumps, Δa_k, otherwise the spectral function is constant. Now (20) can be approximated by the matrix equation

$$\mathbf{c} = A \, \Delta\mathbf{a} \tag{22}$$

which, coupled with the condition $\Delta\mathbf{a} \geq 0$ is the classic problem of feasibility of a linear program, described in 4.06. It is easier to treat the case of exact fitting as a special case of the more interesting problem of fitting to within a tolerance. For a given level of approximation we can solve the optimization problem

$$\min_{\Delta\mathbf{a} \geq 0} \| \mathbf{c} - A \, \Delta\mathbf{a} \| . \tag{23}$$

Here we use the Euclidean length for the norm and we treat the real and imaginary parts as separate real quantities. As usual we shall declare a solution acceptable if the minimum norm is less than T. We can imagine systematically improving the approximation by letting the largest sampling interval $\Lambda_{k+1} - \Lambda_k$ tend to zero, at the same time as Λ_L increases without bound. If as the discretization improves we are sure never to delete sampling points, but only to add them, the best misfit can never get worse. The misfit is bounded below (by zero) so as we solve a sequence of optimization problems (23) with ever better approximation, the corresponding sequence of misfits must tend to a limit. We need to show that, if there exists a spectral function whose misfit is less than T, the process we have just described will discover a vector $\Delta \mathbf{a}$ giving a misfit less than or equal to T.

We treat only the case of discrete spectra in the form of (18). The idea of the proof is to show that by appropriate choices of components of $\Delta \mathbf{a}$ it is possible to get the admittance of the finite model as close as we desire to that of the posited true spectral function; hence its misfit will be arbitrarily close and certainly below T. We examine the admittance at one frequency; since the choices we make concerning $\Delta \mathbf{a}$ are independent of frequency, the two responses will converge at each frequency. Let us abbreviate the numerators in (18) by b_n and let the response of the finite model be $\tilde{c}_j$. Then

$$|c_j - \tilde{c}_j| = \left| \sum_{n=1}^{\infty} \frac{b_n}{\lambda_n + 2\pi i f_j} - \sum_{k=1}^{L} \frac{\Delta a_k}{\Lambda_k + 2\pi i f_j} \right|. \tag{24}$$

We denote by $\tilde{\lambda}_n$ the member of the set $\{\Lambda_k\}$ closest to the eigenfrequency λ_n. There are K values in this set and we assume the sampling is dense enough that they are all distinct; it follows that the sequence $\tilde{\lambda}_1, \tilde{\lambda}_2, \tilde{\lambda}_3, \cdots$ is increasing. We shall choose Δa_k to be zero except for those k that correspond to elements $\tilde{\lambda}_n$ just defined, where we set $\Delta a_k = b_n$. The optimization program is free to choose Δa_k and its choice may be better than ours in reducing the misfit, but it can be no worse. With these choices we have

$$|c_j - \tilde{c}_j| = \left| \sum_{n=1}^{K} \frac{b_n}{\lambda_n + 2\pi i f_j} + \sum_{n=K+1}^{\infty} \frac{b_n}{\lambda_n + 2\pi i f_j} \right.$$
$$\left. - \sum_{n=1}^{K} \frac{b_n}{\tilde{\lambda}_n + 2\pi i f_j} \right| \tag{25}$$

$$\le \sum_{n=1}^{K} b_n \left| \frac{1}{\lambda_n + 2\pi i f_j} - \frac{1}{\tilde{\lambda}_n + 2\pi i f_j} \right| + \left| \sum_{n=K+1}^{\infty} \frac{b_n}{\lambda_n + 2\pi i f_j} \right|. \tag{26}$$

Because the series (18) is convergent, the second term in (26) must vanish as K tends to infinity. Let us examine the first term:

$$\sum_{n=1}^{K} b_n \left| \frac{\tilde{\lambda}_n - \lambda_n}{(\lambda_n + 2\pi i f_j)(\tilde{\lambda}_n + 2\pi i f_j)} \right|$$

$$\leq |\tilde{\lambda}_n - \lambda_n|_{max} \sum_{n=1}^{K} \frac{1}{|\tilde{\lambda}_n + 2\pi i f_j|} \left| \frac{b_n}{\lambda_n + 2\pi i f_j} \right|. \tag{27}$$

Since the sequence $\tilde{\lambda}_1, \tilde{\lambda}_2, \tilde{\lambda}_3, \cdots$ is increasing

$$|\tilde{\lambda}_1 + 2\pi i f_j| < |\tilde{\lambda}_n + 2\pi i f_j|, \quad n > 1 \tag{28}$$

and so if s is the sum in (27)

$$s < \frac{1}{|\tilde{\lambda}_1 + 2\pi i f_j|} \sum_{n=1}^{\infty} \left| \frac{b_n}{\lambda_n + 2\pi i f_j} \right|. \tag{29}$$

It is easy to show that if (18) is convergent it is absolutely convergent and so the sum in (29) is finite; therefore s is always bounded above by some constant no matter how large K becomes. Furthermore, in (27) as the approximation gets better, the difference $|\lambda_n - \tilde{\lambda}_n|$ approaches zero. Thus both terms in (26) tend to zero and this proves that in the limit of ever better approximation, the optimum solution of (23) can achieve a misfit as small as that of any conductivity profile fitting the data.

The result we have just obtained leaves us with a *necessary* condition for the compatibility of a data set and the one-dimensional model assumptions: if a physical conductivity profile can satisfy the admittance data to tolerance T, the optimization process will yield a misfit less than or equal to T. Conversely, if the optimization produces too large a misfit, we are certain there can be no models in accord with the observations. But it is possible that optimization over $a(\lambda)$ will do a *better* job of fitting than the physical system can and then it may appear there are acceptable solutions when in fact none exists. To study this question we must discover the kind of spectral function the optimization process generates: not every sequence of λ_n and b_n is connected with the eigensystem of a differential equation or a physically realizable model. We shall show that the optimizing process does indeed generate physical models, but most unusual ones.

We propose solving the optimization problem (23) with an algorithm like Lawson and Hanson's NNLS. Recall from chapter

4 that this procedure constructs an optimum vector in which the number of positive components is never more than the number of rows of A; the rest are exactly zero. There are N frequencies in the data set and this translates into $2N$ rows of A, which has L columns. As we improve the approximation, it is L that increases, while N, of course, is constant. Thus no matter how superb the degree of approximation, the total number of jumps in the spectral function is $2N$ or fewer. This means the physical system has a finite number of resonances. Obviously, if there is a corresponding physical picture, it cannot be a strictly positive, continuous σ for we have seen that every such function possesses infinitely many eigenvalues. Krein (1952) first studied a physical system with this unusual property, the vibrating string mentioned earlier. If all the mass of the string is concentrated into a number of point masses, joined by massless string, there are only finitely many degrees of mechanical freedom and only finitely many frequencies of free vibration. The electrical analogy is an insulating halfspace with finitely many concentrations of conductivity in thin sheets, delta functions of conductivity. We shall study such a system, called a model in the class D^+, and obtain its admittance in the complex f plane; it would be more satisfactory to start over and show that the spectral representation (20) is valid for a linear vector space of conductivities large enough to include delta functions and other ordinary functions as well. The reader is referred to Parker (1980) for this development, for which we have no room.

We compute the admittance of the conductivity profile

$$\sigma(z) = \sum_{m=1}^{M} \tau_m \, \delta(z - z_m) \tag{30}$$

with $0 \le z_1 < z_2 < \cdots < z_M < H$ and where $\tau_m > 0$ is the total conductance of the m th layer. We must solve the electric field equation 5.06(1) which we restate for convenience:

$$\frac{d^2 E}{dz^2} = 2\pi i f \, \mu_0 \sigma(z) E(z) . \tag{31}$$

We set $E(H) = 0$ as always. Consider integrating (31) across one of the conducting layers:

$$\int_{z_m - \varepsilon}^{z_m + \varepsilon} \frac{d^2 E}{dz^2} \, dz = \int_{z_m - \varepsilon}^{z_m + \varepsilon} 2\pi i f \, \mu_0 \sigma(z) E(z) \, dz . \tag{32}$$

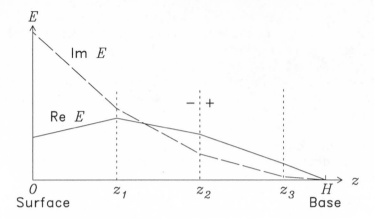

Figure 5.08b: The complex electric field for a model in D^+ with three conducting layers.

Then letting ε go to zero we have:

$$D_m^+ - D_m^- = 2\pi i f \mu_0 \tau_m E_m \tag{33}$$

where D_m^+ is the limit as we approach z_m from below, of dE/dz, D_m^- is the limit from above (because z increases downwards, the notation is the reverse of one's immediate expectations; see figure 5.08b), and E_m is just $E(z_m)$, which is defined since E is continuous. Now

$$\frac{d^2E}{dz^2} = 0, \quad z_m < z < z_{m+1} \tag{34}$$

and therefore E varies linearly between the layers, experiencing a jump in gradient at each conductor. It follows that

$$D_m^+ = D_{m+1}^- \tag{35}$$

and taking E across an insulating gap:

$$E_{m+1} = E_m + (z_{m+1} - z_m)D_{m+1}^- . \tag{36}$$

Define the admittance just above the m th conductor by

$$C_m = -E_m/D_m^- . \tag{37}$$

Now we eliminate the electric fields and their gradients and connect C_{m+1} to C_m using (33), (35), and (36):

$$C_m = \cfrac{1}{2\pi i f \mu_0 \tau_m + \cfrac{1}{z_{m+1} - z_m + C_{m+1}}} . \tag{38}$$

We combine these relations from the bottom, where $E = 0$, to the top and thereby obtain a grand *continued fraction* expression for the admittance at $z = 0$:

$$c(f) = H_0 + \cfrac{1}{2\pi i f\,\mu_0\tau_1 + \cfrac{1}{H_1 + \cfrac{1}{2\pi i f\,\mu_0\tau_2 + \cdots \cfrac{1}{H_M}}}} \tag{39}$$

where we name the separations between the conductors $H_m = z_{m+1} - z_m$, $H_0 = z_1$, and $H_M = H - z_M$. The term H_0 in (39) arises from a possible surface insulator. If we think about simplifying (39) we conclude that some algebraic manipulation will rearrange the continued fraction into a ratio of two polynomials in frequency:

$$c(f) = \frac{p_0 + p_1(2\pi i f) + p_2(2\pi i f)^2 + \cdots + p_M(2\pi i f)^M}{1 + q_1(2\pi i f) + q_2(2\pi i f)^2 + \cdots + q_M(2\pi i f)^M} \tag{40}$$

where p_k and q_k are real. Equation (40) in turn can be written as a partial fraction, in the manner invented to calculate elementary integrals:

$$c(f) = b_0 + \sum_{m=1}^{M} \frac{b_m}{\lambda_m + 2\pi i f}. \tag{41}$$

Note (39), (40), and (41) are identical functions of f, just written differently. Equation (41) shows something we had predicted: the admittance has a finite number of poles on the imaginary axis. The term b_0 is again due to a surface insulator, which is not allowed in the earlier treatments. But the relationship to (39) can be reversed: if the coefficients in (41) are discovered by the optimization process, they can be converted into separations and conductances by building the continued fraction corresponding to (39); see exercise 5.08(ii) for a simple scheme. The transformation procedure can be automated in several ways without the need of an intermediate step to create a rational function in (40), but many of them are numerically ill-conditioned; Parker and Whaler (1981) describe a stable algorithm in detail.

At this point we have shown that every system with finitely many eigenvalues can be interpreted in terms of a model in D^+, which by stretching things a little, can be regarded as a conductivity model. But one point remains to be cleared up if the output

Table 5.08A: A Sequence of Three Optimizations

$L = 54$			$L = 102$			$L = 150$		
k	Λ_k	Δa_k	k	Λ_k	Δa_k	k	Λ_k	Δa_k
1	0	0.1294	1	0	0.1265	1	0	0.1271
15	0.0116	0.2366	17	0.0116	0.2475	19	0.0116	0.1301
19	0.0457	0.2268	26	0.0494	0.1129	20	0.0118	0.1184
23	0.1796	1.542	27	0.0568	0.1638	34	0.0524	0.0215
24	0.2529	0.3335	34	0.1943	1.134	35	0.0539	0.2525
26	0.5016	1.377	35	0.2089	0.7874	49	0.1972	0.4231
27	0.7063	0.0892	44	0.5016	0.1337	50	0.2001	1.489
37	21.67	1.687	45	0.5425	1.237	70	0.5343	1.019
38	30.52	3.006	66	26.98	3.16	71	0.5425	0.3601
42	120.0	24.51	67	28.75	1.887	97	27.33	2.533
43	169.0	15.2	77	129.8	7.932	98	27.68	2.484
49	1318	103.5	78	139.6	31.42	118	135.7	8.028
50	1857	269.6	94	1641	360.6	119	137.7	31.3
			95	1749	5.667	140	1641	337.2
						141	1663	29.19

of the optimization program is to be accepted as the smallest possible misfit: are the parameters τ_m and H_m resulting from the transformation of the finite spectral function always positive? The answer is yes. A proof can be found in the classical theory of continued fractions in the book by Herbert Wall (1948, chapter 5). If we set $\zeta = 2\pi i f$, then the fraction (39) is equivalent to what is known as a *real J-fraction*. If the constants are positive, Wall shows by mathematical induction that the imaginary part of the value of the fraction is negative when $\text{Im}\,\zeta > 0$; conversely it can be seen that whenever negative constants occur, the imaginary part will be positive in a part of the upper half plane of ζ. But this is incompatible with (41) as we can see from expanding into real and imaginary parts as we did with (19). Therefore, only fractions with positive constants are generated from partial fractions like (41) and models with positive conductances always exist achieving the minimum misfit. And so we have shown that the necessary condition is also sufficient.

If the association of the smallest misfit with a series of delta functions in conductivity is dismaying, Parker and Whaler (1981) show that layered or smooth profiles can always be constructed that are arbitrarily close in misfit to a model in D^+. The process we have described is the basis for an effective practical scheme for

Table 5.08B: Optimal Admittance

n	λ_n	b_n	n	λ_n	b_n
0	–	0	5	0.5365	1.380
1	0	0.1271	6	27.51	5.018
2	0.01170	0.2484	7	137.3	39.33
3	0.05375	0.2741	8	1643.	366.4
4	0.1995	1.912			

finding the least misfit and corresponding D^+ models. Naturally, we cannot realize the limit of arbitrarily fine subdivision of the λ axis, nor can we allow Λ_L to be infinitely great, but empirically the convergence rate as the discretization is improved appears to be satisfactory: not only does X, the misfit, converge rapidly, but the admittance also tends a particular rational function, which is something we have not discussed. Let us illustrate these ideas in their application to the MT sample data set studied throughout this chapter. Two slight modifications are needed: first, we introduce a weight into the misfit measure (23) to allow for unequal uncertainties in the data as in 5.06(37); second, a component is added to $\Delta\mathbf{a}$ to accommodate the constant term b_0 in (41).

We solve the modified version of (23) with NNLS. The positive λ axis is sampled three times with increasing density, taking $\Lambda_L = 5185 \, \text{s}^{-1}$; nonzero elements of $\Delta\mathbf{a}$ (in km s^{-1}) are tabulated together with the corresponding Λ_k. The misfits X for each optimization are 4.647, 4.603, and 4.6024. It will be noted that after the third refinement, the positive components, except for the first, are grouped in pairs. This is taken to mean that the proper sample point for each pair lies between the two on the λ axis, and the size of Δa_k indicates the relative weight in a linear interpolation. If these interpolated points are added, the final optimization yields a misfit of 4.6023 with a total of only eight positive components, although it will be recalled a maximum of thirty-four is permitted by the theory. Table 5.08B gives the final admittance as described by (41). This translates into a model in D^+ with eight conductors, tabulated in table 5.08C and plotted in figure 5.08c. Note that $H = \infty$, which follows from the presence of a pole in c at zero frequency.

We are naturally interested in comparing the solution with the smallest misfit with those found by iterative regularization. The singular nature of the delta functions makes a direct

Table 5.08C: Optimal Model in D^+

k	z_k (km)	τ_k (S)	H_k (km)
1	0	1.9189	0
2	0.28307	17.122	0.28307
3	0.65441	94.538	0.37134
4	1.2124	93.589	0.55804
5	14.483	462.94	13.270
6	22.471	1094.0	7.9880
7	44.372	1473.1	21.902
8	78.320	3023.4	33.948

comparison between figures 5.08c and 5.06d difficult, except for broad features, like the low conductivity zone around 10 km depth and the concentration of conductance just below 1 km. We can register the two kinds of solution if instead of conductivity we plot integrated conductivity to a given depth:

$$\tau(z) = \int_0^z \sigma(s)\,ds \ . \tag{42}$$

This quantity is plotted in figure 5.08d; the best-fitting solution appears as the step function, while the smooth curve is the regularized model found by Occam's process in W_2^1 with misfit 5.788. The similarity between the two is quite remarkable.

We can exploit our ability to find a best-fitting solution to answer a question about the information contained in a given data set. From our discussion in 5.02 of the penetration of the electromagnetic field into a uniform conductor and the analysis of resolution given in the previous section we can conclude that knowledge of the conductivity gets more and more imprecise as we look deeper in the Earth. Is it the case that there is something to be learned from the observations at any depth however great, or is there a definite boundary beyond which nothing whatever can be deduced? The answer, perhaps surprisingly, is the latter. To understand this consider what happens in a system where there is a layer of extremely high conductivity: the electric and magnetic fields are attenuated to vanishing point in the layer so that no energy penetrates into the region below. Then whatever the behavior of the conductivity beneath the layer, no information about it is returned to the surface: the layer "shields" the deeper region. Therefore, if there is a model fitting the measure-

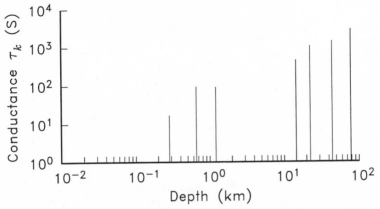

Figure 5.08c: Optimal model in D^+: vertical lines indicate position of thin conducting layers, with conductances as shown. Surface layer omitted because of logarithmic depth scale.

ments with a perfectly conducting layer a $z = H$, everything below H is in principle unknowable because we can put any conductivity structure we like there and still satisfy observation. If one is willing to place an upper bound on the conductivity in the ground, it

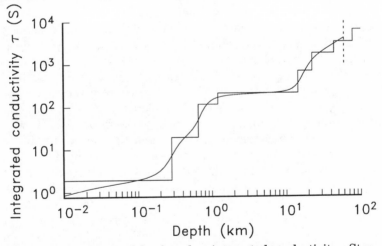

Figure 5.08d: Two models plotted as integrated conductivity. Step function: optimum model in D^+ shown in the previous figure. Smooth curve: integrated model in W_2^1 with $X = 5.788$ shown as solid line in figure 5.06b; this model terminates at $z = 60$ km.

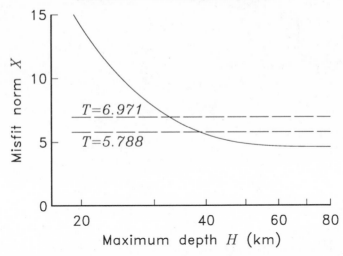

Figure 5.08e: Smallest misfit norm as a function of layer thickness H. The two dashed lines show misfit at the expected value and the 95 percent level.

would then be possible to obtain information from deeper in the section: but our approach is intended to provide the depth range intrinsic to the data set, without outside assumptions. We already know how to construct models within a finite layer terminated with a perfect conductor because that corresponds precisely to the boundary condition $E(H)=0$ we have applied all along. To force the layer thickness to be H in the optimization (23) we include the "datum" $c(0)=H$ with zero (or tiny) uncertainty. Now we solve a series of problems, sweeping through H and examining the misfit. There is a trade-off between how well we believe the data must be matched and the depth of ignorance, as we see in figure 5.08e. If we demand agreement up to the expected value of the norm of the noise, the limiting depth in the example is 38.6 km; to be more cautious we might want the misfit norm to be exceeded by chance only 5 percent of the time: then we would say everything below 32.7 km is unreachable with these data.

The existence question has been settled for a handful of other nonlinear inverse problems besides the magnetotelluric inverse problem. Solutions have been obtained for the seismic τ-p and X-p inverse problems (Garmany et al., 1979) and the resistivity sounding problem (Parker, 1984). In every case a secondary linear inverse problem is introduced in which positivity of the unknown function plays a central role. This is unsurprising for the resistivity problem since the solution is closely modeled on the one for the MT problem. The seismic problems are completely different in character, but lead nonetheless to the same kind of solution strategy. The author suspects that this is not a fundamental insight about the nature of nonlinear inverse problems, but rather a reflection of our limited knowledge.

Exercises

5.08(i) Calculate and plot the spectral function of a uniform conducting layer terminated with a perfect conductor. Examine the limit as H, the layer thickness, tends to infinity.

5.08(ii) Consider the admittance

$$c(f) = 1 + \frac{2}{1 + 2\pi i f} + \frac{2}{4 + 2\pi i f} .$$

By elementary algebra translate this into the ratio of two quadratic polynomials, in the form of (39). Study the following scheme for converting a ratio of polynomials into continued fraction in the form of (38) then apply it to your function. We start with two degree-n polynomials: $P_n(z), Q_n(z)$. Perform the symbolic division (a); rewrite by taking the reciprocal of the rational function (b); perform another division (c) and now repeat step (a) on the ratio in the denominator, and so on.

$$\frac{P_n(z)}{Q_n(z)} = a + \frac{\breve{P}_{n-1}(z)}{Q_n(z)} \tag{a}$$

$$= a + \frac{1}{Q_n(z)/\breve{P}_{n-1}(z)} \tag{b}$$

$$= a + \frac{1}{bz + \dfrac{\breve{Q}_{n-1}(z)}{\breve{P}_{n-1}(z)}} \tag{c}$$

Hence calculate the parameters of the model in D^+.

5.08(iii) Show that the curve of misfit X against H, for example, figure 5.08e, is monotonically decreasing or may exhibit a single minimum. How is the penetration theory interpreted for the case with a minimum? What are the values of X at $H = 0$ and $H = \infty$?

5.09 Coda

Our treatment of nonlinear inverse problems is of necessity very incomplete. Throughout the book we have generally avoided the more mathematical kinds of problems that deal with uniqueness, existence, and construction of solutions based on exact, complete data. Experience suggests that those often elegant theories are hardly ever useful when one is faced with geophysical field measurements. And yet the powerful technique of Gel'fand and Levitan (1956) almost deserves a place here. Invented to solve an inverse problem in quantum scattering theory, the original method constructs the potential in a Schroedinger equation from its spectral function. This remarkable method sets out by assuming there is a most unusual integral transform relating the solution of the Schroedinger equation to a cosine function. The kernel in the transformation is unknown, but by exploiting the completeness and orthogonality of the eigenfunctions, Gel'fand and Levitan obtain a Fredohlm integral equation of the second kind for the unknown kernel. Once the transformation is known so is the solution to the Schroedinger equation and the unknown potential seemingly appears as a by-product. This is an analytic solution that maps one function to another and therefore is not directly applicable to finite data sets. It has been transformed in many ways to solve a variety of inverse problems in geophysics and other fields. The reader who has followed our treatment of the magnetotelluric inverse problem to the end is adequately prepared for Weidelt's (1972) marvelous adaptation of the theory. We note as an aside that the one-dimensional magnetotelluric problem is one of the most thoroughly studied; the reader can get a sketch of the different approaches this problem has attracted from the book by Whittall and Oldenburg (1992), who provide an up-to-date survey and much insight. Another important modification of the Gel'fand and Levitan theory is its transformation into the time domain for application to inversion of vertical incidence, seismic-reflection data; see Bube and Burridge (1983) for a review of this inverse problem.

A difficulty peculiar to nonlinear problems concerns the possible existence of multiple minima in misfit when one attempts to match the measurements. In the linear theory we optimize convex functionals (like norms of linear functionals) over convex sets; the functional value at a local minimum is always the best one can do, the global minimum. There is no such guarantee in the case of nonlinear problems. This is the fundamental reason why a numerical search like Occam's process can never be regarded as truly satisfactory since the procedure may arrive at a local minimum which is in fact not optimal; the existence of a better solution with a smaller value of roughness or data misfit would then go undetected. The two numerical examples we studied were apparently quite well behaved in this respect, although we have no proof of the optimality of the iterative solutions we obtained in 5.05 and 5.06. One of the most troublesome geophysical inverse problems because of multiple minima is seismic waveform inversion: here one attempts to reproduce detailed seismograms from a velocity model. The vertical-incidence problem mentioned earlier is a special case of this more general problem, where sources and receivers are spread out horizontally. When a seismogram exhibits almost sinusoidal oscillations, the predicted waveforms may lead or lag behind the observed one by a cycle and still fit almost perfectly; but the match will be poor for every half-cycle delay. This effect causes a series of highs and lows in misfit as model travel time varies. Methods based upon the local linearization, like those we have developed, are incapable of dealing with this kind of difficulty; all they can do is home in on a local minimum.

Current research on the question centers on exploring the model space with the aid of random fluctuations. A purely random search would be hopelessly inefficient, even if the dimension of the space is only a hundred or so. Two promising ideas have been designed in analogy with natural processes: slow cooling and natural selection. The first, called *simulated annealing*, is based on the way in which a thermal system comes to its lowest energy state despite the existence of metastable states, local minima in the energy functional. Random fluctuations in the system, because they are not always heading downwards in energy, allow the search to examine a larger region of model space. As "time" advances and the "temperature" falls, the size of the random excursions is decreased. Press et al. (1992) give a numerical algorithm; see Johnson and Wyatt (1994) for a geophysical application. Natural selection in biology may be viewed as a process that optimizes a functional called fitness. So-called *genetic algorithms* assemble a population of models whose features are treated as "genes"; the models within the family "mate" pair-wise and exchange genes, but a small proportion of random "mutations" is also introduced. The offspring are tested for fitness, which in our application means smallness of the objective functional, and the best go forward to mate once more, although again a random element appears here too, occasionally allowing a "lucky" model to survive despite a poor performance. This approach promises to be more effective than simulated annealing and has been applied with considerable success to the seismic waveform problem (Sambridge and Drijkoningen, 1992). Both the genetic algorithm and simulated annealing tend to require a great deal of computer time. Unfortunately, just as with Occam's process for simpler problems, there is no mathematical proof as yet that the final solution obtained truly is optimal.

THE DILOGARITHM FUNCTION

The special function arising in spherical spline interpolation in section 2.07 has been neglected by the compilers of approximations for computers. Abramowitz and Stegun (chapter 27, 1968) do not give a suitable algorithm, nor do the references quoted by them. Here is my effort, should anyone wish to compute the function. The definition, equation 2.07(10), naturally gives rise to the power series

$$\operatorname{dilog}(1-x) = x + \frac{x^2}{2^2} + \frac{x^3}{3^2} + \cdots \qquad |x| \le 1. \qquad (A1)$$

But this is too slowly convergent to be much use for x in the interval $(0,1)$ as we need it. (Can the reader estimate how many terms would be needed to obtain four significant figures of accuracy at $x=1$?) One approach is to use a symmetry property given by Abramowitz and Stegun to map the function with argument in $(\frac{1}{2}, 1)$, where (A1) converges moderately well, into $(0, \frac{1}{2})$ like this:

$$\operatorname{dilog}(x) = \frac{1}{6}\pi^2 - \operatorname{dilog}(1-x) - \ln x \, \ln(1-x). \qquad (A2)$$

Up to 41 terms are needed if computer double precision (say, 1 in 10^{14}) is to be attained by raw use of (A1) on the smaller interval $(\frac{1}{2}, 1)$. The power series can and should be telescoped (see below) but this approximation is somehow still unsatisfying.

    A simple change of variable fixes the problem: in the integral 2.07(10) let $t = e^{-s} - 1$ and then

$$
\begin{aligned}
\operatorname{dilog}(e^{-y}) &= \int_0^y \frac{s \, ds}{e^s - 1} \\
&= \sum_{n=0}^{\infty} B_n \frac{y^{n+1}}{(n+1)!}. \qquad |y| \le 2\pi \\
&= y - \frac{y^2}{4} + \frac{y^3}{36} - \frac{y^5}{3600} + \frac{y^7}{211680} - \cdots \qquad (A3)
\end{aligned}
$$

where we have expanded the integrand in its classical Taylor series and B_n are the Bernoulli numbers (chapter 23, Abramowitz and Stegun, 1968); notice that, except for the quadratic term, the function is expressed in odd powers of y. Mapped into the original variable the power series (A3) converges for x in the interval $(e^{-2\pi}, e^{2\pi})$ or about $(0.002, 540)$.

A good computer algorithm for the dilogarithm can now be designed. For $\frac{1}{2} \le x \le 1$, which maps into $-\ln 2 \le y \le 0$, we use (A3) in telescoped form; for $0 \le x < \frac{1}{2}$ we employ (A2). In the interests of completeness, we note that (A3) works up until $x = 2$ and for $x > 2$ there is the relation

$$\text{dilog}(x) = -\text{dilog}(1/x) - \frac{1}{2}(\ln x)^2 .$$

Telescoping or polynomial economization (see chapter 12, Acton, 1970) recognizes the fact that the approximation obtained by terminating the Taylor series is most accurate near $x = 0$ and least accurate at the end of the interval on which the function is required. It is possible to adjust the coefficients of the polynomial to redistribute the error evenly on the interval, thus lowering the maximum error committed by a fixed-degree polynomial on a given interval. Using Remes's algorithm (Acton, 1970) I calculated the following economized polynomial approximations for my 32-bit computer. A four-term, degree-five approximation is sufficient for single-precision arithmetic:

$$\text{dilog}(e^{-y}) = a_1 y + a_2 y^2 + a_3 y^3 + a_5 y^5 + y\,\varepsilon_1(y) .$$

With the values in the table below. when $|y| \le \ln 2$ we find $|\varepsilon_1| < 8.3 \times 10^{-8}$.

$$a_1 = 1.00000008 \qquad a_3 = 0.02777598259$$
$$a_2 = -0.25 \qquad a_5 = -0.0002719883$$

For double-precision accuracy we need take only three more terms in the expansion:

$$\text{dilog}(e^{-y}) = a_1 y + a_2 y^2 + a_3 y^3 + a_5 y^5 + a_7 y^7 +$$
$$a_9 y^9 + a_{11} y^{11} + y\,\varepsilon_2(y) .$$

The corresponding numerical values are given below. On the same interval as before the error function satisfies the bound $|\varepsilon_2| < 6.6 \times 10^{-15}$.

$$a_1 = 1 \qquad\qquad\qquad a_7 = 4.724071696 \times 10^{-6}$$
$$a_2 = -0.25 \qquad\qquad a_9 = -9.1764954 \times 10^{-8}$$
$$a_3 = 0.027777777777213 \qquad a_{11} = 1.798670 \times 10^{-9}$$
$$a_5 = -2.7777776990 \times 10^{-4}$$

TABLE FOR 1-NORM MISFITS

In section 3.01 we considered the 1-norm of the misfit vector, which is the vector of differences between the true noise-free observation and the actual noisy value. We suppose the noise consists of N statistically independent, zero-mean Gaussian random variables with the same standard error, σ. Then we can compute the probability that a specific value of $\|\mathbf{X}\|_1$ would be exceeded by chance; see Parker and McNutt (1980) for the method of computation.

A table is provided for small N. Suppose the given value of the 1-norm is S. The table gives the critical values of S/σ for a set of probabilities p listed in the first row and N in the first column. Thus if $S/\sigma = 2.756$ when $N = 2$, we see from the table that the chance that the 1-norm is this large or larger is 0.10. The author has reason to believe the results in this table are slightly more accurate than those in the one given by Parker and McNutt.

When $N > 10$ one can use the following approximation. Let x be the standardized value of S, that is,

$$x = (S - \mu)/v^{\frac{1}{2}}.$$

Here the mean μ and the variance v are those of the distribution of the 1-norm, namely,

$$\mu = (2/\pi)^{\frac{1}{2}}\sigma N \quad \text{and} \quad v = (1 - 2/\pi)\sigma^2 N.$$

Percentage Points for 1-Norm Misfit

N	0.99	0.95	0.90	0.75	0.50	0.25	0.10	0.05	0.01
2	0.178	0.402	0.576	0.954	1.487	2.119	2.756	3.163	3.969
3	0.497	0.871	1.124	1.626	2.289	3.046	3.798	4.276	5.219
4	0.901	1.398	1.714	2.317	3.088	3.953	4.802	5.339	6.398
5	1.357	1.958	2.329	3.020	3.887	4.847	5.782	6.371	7.530
6	1.848	2.543	2.961	3.732	4.685	5.731	6.745	7.381	8.630
7	2.366	3.145	3.607	4.451	5.483	6.609	7.694	8.374	9.705
8	2.903	3.760	4.264	5.175	6.281	7.481	8.633	9.353	10.761
9	3.457	4.387	4.929	5.903	7.079	8.348	9.563	10.321	11.800
10	4.025	5.023	5.601	6.635	7.877	9.212	10.486	11.280	12.827

Then by means of the asymptotic moment expansion it can be shown that the probability that the norm of a set of random variables should exceed the given value S is

$$\Pr(S \le \|\mathbf{X}\|_1) \approx P(x) - 0.0662 N^{-\frac{1}{2}}(x^2 - 1)e^{-x^2/2}$$

where $P(x)$ is the normal probability integral; see Abramowitz and Stegun, chapter 26, 1964.

REFERENCES

Acton, F. S. *Numerical Methods that Work*. New York: Harper and Row, 1970.

Abramowitz, M., and I. A. Stegun. *Handbook of Mathematical Functions*. New York: Dover, 1968.

Backus, G. E. "Inference from inadequate and inaccurate data," I. *Proc. Nat. Acad. Sci. USA* 65:1–7, 1970a.

———. "Inference from inadequate and inaccurate data," II. *Proc. Nat. Acad. Sci. USA* 65:281–87, 1970b.

———. "Confidence set inference with a prior quadratic bound." *Geophys. J.* 97:119–50, 1989.

Backus, G. E., and J. F. Gilbert. "Numerical applications of a formalism." *Geophys. J. Roy. Astron. Soc.* 13:247–76, 1967.

———. "The resolving power of gross Earth data." *Geophys. J. Roy. Astron. Soc.* 16:169–205, 1968.

———. "Uniqueness in the inversion of inaccurate gross Earth data." *Phil. Trans. Roy. Soc.* A 266:187–269, 1970.

Bard, Y. *Nonlinear Parameter Estimation*. New York: Academic Press, 1974.

Barnett, V., and T. Lewis. *Outliers in Statistical Data*. New York: John Wiley & Sons, 1984.

Bazaraa, M. S., and C. M. Shetty. *Nonlinear Programming—Theory and Algorithms*. New York: John Wiley & Sons, 1979.

Bracewell, R. N. *The Fourier Transform and its Applications*. New York: McGraw-Hill, 1978.

Bube, K. P., and R. Burridge. "The one-dimensional inverse problem of reflection seismology." *SIAM Rev.* 25:497–559, 1983.

Caignard, L. "Basic theory of the magneto-telluric method of geophysical prospecting." *Geophysics* 18:605–35, 1953.

Caress, D. W., and R. L. Parker. "Spectral interpolation and downward continuation of marine magnetic anomaly data." *J. Geophys. Res.* 94:17393–407, 1989.

Coddington, E. A., and N. Levinson. *Theory of Ordinary Differential Equations*. New York: McGraw-Hill, 1955.

Constable, C. G., and R. L. Parker. "Statistics of the geomagnetic secular variation for the past 5 m.y." *J. Geophys. Res.* 93:11569–81, 1988a.

———. "Smoothing, splines and smoothing splines; their application in geomagnetism." *J. Comp. Phys.* 78:493–508, 1988b.

Constable, C. G., R. L. Parker, and P. B. Stark. "Geomagnetic field models incorporating frozen flux constraints." *Geophys. J. Int.* 113:419–33, 1993.

Constable, S. C., R. L. Parker, and C. G. Constable. "Occam's inversion: a practical algorithm for generating smooth models from electromagnetic sounding data." *Geophysics* 52:289–300, 1987.

Cox, A. "A stochastic approach towards understanding the frequency and polarity bias of geomagnetic reversals." *Phys. Earth Planet. Int.* 24:178–90, 1981.

de Boor, C. *A Practical Guide to Splines*. New York: Springer-Verlag, 1978.

Dorman, L. M. "The gravitational edge effect." *J. Geophys. Res.* 80:2949–50, 1975.

Dym, H., and H. P. McKean. *Fourier Series and Integrals*. New York: Academic Press, 1972.

Garmany, J. D., J. A. Orcutt, and R. L. Parker. "Travel time inversion: a geometrical approach." *J. Geophys. Res.* 84:3615–22, 1979.

Gass, S. I. *Linear Programming: Methods and Applications*. New York: McGraw-Hill, 1975.

Gel'fand, I. M., and R. M. Levitan. "On the determination of a differential equation from its spectral function." *Am. Math. Soc. transl. ser. 1*, 253–304, 1956.

Gilbert, F., and A. M. Dziewonski. "An application of normal mode theory to the retrieval of structural parameters and source mechanisms from seismic spectra." *Phil. Trans. Royal Soc. Lond.* A 278:187–269, 1975.

Gill, P. E., W. Murray, and M. H. Wright. *Practical Optimization*. New York: Academic Press, 1981.

Golub, G. H., and C. F. Van Loan. *Matrix Computations*. Baltimore: Johns Hopkins University Press, 1983.

Grant, F. S., and G. F. West. *Interpretation Theory in Applied Geophysics*. New York: McGraw-Hill, 1965.

Harland, W. B., R. L. Armstrong, A. V. Cox, L. E. Craig, A. G. Smith, and D. G. Smith. *A Geologic Time Scale*. Cambridge: Cambridge University Press, 1989.

Henry, M., J. A. Orcutt, and R. L. Parker. "A new approach for slant stacking refraction data." *Geophys. Res. Lett.* 7:1073–76, 1980.

Jackson, A. "Accounting for crustal magnetization in models of the core magnetic field." *Geophys. J. Int.* 103:657–74, 1990.

Jeffreys, H. *Theory of Probability*, 3d ed. Oxford: Clarendon Press, 1983.

Jenkins, G. M., and D. G. Watts. *Spectral Analysis and its Applications*. London: Holden-Day, 1969.

Johnson, H. O., and F. K. Wyatt. "Geodetic network design for fault-mechanics studies." *Manuscripta Geodetica*, submitted 1994.

Jones, G. J., A. D. Chave, G. Egbert, D. Auld, and K. Bahr. "A comparison of techniques for magnetotelluric response function estimation." *J. Geophys. Res.* 94:14201–14, 1989.

Klitgord, K. D., S. P. Huestis, J. D. Mudie, and R. L. Parker. "An analysis of near-bottom magnetic anomalies: sea-floor spreading and the magnetized layer." *Geophys. J. Roy. Astron. Soc.* 43:387–424, 1975.

Korevaar, J. *Mathematical Methods*. New York: Academic Press, 1968.

Krein, M. G. "On inverse problems for inhomogeneous strings." *Dolk. Akad. Nauk SSSR* 182:669–72, 1952.

Lanczos, C. *Linear Differential Operators*. New York: Van Nostrand, 1961.

Lawson, C. L., and R. J. Hanson. *Solving Least Squares Problems*. Englewood Cliffs, NJ: Prentice-Hall, 1974.

Luenberger, D. G. *Optimization by Vector Space Methods*. New York: John Wiley and Sons, 1969.

———. *Introduction to Linear and Nonlinear Programming*. Reading, Pa.: Addison-Wesley, 1984.

McBain, J. "On the Fréchet-differentiability of the one-dimensional magnetotellurics problem." *Geophys. J. Roy. Astron. Soc.* 86:669–72, 1986.

Macdonald, K. C., S. P. Miller, S. P. Huestis, and F. N. Spiess. "Three-dimensional modeling of a magnetic reversal boundary from inversion of deep-tow measurements." *J. Geophys. Res.* 85:3670-80, 1980.

Masters, G., and F. Gilbert. "Attenuation in the earth at low frequencies." *Phil. Trans. Royal Soc. Lond.* A 308:479–522, 1983.

Moritz, H. *Advanced Physical Geodesy*. Tunbridge: Abacus Press, 1980.

Morse, P. M., and H. Feshbach. *Methods of Theoretical Physics*. New York: McGraw-Hill, 1953.

Noble, B., and J. W. Daniel. *Applied Linear Algebra*. Englewood Cliffs, NJ: Prentice-Hall, 1977.

Oldenburg, D. W. "Funnel functions in linear and nonlinear

appraisal." *J. Geophys. Res.* 88:7387–98, 1983.

Parker, R. L. "Inverse theory with grossly inadequate data." *Geophys. J. Roy. Astron. Soc.* 29:123–38, 1972.

―――. "Best bounds on density and depth from gravity data." *Geophysics* 39:644–49, 1974.

―――. "The theory of ideal bodies for gravity interpretation." *Geophys. J. Roy. Astron. Soc.* 42:315–34, 1975.

―――. "Linear inference and underparameterized models." *Rev. Geophys. Space Phys.* 15:446–56, 1977.

―――. "The inverse problem of electromagnetic induction: Existence and construction of solutions based upon incomplete data." *J. Geophys. Res.* 85:4421–28, 1980.

―――. "The existence of a region inaccessible to magnetotelluric sounding." *Geophys. J. Roy. Astron. Soc.* 68:165–70, 1982.

―――. "The inverse problem of resistivity sounding." *Geophysics* 49:2143–58, 1984.

Parker, R. L., and M. K. McNutt. "Statistics for the one-norm misfit measure." *J. Geophys. Res.* 85:4429–30, 1980.

Parker, R. L., and L. Shure. "Efficient modeling of the Earth's magnetic field with harmonic splines." *Geophys. Res. Lett.* 9:812–15, 1982.

Parker, R. L., and K. Whaler. "Numerical methods for establishing solutions to the inverse problem of electromagnetic induction." *J. Geophys. Res.* 86:9574–84, 1981.

Parker, R. L., L. Shure, and J. A. Hildebrand. "The application of inverse theory to seamount magnetism." *Rev. Geophys.* 25:17–40, 1987.

Parkinson, W. D. *Introduction to Geomagnetism.* New York: Elsevier, 1983.

Peterson, J., C. R. Hutt, and L. G. Holcomb. "Test and calibration of the seismic research observatory." Open file Report 80–187. *U.S. Geological Survey*, Albuquerque, NM, 1980.

Press, W. H., B. P. Flannery, S. A. Teukolsky, and W. T. Vettering. *Numerical Recipes in Fortran—The Art of Scientific Computing.* Cambridge: Cambridge University Press, 1992.

Priestley, M. B. *Spectral Analysis and Time Series.* New York: Academic Press, 1981.

Ralston, A. *A First Course in Numerical Analysis.* New York: McGraw-Hill, 1965.

Reinsch, C. H. "Smoothing by spline functions." *Numer. Math.* 10:177–83, 1967.

Sambridge, M., and G. Drijkoningen. "Genetic algorithms in

seismic waveform inversion." *Geophys. J. Int.* 109:323–42, 1992.

Schouten, H., and K. McCamy. "Filtering marine magnetic anomalies." *J. Geophys. Res.* 77:7089–99, 1972.

Seber, G. A. *Linear Regression Analysis.* New York: John Wiley, 1977.

Shen, P. Y., and A. E. Beck. "Determination of surface temperature history from borehole temperature gradients." *J. Geophys. Res.* 88:7485–93, 1983.

Shure, L., R. L. Parker, and G. E. Backus. "Harmonic splines for geomagnetic modelling." *Phys. Earth Planet. Int.* 28:215–29, 1982.

Smith, D. R. *Variational Methods in Optimization.* Englewood Cliffs, NJ: Prentice-Hall, 1974.

Smith, R. A. "A uniqueness theorem concerning gravity fields." *Proc. Camb. Phil. Soc.* 57:865–970, 1961.

Stacey, F., D. *Physics of the Earth.* New York: John Wiley and Sons, 1977.

Stark, P. B., and R. L. Parker. "Velocity bounds from statistical estimates of $\tau(p)$ and $X(p)$." *J. Geophys. Res.* 92:2713–19, 1987.

———. "Bounded-variable least-squares: an algorithm and applications." *Comp. Stat.* submitted 1994.

Stark, P. B., R. L. Parker, G. Masters, and J. A. Orcutt. "Strict bounds on seismic velocity in the spherical earth." *J. Geophys. Res.* 91:13892–902, 1986.

Strang, G. *Linear Algebra and its Applications.* New York: Academic Press, 1980.

———. *Introduction to Applied Mathematics.* Wellesley, Mass.: Wellesley-Cambridge Press, 1986.

Tarantola, A. *Inverse Problem Theory: Methods for Fitting and Model Parameter Estimation.* New York: Elsvier, 1987.

Vanmarcke, E. *Random Fields: Analysis and Synthesis.* Cambridge, Mass.: MIT Press, 1983.

Wahba, G. *Spline Models for Observational Data.* Philadelphia: Society for Industrial and Applied Mathematics, 1990.

Wall, H. S. *Analytic Theory of Continued Fractions.* New York: Chelsea Publishing Co., 1948.

Watson, G. N. *A Treatise on the Theory of Bessel Functions.* Cambridge: Cambridge University Press, 1966.

Weidelt, P. "The inverse problem of geomagnetic induction." *Zeit. Geophys.* 38:257–89, 1972.

———. "Conduction of conductance bounds from magnetotelluric impedances." *Zeit. Geophys.* 57:191–206, 1985.

Whaler, K. A., and D. Gubbins. "Spherical harmonic analysis of the geomagnetic field: an example of a linear inverse problem." *Geophys. J. Roy. Astron. Soc.* 65:645–93, 1981.

Whittall, K. P., and D. W. Oldenburg. *Inversion of Magnetotelluric Data for a One-dimensional Conductivity.* SEG Monograph Ser. 5, 1992.

Widmer, R., G. Masters, and F. Gilbert. "Spherically symmetric attenuation within the earth from normal mode data." *Geophys. J. Int.* 104:541–53, 1991.

Wilkinson, J. H. *The Algebraic Eigenvalue Problem.* Oxford: Clarendon Press, 1965.

Woodhouse, J. H. "On Rayleigh's principle." *Geophys. J. Roy. Astron. Soc.* 46:11–22; 1976.

Young, C. T., J. R. Booker, R. Fernandez, G. R. Jiracek, M. Martinez, J. C. Rogers, J. Stodt, H. S. Waff, and P. E. Wannamaker. "Verification of five magnetotelluric systems in the mini-EMSLAB experiment." *Geophysics* 53:553–57, 1988.

INDEX

Abramowitz, M., 123, 124, 129, 288, 371, 374
Acton, F. S., 372
Addition Theorem for spherical harmonics, 95, 98, 232
annihilator, 66
apparent resistivity, 299–300, 302
approximate linear independence, 108, 116–17
associated Legendre function, 94
autocovariance function, 281–82, 288, 290
averaging kernel, 208–13
averaging length, 209–14, 276

Backus, G. E., 208, 222, 233, 308, 309, 325
Backus-Gilbert Theory, 200, 206–14, 221, 285, 286, 289, 294, 326, 328, 348
Banach space, 6, 21, 32, 253–54
Bard, Y., 58
Barnett, V., 130
basic feasible solution, 256
basic solution, 254–55, 259–62, 264
basis, 8–10; depleted, 189–92, 194–97; orthogonal, 95, 148, 153, 158, 171, 195–96
Bazaraa, M. S., 48
Beck, A. E., 87
Bernoulli numbers, 371
Bessel function, 291, 304
bottomless valley, 321, 348–49
bounded linear functional, 32, 58, 62, 86–89, 97, 132, 214, 254, 256, 278, 305–6
bounded-variable least squares. See BVLS
Bracewell, R. N., 70
Bube, K. P., 368
Burridge, R., 368
BVLS, 267, 274, 278

Caignard, L., 296
Caress, D. W., 280
Cauchy sequence, 20–26, 32, 202, 208
Chain rule for Fréchet derivatives, 306, 314
Cholesky factorization, 45, 127, 141, 152, 162, 167, 174, 179, 190, 205, 291
closed subspace, 32–35, 43
Coddington, E. A., 303, 352
collocation, 78, 93, 101; least-squares, 191, 233, 280; non-linear, 313
complete set of functions, 95, 352–54
complete space, 24–27, 31, 37, 78, 92, 95–96, 195
completion of a space, 18, 20, 24–26, 89, 208
condition number, 16–17, 40–42, 44–45, 108–11, 116–17, 149, 166, 179–82, 239, 338, 340, 361
Constable, C. G., 74, 191, 279
Constable, S. C., 317
constrained optimization, 46–50, 134, 258
continued fraction, 288, 361–62, 367
convergence, 9, 18, 20–26, 67, 95, 116, 141, 163, 257, 260
convexity, 51–53, 136, 224, 257, 264, 294, 316, 368
convolution, 68, 71, 158, 162, 187–88, 282–84
Convolution Theorem, 71, 188
covariance, 127, 132
covariance matrix, 64, 127, 141, 150, 291
Cox, A., 279, 282
creeping algorithm, 311–12, 314, 319, 326–28, 330–32, 340–42, 345–46